CAPTAIN COOK REDISCOVERED

CAPTAIN COOK REDISCOVERED

Voyaging to the Icy Latitudes

DAVID L. NICANDRI

UBCPress · Vancouver · Toronto

29 28 27 26 25 24 23 22 21 20   5 4 3 2 1

Printed in Canada on FSC-certified ancient-forest-free paper
(100% post-consumer recycled) that is processed chlorine- and acid-free.

Library and Archives Canada Cataloguing in Publication

Title: Captain Cook rediscovered : voyaging to the icy latitudes / David L. Nicandri.
Names: Nicandri, David L., author.
Description: Includes bibliographical references and index.
Identifiers: Canadiana (print) 20200289799 | Canadiana (ebook) 20200289969 |
   ISBN 9780774862226 (hardcover) | ISBN 9780774862240 (PDF) |
   ISBN 9780774862257 (EPUB) | ISBN 9780774862264 (Kindle)
Subjects: LCSH: Cook, James, 1728-1779—Travel—Polar regions. | LCSH: Polar
   regions—Research—History—18th century. | LCSH: Polar regions—Discovery
   and exploration.
Classification: LCC G246.C7 N53 2020 | DDC 910.92—dc23

UBC Press gratefully acknowledges the financial support for our publishing program of the Government of Canada (through the Canada Book Fund), and the British Columbia Arts Council.

This book has been published with the help of a grant from the Canadian Federation for the Humanities and Social Sciences, through the Awards to Scholarly Publications Program, using funds provided by the Social Sciences and Humanities Research Council of Canada.

UBC Press
The University of British Columbia
2029 West Mall
Vancouver, BC V6T 1Z2
www.ubcpress.ca

FOR CHRIS

Contents

Part 4: Sequels

Illustrations

Acknowledgments

First, I want to thank my friend Robin Inglis for the encouragement, insights, and assistance that made this book possible. Robin offered suggestions on reading, provided a helpful review of my manuscript at an early stage, and, as North America's premier expert on Cook cartography and related illustrative material, identified and secured permission for the many maps and images that illuminate this book.

I have long held the work of Ian MacLaren in high regard for the true enlightenment he has offered on how to read and interpret travel literature. Ian is the most assiduous scholar I know, and I have attempted to replicate the thoroughness and innovation that he brings to his research and writing, skills that he brought to his review of my manuscript for UBC Press. This book has also profited immensely from the many helpful comments and suggestions of Michael F. Robinson.

I would also like to acknowledge the technical analysis of sea ice science provided by Harry Stern and Bonnie Light from the University of Washington. Similarly, the late Les Eldridge, one of Puget Sound country's leading maritime historians, was very helpful in identifying the mistakes in nautical nomenclature that a landsman such as myself is prone to.

I am especially indebted to the editorial team at UBC Press for seeing this book into print. At the outset, James MacNevin took interest in an unorthodox interpretation of a somewhat controversial subject. Megan Brand and Lesley Erickson guided the editorial process and smoothed the rough seas enveloping some of my text. Thank you.

It has been said that people like to have written, but no one likes the rigours of writing. For me, at least, the ten-year effort to get this book into print could only have been sustained by the reassurance of dear colleagues and friends. First among these is Clay Jenkinson, the editor of my *River of Promise: Lewis and Clark on the Columbia.* I will never be able to thank him enough for that opportunity, which opened the window to this book. I meet weekly with a self-styled writers group, consisting of Ken Balsley, Joe Illing, and Dick Pust, and appreciate their support and their reading of a few chapters. And then there is the group at Tumwater Valley Athletic Club, led by Fred Cook (no relation to the great navigator), who scolded me for even thinking of taking on a project other than this one.

And lastly, to my best friend Chris, thanks for putting up with the dislocations this book and other writing projects have brought to our household.

CAPTAIN COOK REDISCOVERED

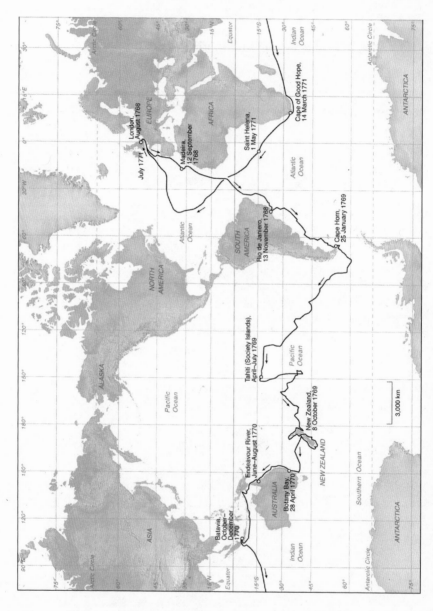

Cook's First Voyage, *Endeavour*, 1768–71. Map by Eric Leinberger

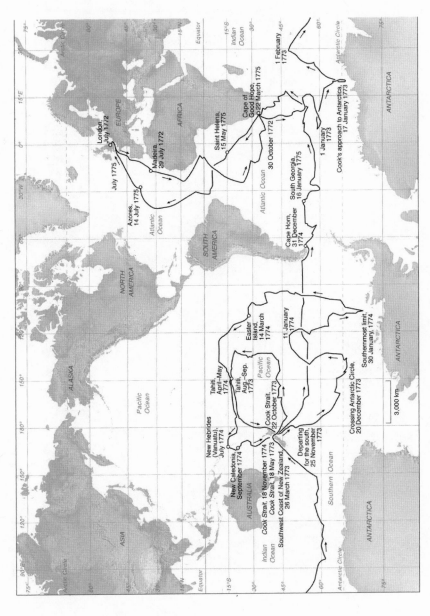

Cook's Second Voyage, *Resolution*, 1772–75. Map by Eric Leinberger

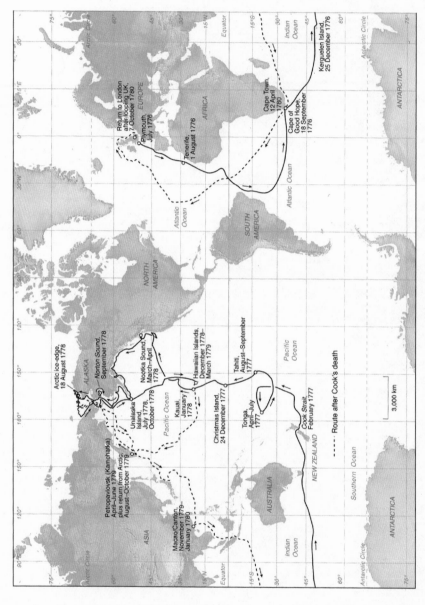

Cook's Third Voyage, *Resolution*, 1776–80. Route of exploration and return journey to England after Captain Cook's Death. Map by Eric Leinberger

Introduction

On January 30, 1774, in the mid-summer sun, James Cook's *Resolution* stretched southward at an unprecedented rate. Cook was on his second voyage, a quest for Terra Australis Incognita, the hypothetical southern continent that mirrored the Eurasian land mass. Cook had just crossed the Antarctic Circle (66° 33' S) for the third time, once in the Indian Ocean and earlier that same season in the Pacific. In the previous instances, after crossing the line, he had encountered the ice pack, which prevented him from sailing farther south. Before *Resolution* had taken to sea, Joseph Banks, a nobleman-naturalist and the most celebrated figure from Cook's first voyage, had joked about the prospect of cruising directly to the South Pole. In a fit of egotism, Banks talked himself out of accompanying Cook on the second voyage. But to all aboard *Resolution* in that day's long light and remarkably mild weather, heretofore unprecedented at that or any near latitude, it seemed Banks's quip was about to be realized.

Nature was only teasing, of course, because Cook soon detected the blink, the sun's reflection off the impenetrable ice pack guarding Antarctica's shore. Cook reached 71° 10' S, nearly four degrees of latitude closer to the pole than his previous high mark. At this juncture, southwest of Cape Horn, Cook inscribed in his journal the most famous line of text that he or any other explorer has ever committed to writing: "I whose ambition leads me not only farther than any other man has been before me, but as far as I think it possible for man to go, was not sorry at meeting with this interruption, as it in some measure relieved us from the dangers and hardships, inseparable with the Navigation of the Southern Polar regions."[1]

Cook turned north at that point to winter in the tropics before resuming his search for Terra Australis in the South Atlantic the next year. But this passage was later immortalized. In the space age of the 1960s, Gene Roddenberry adapted it into the epigram for his *Star Trek* series. Stylistically, the passage also prefigured Neil Armstrong's famous "great leap for man" exclamation, made when he first set foot on the moon, two hundred years to the month after Cook left Tahiti, his defining destination for the *Endeavour* voyage, in 1769. More recently, Tony Horwitz adapted Cook's text for the subtitle of his popular book *Blue Latitudes: Boldly Going Where Captain Cook Has Gone Before* (2002).[2]

It is ironic that the forceful imagery and narrative expressiveness that Cook employed on reaching the farthest south became the emblematic expression for his career. It is virtually the only aspect of his voyages into the icy latitudes that students of his career are intimately familiar with. The only incident that comes close came during Cook's third voyage in search of a Northwest Passage across the top of North America. In the summer of 1778, north of the Alaskan subcontinent that he would be the first to delineate cartographically, Cook and his men on *Resolution* saw another blink, presaging that the Arctic ice pack would stymie their progress to the northeast and Baffin Bay. This time, Cook turned west, hoping to flank the ice. He then sent some of the crew out in the ship's small boats to hunt walruses to supplement the provisions stored on board. There was some grumbling about this unappetizing meat recorded in the journals of a few midshipmen. Historians later conflated these remarks into a larger narrative that Cook had, by this point, lost his touch as a commander, a mere six months before he would be killed in Hawaii. Yet this same community of historians has long recounted a similar story from the *Endeavour* voyage to favourably illustrate how Cook implemented dietary controls in his legendary battle against scurvy. During that voyage, the crew rebelled over having to eat sauerkraut until Cook cagily directed that the officers be seen eating it. Cook recorded: "Altho it be ever so much for their good yet it will not go down with them and you will hear nothing but murmurings gainest the man that first invented it; but the Moment they see their Superiors set a Value upon it, it becomes the finest stuff in the World and the inventer an honest fellow."[3]

These vignettes underscore the two major revisions to the Cook story presented in this book. First, Horwitz's travelogue falls comfortably within what I call the palm-tree paradigm. Notwithstanding that Horwitz subtitled his book after Cook's legendary statement from the edge of the Antarctic ice pack, he focuses on the sun-drenched beaches of Hawaii, Tahiti, and other South Pacific islands – where Cook's famous cross-cultural encounters occurred. But Horwitz largely ignores Cook's travels to those parts of the world that are of ever-increasing significance in the twenty-first century: the icy latitudes of the Indian, Pacific, Atlantic, and Arctic oceans.

Horwitz dismissed reading about Cook's Antarctic probes as "the literary equivalent of chewing on ice cubes."[4]

Most historians, indeed Cook's contemporaries, seized on enchanting island venues as the essential setting for understanding his expeditions. The icy latitudes and their cold temperatures never generated comparable interest in the literature, in the eighteenth century or since. Simon Winchester argues that palm trees became "central to Pacific imagery" because they provide "a picture-perfect and theatrically green backdrop for every beach scene."[5] That the polar zones are lightly inhabited and infrequently visited should not make them less relevant to the study of Cook. Given the current global climate crisis, the opposite could be true.

The anthropological perspective that dominates Cook discourse comes at the cost of understanding the full geographic scope of his endeavours, including their new climatological relevance.[6] In most books about Cook, the story is largely confined to the following formula: no encounter, no voyage. But taken as environmental history, Cook's experience in frigid seas can be considered a compelling indicator of the pace of global warming. This perspective is particularly true of his final voyage in search of the Northwest Passage. If Cook had sailed through the Bering Strait in the conditions of August 2020 instead of August 1778, he might have passed eastward through the northern Canadian archipelago, emerged at Baffin Bay, and headed home to England. In that sense, Cook did not fail to discover the Northwest Passage: he was merely ahead of his time.

The Cook we think we know, the tropical Cook, is a narrative construct – he is largely the product of other writers, including the editors of his accounts. The modern literature analyzing his career, though voluminous, is remarkably orthodox. The double standard evident in the walrus meat and sauerkraut stories highlights the most constant assertion in contemporary Cook historiography and the second revisionist theme of this book: that Cook never should have conducted his fatal third voyage because he was exhausted after piloting the first two and fatally overextended himself by overseeing another. The most salient sub-elements of this view are 1) that Cook had become complacent, perhaps careless or cruel, in his relations with Indigenous peoples; 2) that he lacked his customary professional detachment, resulting in a more fractious relationship with his crew; and 3) that he was not as geographically curious on this voyage as he had been during his first two expeditions. It is routinely observed that these presumptive failings prefigure his inevitable demise at Kealakekua Bay in February 1779.

The interpretive homogeneity applied to Cook's third expedition is a function of the oversized influence of John Cawte Beaglehole, editor of *The Journals of Captain James Cook.* Few historians have had such sway over a subject. His summative biography of Cook, which grew out of his editorial work, created such an indelible image that it has become difficult to see Cook outside of Beaglehole's lens. Historians Robin Fisher and Hugh Johnston asserted a generation ago that Beaglehole "dominated the field of Cook studies in a way that no individual now can or, perhaps, ought to do." In the introduction to their 1979 edited volume, *Captain James Cook and His Times,* they maintained that the best scholarship emanated from the South Pacific and that no figure exemplified "antipodean domination" more than Beaglehole, a native New Zealander. Fisher and Johnston's goal was to bring geographic balance to the interpretation of Cook's career. One of the contributors to the volume, Michael Hoare, confidently claimed that "the pendulum of Cook scholarship is moving back to Europe, to the north Pacific, its islands and coasts."[7]

Yet this shift never happened. What Fisher and Johnston could not have anticipated was the academic dust-up between Marshall Sahlins, Gananath Obeyesekere, and scholarly book reviewers that raged in the 1990s. The Cook-Lono debate – on whether Hawaiians treated Cook as a deity and how related circumstances precipitated his death – solidly reinforced Cook studies within the palm-tree paradigm. Although this literary intensity has ebbed in the quarter century since, one consequence endures: Cook's story in the icy latitudes is still relatively unknown.

Cook's fastidiousness as a navigator is oft remarked on, but one facet of his style has been overlooked – his fidelity to mission. His strict adherence to the strategic purpose of the third voyage is a probative example. Historians of the Pacific Northwest commonly disparage Cook's competence by noting that he missed the outfall of the Columbia River and the Straits of Juan de Fuca when he sailed up the Pacific Coast in 1778. But, as stipulated by Admiralty instructions, he was not to look for the Northwest Passage until he reached "the Latitude of 65°, or farther, if you are not obstructed by Lands or Ice." This specification had been informed by Samuel Hearne's terrestrial exploration northwest of Hudson Bay earlier that decade. Cook was cautioned "not to lose any time in exploring Rivers or Inlets" until he got to 65° N. Only then was he to search for those openings "as may appear to be of a considerable extent and pointing towards Hudsons or Baffins Bay."[8]

Cook scrupulously adhered to this guidance. But because he was occasionally out of sight of land, he never recorded those mid-latitude apertures, exposing himself to second-guessing by maritime fur traders who followed his track more minutely. Historians conventionally posit George Vancouver's expedition as a corrective to Cook's supposed inadequacies, but Vancouver's and Cook's missions differed. Vancouver was looking for a different version of the Northwest Passage – the one pelt merchants and hypothetical geographers had conjured in Cook's wake. Ironically, Cook's faithful adherence to the specifications of his third voyage – including avoidance of attractive nuisances such as rivers and inlets – caused his thoroughness to be called into question.

Stories of Cook's supposed nonfeasance along the Northwest Coast are a regional extension of Beaglehole's notion that the Cook of his first and second voyages would not have let slip the opportunities for exploration that the third afforded. Before reaching North America, Cook passed on chances to survey dozens of South Pacific islands. Many were mere reefs and sandy islets, but Beaglehole was shocked that even when it came to Samoa and Fiji, the great Cook was "content to enquire into them no further." Seeming to take Cook's alleged indifference to the southwest Pacific as a regional slight, Beaglehole then put forward his defining proposition: "Can there be any doubt that Cook on his second voyage, if he had heard of their existence ... would have been after them, fastened them down securely on his Pacific chart, even at the cost of minor disorganization to his time plan?" Beaglehole followed this suggestion with the most influential question ever asked about Cook's career and certainly about his execution of the third voyage: "Is it possible that, just as unsuspected strain on his mind was beginning to affect his attitude to the human situation, so, in relation to unexpected geographic possibilities, he was beginning to experience a certain tiredness?"[9]

In *Cook: The Extraordinary Voyages of Captain Cook*, Nicholas Thomas highlights that this single rhetorical question led to the conventional view that Cook should have quit after his first two expeditions. From its careful, tentative birth in Beaglehole's introduction to the journals of the third voyage, the notion that Cook was experiencing fatigue became the fundamental premise for understanding his last expedition. The idea was especially favoured and expanded on, Thomas argues, by postcolonial authors whose allegiance lay with the aggrieved Indigenous peoples whom Cook visited. In these historians' hands, Beaglehole's merely fatigued Cook becomes a violent and irrational man whose compromised judgment led to his death.[10]

Beaglehole contended that the third voyage differed from the others in the obvious sense of geographic scope but more critically, if elusively,

in feeling. Like most hypotheses, Beaglehole said his could be contro-
verted, but no one has tried. Books published this century still habitually
posit the axiomatic James Cook – that is, the diminished-third-voyage-
explorer trope stipulating that his "behaviour had shifted significantly"
or that he was acting "out of character." Usually, such assertions have a
teleological purpose; one author noted that Cook "had become a tired
and sick man, and his condition may have contributed to his death."[11]

Most authors treating Cook's career follow the narrative convention
of disentangling his life in chronological order. This book is sequential,
too, but it deviates from the norm in not privileging Cook's first voyage
(briefly treated in Part 1) or, more generally, the time he spent in the
tropics during all three voyages. In these pages, the emphasis is on his
second and third voyages, particularly in the icy latitudes. Though he has
been cemented in the popular and scholarly imaginations within the
tropics, Cook was a polar explorer of the first rank. Even less appreciated
is that he was a pioneering ice scientist. In the Arctic, that honour is
sometimes bestowed on William Scoresby – a whaler who studied the
natural history of the region, including sea-ice formation – based on a
paper he delivered at a scientific meeting in Scotland in 1815. Others
credit the better-known Fridjof Nansen, whose ice-embedded voyage in
the *Fram* (1893–96) gained worldwide attention. James Eights, the naturalist aboard Nathaniel Palmer's 1829 sealing and exploratory voyage is
often acclaimed as the first Antarctic scientist. Turning the palm-tree
paradigm on its head, I argue that James Cook and Johann Forster, chief
naturalist on the second voyage's circumnavigation of Antarctica, were
the true originators of polar climatology.

Any discussion of Cook in the icy latitudes must take into account the
prevailing theory that deep saltwater did not freeze. Cook's contemporaries believed that icebergs and packed ice were frozen masses that had
emanated from rivers. This was an ancient idea, popularized in Cook's
time by Daines Barrington, a member of the Royal Society with connections
to the British Admiralty (though the foremost contemporary theoretician
was the Swiss bibliophile Samuel Engel). The now preposterous notion
that seawater did not freeze fed a corollary proposition almost more incredible to modern sensibilities – that the North Pole was altogether free
of ice because no land was thought to be proximate to it. As shown in
Part 2, the great masses of ice that Cook and Forster encountered while
criss-crossing the Southern Hemisphere's empty high latitudes, juxtaposed
with the shrinking size of any putative southern continent at or near the
South Pole, informed their skepticism of reigning glaciological theory.

Cook then refined their scientific breakthroughs during his subsequent voyage to the Arctic. Our modern understanding of polar hydrology owes much to his observations.

Another common practice in Cook historiography since Beaglehole has been to view the alleged shortcomings of the third voyage through the gauzy lens that was turned on the first two – in other words, to emphasize supposed deviations from a previously exemplary pattern. I challenge that perspective by documenting the consistency of Cook's deportment across all three voyages. In doing so, I highlight activities on the earlier voyages that are typically unimpeached in the Cook literature but would not be if they had occurred during the third. Cook's last expedition is usually characterized as an anticlimactic quest for the Northwest Passage, as a mere prologue to his undoing in Hawaii. Here, I invert that model, for, if studied within the context of Cook's mission and not his death, the northern voyage was the most ambitious and consequential in terms of geographic comprehension.

In Part 3, I argue that Cook the navigator and geographic problem solver was as conscientious during the third voyage as he was during the first two. I present evidence that controverts the common supposition that Cook's abilities had been stretched too far by analyzing his time in the high northern latitudes on its own terms, not as an extension of the southern voyages nor as an interlude before his inevitable death in Hawaii. Cook was always conscious of the true mission of the final voyage, even if some of his shipboard contemporaries, and many modern authors, fault the way he executed it. After he completed the second expedition circumnavigating Antarctica, Cook considered himself "done" with the (south) Pacific.[12] Accordingly, as is documented in Chapter 8, prior to striking out for North America's Pacific Coast, Cook had no intention of making discoveries in Polynesia. The region was merely a staging area for the sail north.

Cook's Arctic campaign reached its crescendo in August 1778, when, off the Alaskan coast at 70° 44' N, he was stymied by a wall of ice twelve feet high. This was as far north as he would get, not quite matching the southern extremity reached on the second voyage. At his northern apex, twenty-five months after the expedition's launch, with cold and fatigue settling into the bodies and minds of his crew, Cook diligently probed westward along the ice edge for eleven more gruelling days. He exhausted every prospect for an opening through or around the ice pack and rarely had a clear view of his surroundings because of the Arctic fog. He relied on navigational guidance from the incessant barking of the walruses abounding on the ice edge. Contrary to the tired-voyager hypothesis, it was the most vigorous sailing of his career, a mere six months before his death.

When Cook left Alaska in October 1778, the expedition, according to the original timeline, should have been coming to a close. But having come so far and unsatisfied with his first attempt, he announced a plan to extend the voyage into an unprecedented fourth year. Cook had so thoroughly inculcated a culture of diligent exploration and fidelity to mission that even after his death the expedition's demoralized crew, now commanded by Charles Clerke, returned to the Arctic in the summer of 1779. Most books treat Clerke's return and subsequent events in China as an afterthought; many ignore it completely. To an extent, this is to be expected; a biographical portrait can only extend to the duration of a subject's life. But this tendency need not apply to the history of an expedition, as opposed to a man. The interpretive pattern that presumes a supposedly lesser figure such as Clerke does not merit much attention has damaged our understanding of Cook's final voyage and its relationship to Arctic environmental history. In Part 4, we see an expedition still guided by Cook's logic model and ethos. Even after the second fruitless attempt in the Arctic, and Clerke's own death shortly thereafter, the surviving leadership team dedicated itself to making further contributions to Europe's understanding of East Asian geography. On its way home, Cook's expedition inadvertently seeded the maritime fur trade along the Northwest Coast, the one aspect of Cook's execution of the third voyage for which historians have given him more credit than he deserves. This mercantile development spurred a new vision for the Northwest Passage, one that culminated in the clarity that George Vancouver brought to regional geography.

Cook's final voyage was not a continuation of his earlier expeditions in the South Pacific, nor a fatal mistake, but a crowning navigational achievement. More largely put, by emphasizing Cook's work in the icy latitudes, where he spent more time under sail than in the tropical zones to which he is usually consigned (of necessity by anthropologists; for historians, by their choice), we can discover a new Captain Cook. In the twenty-first century, an age whose hallmark will be massive climate change, perhaps it is time to acknowledge that the environmental backdrop for a newly relevant Cook is not a warm sandy beach, nor even the ocean blue, but a cool summer along that Alaska coastline that leads to the Arctic ice pack.

PART ONE *Prequels*

ONE

The North Sea and Canada

Little in James Cook's humble beginnings suggested a momentous life was in store for him. Born the son of a day labourer in 1728, Cook was raised on Aireyholme Farm in the Yorkshire township of Great Ayton, where his yeoman father was overseer for Thomas Skottowe, lord of the manor. Young James probably helped on the farm by working in the fields or tending to animals. Though no one could have predicted a nautical application of these agrarian experiences, during the course of his voyages Cook occasionally supplemented his ships' provisions by procuring fresh grasses and other wild plants and berries to fend off scurvy. Animal husbandry was another important part of these expeditions, as goats and cattle were transported to the South Pacific as gifts from himself and the King.

Cook attended school in Great Ayton, where he learned the rudiments of letters and numbers, skills that facilitated his escape from the terrestrial clutches of rural existence. This separation commenced at age seventeen, when he was introduced to the worlds of commerce and maritime life in neighbouring Staithes, a fishing village that he moved to in 1745 to work as an assistant in William Sanderson's general store. A bigger step came a year later, when he started a three-year apprenticeship in the busy Yorkshire port city of Whitby under Captain John Walker, a ship owner and a major figure in marine affairs.

Whitby today, in the estimation of local historian Sophie Forgan, is "a charming town, rather off the beaten track, beloved of tourists ... but it is a town whose connection with the sea is largely that of leisure."[1] But in the mid-eighteenth century, Whitby was a prominent centre for ship-building, notably colliers, which were essential to the coastal coal-carrying trade from nearby Newcastle to London. Some four hundred ships made this run over the course of a year. They typically made ten voyages a season and in some years transported 30 percent of London's coal supply. Whitby's commercial fleet carried other cargoes and served other ports. For example, up to 50 percent of the timber imported from Norway passed over

Whitby's docks, and 20 percent of Britain's general trade from Baltic ports did likewise.

As a trainee, Cook started as Walker's servant, running errands and stocking ships with food and chandlery supplies. These duties were an extension of his work for Sanderson's store in Staithes, but the expectation inherent in his apprenticeship was that by its end he would become a sailor. Few internships have been more productive. Cook was exposed to shipboard responsibilities through and beyond his apprenticeship for a total of nine years employment aboard Walker's vessels. This service laid the foundation for his future greatness in pilotage and ship management. During the course of commercial voyages in the North and Baltic Seas, Cook learned key maritime skills, such as casting a lead line to measure depth and ascertain the composition of the sea bottom (to get a read on shifting sands, submerged rocks, or shoal water). He also mastered observation of shorelines and the movement of waves, currents, and tides.

Cook also learned how to read a ship's speed via a log line, a piece of wood tied to a rope of fixed length that floats behind the ship. Using a sandglass, he'd determine the time it took for the line to unwind from its spool, which would yield the rate. Of greater importance was being able to determine latitude by using a quadrant to measure the sun's height above the horizon. The most difficult navigational skill to master in this era was calculating longitude, an aptitude less important to Cook early in his career but a major theme during his Pacific voyages. Central to the high-latitude orientation of this book, Cook probably had his first experience sailing near sea ice as a master's mate conducting Baltic voyages to such harbours as Riga.

Walker had less tangible influences on Cook's life and professional outlook. The Society of Friends, or Quakers, were a major presence in Whitby, and Walker was one of their number. Cook boarded with Walker's family for the duration of his time with the company, and though he never became a Quaker, a case has been made that some of the central tenets of Quakerism – moderation in all things, self-discipline bordering on austerity, work as an end in itself, an aversion to violence, and the stewardship of interpersonal relationships through fair treatment and care for colleagues – are reflected in the broad pattern of Cook's career.[2] This line of thought extends even to Cook's path-breaking open-mindedness toward Indigenous peoples and their customs, though this temperament was influenced more by the Enlightenment's intellectual currents.

Cook was fortunate to have many mentors, but the near decade he spent under Walker's tutelage was formative and long-lasting. Cook wrote

Walker after returning from his first two circumnavigations, and he once visited him in Whitby. The letters were prosaic in some respects, describing his ships' state of repair and alluding to completed or projected itineraries. But beyond those technical facets, the correspondence is important because after Cook died his wife destroyed all their correspondence. In Forgan's phrasing, "Cook unbent just a little to Walker," and those missives provide the few interior glimpses we have of Cook's state of mind at pivotal junctures in his career.[3]

By 1755, Cook was in line for a promotion to sailing master, the senior petty officer responsible for (under a captain's oversight) the tactical guidance of a vessel's course and sail arrangement. Instead of taking this advancement, Cook forsook nine years seniority as a merchantman under Walker and enlisted as an able-bodied seaman, or AB, in the Royal Navy. An AB could become a petty officer, or reach higher noncommissioned ranks, as Cook's eventual ascension proved. Still, historians have puzzled over his decision to enlist. The most plausible explanations are a patriotic impulse (war broke out with France in 1754) or simple wanderlust. The first seems more convincing since Cook could have sought employment in any of the far-flung Crown-charted trading enterprises, such as the East India Company. Cook's less than lateral move certainly does not appear to be pecuniary repositioning. Nor was Cook avoiding a press gang: as sailing master for a merchantman, he was exempt from induction into the navy.

⚓

War often catapults relatively unknown individuals to the forefront of history, and so it was with Cook. The preliminary skirmishes of the Seven Years' War between England and France occurred in 1754–55, though most of the combat occurred between 1756 and 1763, thus the name. A global conflict, in North American historiography the confrontation is commonly called the French and Indian War. Cook's involvement in the hostilities laid the foundation for his distinguished career in the Royal Navy. The Cook we know from history – the great navigator and cartographer – was formed there. In the words of writer and novelist Victor Suthren, "Canada and its waters were the crucible within which the materials of a promising but as yet undistinguished naval warrant officer were shaped into the form of the naval captain, surveyor, and cartographer that would amount to greatness."[4] Cook saw action in the logistical support of naval warfare on which Britain's success depended. His best-known role was sounding the rapids – known as the Traverse – downriver from Quebec, the site of the pivotal battle in 1759 for the control of North America. War accelerated Cook's professional growth, and by the conflict's end, his proficiency in surveying and cartography had come to the attention of senior Admiralty officials in London.

On enlistment, Cook joined *Eagle,* a sixty-gun warship. Within a month, he was promoted to master's mate, the rank he held when leaving Walker's merchant service. *Eagle* patrolled home waters looking for French marauders. When the ship took on a new captain, Hugh Palliser, another important figure came to assist Cook's maritime maturation. Palliser saw potential in Cook and made sure he received extra training in chart making. For the next two years, *Eagle's* squadron guarded the English Channel, seeing serious action twice. Among the battle casualties in May 1757 was *Eagle's* master. Palliser promoted Cook to the position, the highest rank possible for a noncommissioned officer, and it came with pay equal to a lieutenant's. Thus, two short years after turning down the promotion Walker offered, Cook achieved a comparable rating in the Royal Navy. He was now responsible for the ship's state of supply, including munitions, plus pilotage (setting courses and arranging sails and rigging), duties the master's position shared with commercial service. That summer, Cook was assigned to *Solebay,* a frigate patrolling for smugglers off the east coast of Scotland. In September 1757, he joined *Pembroke,* captained by John Simcoe, the next mentor in line. As the senior noncommissioned officer on *Pembroke,* Cook came to Canada, an experience that transformed his career.

British war planners in London and Halifax knew that victory in North America hinged on displacing the French from their two great fortresses: Louisbourg on Cape Breton Island, which protected the mouth of the St. Lawrence River, and Quebec, the colonial capital located high above a narrowing of the river and the key to the defence of the interior empire drained by the Great Lakes. Quebec sits on estuarial waters hundreds of miles inland, and the river below the city is so wide that to this day there are no bridges spanning its width. Ground forces would fight the battle, but the campaign's success hinged on the Royal Navy getting the army in the right place. *Pembroke's* squadron of eight ships sailed for North America in February 1758, part of a larger fleet numbering 157 that would attack New France. This was the first long voyage of Cook's career, and it was on this passage that he first witnessed the effects of scurvy. *Pembroke* lost twenty-six men during the crossing.

Louisbourg capitulated on July 26, 1758. The very next day, as the clearing of wreckage in the harbour commenced, the most significant personal encounter in Cook's professional life occurred. At Kennington Cove, southwest of Louisbourg, Cook's curiosity was aroused when he saw Samuel Holland conducting a terrestrial survey of the conquered domain. Holland had charted Fort Ticonderoga in the Hudson River–Lake Champlain theatre the previous year and was now attached to the Atlantic invasion force of Major-General James Wolfe, the eventual hero of Quebec. After securing Simcoe's permission, Cook asked Holland to teach him the science of surveying and map-making. Cook had learned the fundamentals of seamanship in the North Sea coal trade, but the skill that would distinguish his career was the ability to conduct triangulated surveys, a specialized craft he learned from Holland. The Cook brand was born here. Watching Holland at work was a cognitive opening for him. Over a few days, including some after-hours tutorials aboard *Pembroke* at night, Cook acquired the expertise that would make him the paragon of scientific exploration. Years later, Holland told Simcoe's son, John Graves Simcoe, the future founder of York (Toronto) and Upper Canada (Ontario) altogether, that before departing England on his final voyage, Cook had confessed that the training aboard *Pembroke* "had been the sole foundation of the services he had been enabled to perform."[5]

Surveying – diagramming disparate topographic features with precision – was more commonly practised by army men such as Holland (or George Washington) than by navigators. The core of the practice as Cook learned it involved a plane table; a flat, square surface stabilized by a tripod. The table held a sheet of paper above which the surveyor observed distant stations through a rotating alidade, basically, a brass telescope mounted on a protractor that indicated the bearing of the viewed object, which was flagged using bright coloured cloth. With a straight edge, a line was drawn on the incipient chart in the direction of the flag. After a minimum of three points had been established and lines drawn to them from the point of observation, trigonometry was applied to the problem. By subtending, or turning, the angles of a triangle, it is possible to measure the distances on all three sides, provided the length of one side of the triangle, the baseline, is known. This was accomplished by moving the table (Cook used a more mobile theodolite for the balance of his career) a gauged distance, which was computed by using fixed-length rods or chains, early versions of modern tape measures. At the next station, a new apex, or set of bearings taken on flags planted at distant points, was established. Replicated many times over, a kind of geometrical network was created.

The key point about Holland's lessons was the mathematic exactitude triangulation could bring to chart making on scales both large and small.

If you could scientifically plot Kennington Cove, you could theoretically map the entire world with similar precision. By the time of his death, Cook had made a substantial contribution to that very project, single-handedly overseeing the production of charts for the Pacific basin and contributing one-third of the content that led to our normative understanding of the globe's geography. Holland's technique could be and was applied many times by Cook (or men under his direction) in the fashion he first saw demonstrated – by drafting the outline of a harbour and its coastline from a set of land-based vantages.

On water, though, establishing a baseline was procedurally problematic and literally impossible in a physical sense, so an adaptation needed to be made to approach cartographic precision. This method was called a running survey. Cook did not perfect this practice until his second voyage, when it became possible to calculate longitude with ease and accuracy through the use of chronometers that kept Greenwich Time. A baseline was established astronomically by fixing the ship's starting latitude and longitude then pulling up anchor and moving to a new station to discern the difference in latitude and longitude between the two points. Any variation measured by minutes of a degree could be translated into feet, yards, or miles. Best suited for long open shores (such as the Oregon Coast, surveyed during the third voyage), the running survey yielded generally reliable results but never the exactitude of terrestrial surveying because of the vagaries of fixing latitude or longitude at sea, including wave and tidal action on a ship's movement. Since Cook had the rare talent of being able to conduct both land and seaborne surveys, the range of his cartographic talent, from depictions of a harbour in a remote Pacific isle to the whole continental coastline of North America, was unusually wide.

Cook's career simply cannot be understood without an appreciation of his encounter with Holland. Productive outcomes were not long in coming. After Louisbourg, Cook charted the tip of Gaspé Peninsula when his squadron sailed into the Gulf of St. Lawrence in September 1758. Cook's chart, "Bay and Harbour of Gaspee in the Gulf of St. Lawrence," the first of many, would be published in 1759. When the gales of November arrived, *Pembroke*'s unit beat it back to theatre headquarters at Halifax for overwintering in advance of the full fleet's assault on the Citadel at Quebec the next year. Cook had a significant role in these preparations. During the winter of 1758–59, he and Holland arranged provisional charts of the St. Lawrence River patched together from printed sources, a variety of materials captured at Louisbourg, and the feint into the gulf, which scoped the mouth of the river. Their work paid dividends during the forthcoming campaign. Historian D. Peter MacLeod calls this chart "the single deadliest weapon deployed by the British-American forces in the course of the campaign" because "it allowed the invasion fleet to breach Canada's outer defenses and ascend the St. Lawrence without delay and without loss."[6] Cook and Holland made alterations to their draft on the upriver passage to Quebec. Following victory over the French, the chart was finalized in Halifax in spring 1760 and printed later that year in London.

Simcoe encouraged Cook's training in the navigational arts and provided tutelage during the long Halifax winter of 1758–59. For example, Cook prepared his first set of sailing directions for the conquered port of Louisbourg. This nautical form provided narrative guidance on course and distance into a harbour, depth soundings, the location of rocks and other hazards, tidal patterns, variation of compass for taking proper headings and, of course, latitude and longitude. Though Cook's sailing guides were not as well publicized as his maps, he prepared dozens for distant Pacific harbours, and they were incorporated into the Admiralty's accounts of his voyages. But Simcoe's coaching ended when he died during the great fleet's passage to Quebec and was buried at sea.

The definitive campaign up the St. Lawrence commenced on May 5, 1759. Led by Rear-Admiral Philip Durrell, the intended mission of the first squadron, which included *Pembroke* and Cook, as her master, was to arrive on the St. Lawrence early in the navigational season so as to forestall the French fleet's ability to resupply and reinforce the colony. Characteristic of springtime in the North Atlantic, *Pembroke* was forced to manoeuvre through ice floes off the Nova Scotian coast, an encounter with a natural phenomenon that would later become routine in Cook's career. Durrell's squadron sailed into the river behind the last French convoy to make it to Quebec.

Presaging future events, this fleet was led by a rising star in France's army and intellectual life, the twenty-nine-year-old Louis-Antoine de Bougainville. A Parisian of seminoble birth with connections to the Bourbon court, Bougainville had been versed in the classics and mathematics at the University of Paris. He was also conversant in English, which resulted in his being appointed secretary to the French ambassador in London in 1754. That year, he published a treatise on calculus that brought him such intellectual renown he was elected a fellow of the Royal Society of London, Britain's premier learned organization. Bougainville rose through the echelons of the French army as quickly as Cook would in the Royal Navy. In 1756, he was promoted to captain and assigned to New France as aide-de-camp to the marquis de Montcalm, the French general at Quebec. He later conveyed dispatches from Montcalm to Paris, where he was promoted to colonel in 1758. Bougainville, like Cook, would figure prominently in the denouement of the battle for Quebec. So far as we know, the two never met after the French surrender, but in a fashion their paths did cross in Canada, just as they would the following decade in the middle of the Pacific Ocean.

The arrival of Bougainville's convoy in Quebec on May 10 was the very thing Durrell was supposed to prevent, but the British faced a more immediate problem. The river's key feature was a passage beneath the Quebec basin that the fortress overlooked: the Traverse. Influenced by the redirection of currents around the Île d'Orléans, a twenty-one-mile-long island below the Citadel's bluffs, the river's channel here crosses from the north side of the waterway to the south, thus the name. Subject to tidal influence as well as strong river currents, it was a treacherous passage for sailing ships. As J.C. Beaglehole noted, the course of this channel was marked inadequately even in peacetime since "the French pilots knew their business."[7] Those few navigation marks had now all been removed in anticipation of a British invasion. The French strategy was to mystify the river's features. They complacently thought that without proper soundings a British fleet could never gain access to the basin at the foot of their capital.

By June 8, Durrell's squadron had been joined by the entire British fleet. The events of the ensuing seventeen days, in which Cook played a vital role, dictated the future outcome of British victory and French capitulation. The key to British success was re-establishing a system of markers delineating the zigzagging channel. Cook led this effort and was joined by the masters and crewmembers of his own and three other ships. Using longboats, the depth sounding took two days, from June 9 to June 10. The process involved casting lead-weighted lines marked out in fathoms. Underwater rocks and ledges were recorded on charts, but the channel itself was marked with new buoys that replaced those the French had pulled up. On June 11, Durrell recorded that Cook had "returned satisfied with being acquainted with ye Channel."[8]

With the Traverse's course now re-established, *Pembroke* and other ships in Durrell's squadron worked their way upstream directly below Quebec, establishing the route for the rest of the fleet to follow. They did this by mooring their longboats (which flew colour-differentiated flags) to the buoys Cook's team had placed in the river. By June 27, the entire British fleet – including ships of the line, frigates, and transports carrying General Wolfe's troops and supplies – had passed through the channel, a parade concluded by *Neptune*, the flagship of fleet commander Vice-Admiral Charles Saunders. With the two squadrons combined, Saunders now commanded 49 warships plus 119 supply vessels – one-quarter of the entire Royal Navy – that carried twenty thousand sailors and marines, including servicemen from the American colonies. The stage was set for what would prove to be Wolfe's decisive victory over Montcalm on the Plains of Abraham on the fortress's western flank. Suthren states: "Wolfe had been provided the opportunity he sought for conquest, through a consummate demonstration of the mariner's art."[9] This was the apex of Cook's role in the Quebec campaign.

Pembroke joined other ships in ferrying cannons and troops across a relative narrowing of the river and ashore at Pointe Lévis on the south bank of the St. Lawrence opposite the French fortress. This was complicated work in the age of sail and on a river subject to fast tides and strong currents, especially in the season of the freshet, made more difficult by having to dodge musket balls and French cannon shot. The strategic objective was neutralizing the natural superiority of the French position by providing Saunders cover for his ships; they'd provide opposing cannon fire when the ships passed upstream to outflank the fort. The first of these movements took place on the night of July 18, but as the summer wore on, Wolfe deliberated over the prospect of attacking Quebec six miles downstream in the vicinity of Beauport, on the north shore of the St. Lawrence near the falls of the Montmorency River. On the basis of his successful leadership of the Traverse's survey, Cook led a team of surveyors that sounded the Montmorency shore. Wolfe placed great confidence in Cook, telling an adjutant that "the Master of the Pembroke assures the Admiral [Saunders]" that flat-bottomed North Sea "cats" or colliers "can go within less than 100 yards" of the shore.[10]

The Montmorency operation commenced on July 31, but the landing was compromised because British attack vessels grounded farther out than the one hundred yards Cook had predicted, slowing Wolfe's forces as they waded ashore. In the delay and confusion, abetted by the heat of the day and a thunderstorm, Montcalm amassed the French army around Beauport. Wolfe waved off the attack. An even greater disaster was averted when the incoming tide allowed British ships, including Cook's *Pembroke*, to get closer to shore and to the grenadiers who needed rescuing. Assessing whether Cook deserved a share of the blame for this "debacle," Suthren says an explanation for the botched landing might be found in the unreliability of "sounding under attack" and the possibility that "tidal tables were ignored."[11] Cook's reputation survived this setback, which became lost in the haze of both Wolfe's and his own subsequent success.

Wolfe concluded after Montmorency that his only chance to defeat Montcalm would be executing the original plan: landing above Quebec and attacking the fortress from the west. Over the course of August and early September, men and materiel were ferried upstream, proving again that Wolfe's campaign was as much a nautical affair as an army operation. When the final British assault was launched, Cook was far removed from the main action. *Pembroke* was part of a diversionary effort in the Beauport sector, where, based on a combination of wishful thinking, overconfidence, and the earlier skirmish, Montcalm expected the main attack. Instead, the British effected a surprise amphibious night landing at Anse-au-Foulon, a mile and a half upriver from Quebec, in the early morning of September 13, 1759. Samuel Holland, Cook's tutor and Wolfe's confidant, played a central role in this decisive battle, having been positioned across the river to keep an eye on French preparations in that sector.

With a superiority in arms and men and having obviated the fortress's strategic advantage by scaling the bluffs to its rear, Wolfe quickly prevailed over Montcalm in a battle superseded in North American history only by the Union victory at Gettysburg. During the clash, both the British and French commanders received fatal wounds. Indeed, it was Holland who found the mortally injured Wolfe and carried him away from the skirmish line. Wolfe's battlefield death was recorded for posterity in a famous painting by the American artist Benjamin West. Though not included in West's depiction, Holland was among the few actually with Wolfe when he drew his last breath.

Cook might have been a distant observer of the battle, but Bougainville's conduct was central to the disastrous turn of events for France. Montcalm had placed him in charge of a mobile regiment that would shadow British movements on the river above the city and oppose any landing west of Quebec's walls. Bougainville met this objective for several weeks, ranging back and forth as far as ten miles upstream from the fortress to Cap Rouge, where the British lay at anchor. But it is easier to move an army on ships than it is on land. Exhausted by numerous British feints up and down the river over the first two weeks of September, Bougainville's footsore army was, as Suthren phrased it, "nowhere to be found when the thing it had been meant to oppose took place."[12]

The first in a series of French tactical errors, all of which rested on Bougainville, was the failure to notify guard posts along the river that

Bougainville had cancelled a flotilla of small boats with supplies from Montreal intended for the beleaguered French defenders in the fortress of Quebec. From their spies, the British knew that the French intended to replenish the Citadel, so the plan was to pass themselves off as French while riding the tide in their longboats. Thus, when the British landing craft fell down the river in the dark early morning of September 13, no alarms were set off. Tragically for the French, just a few days before Wolfe had settled on scaling the bluffs at Anse-au-Foulon, Bougainville had been stationed at that very spot. Historian C.P. Stacey observed, "100 alert and determined men at the Foulon could have brought Wolfe's scheme to ruin," and Bougainville would have been the saviour of New France. Bougainville compounded this catastrophe by remaining at Cap Rouge after noticing that the British squadron had sailed downstream. Even the noise from the battle on Abraham's farm failed to set him in motion; Montcalm waited in vain. Bougainville arrived after the battle was over and withdrew quickly in the face of the victorious British forces. In Stacey's estimation, Bougainville's "military experience was scarcely equal to the high rank he held and the heavy responsibilities that rested upon him"; "his inefficiency had much to do with the French disaster."[13] Bougainville never publicly acknowledged his share of the blame for the defeat, but with experience he became an accomplished sailor, first by advancing the Enlightenment in the Pacific and then by gaining an element of revenge against the British navy in the Battle of the Cheseapeake, which foreshadowed Cornwallis's fate at Yorktown.

After the French surrender, Cook was transferred to *Northumberland*, part of a squadron that returned to Halifax, while Saunders and much of his command sailed to Great Britain victorious. Over the ensuing winter of 1759–60, Cook finished the definitive chart of the St. Lawrence River that he had begun with Holland. Measuring three by seven feet, in twelve sheets forming the composite view, this chart was printed in London the following summer and complemented the map of Gaspé harbour and the Gulf of St. Lawrence that had been published the previous year. The thirty-two-year-old Cook had risen from obscurity in the Yorkshire countryside and the coal trade to have two superior maps in public circulation. These charts formed a template, oft repeated in the years ahead, marrying "the landward observation methods of the military engineer with the sounding and coastal fixing of the seaman."[14]

Another important but underappreciated aspect of Cook's Canadian career is that we know for certain that it was here that he gained sustained exposure to the defining feature of high-latitude voyaging: sea ice. The

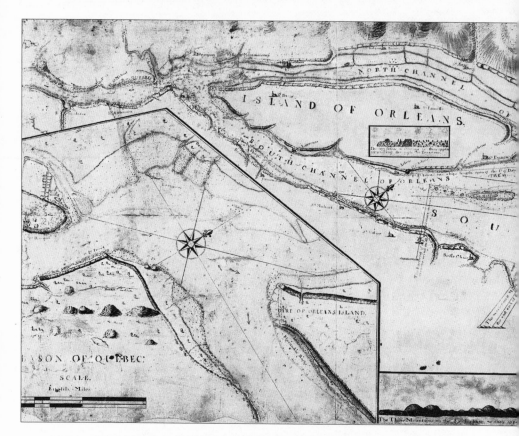

James Cook, *A Plan of the St. Laurence from Green Island to Cape Carrouge,* 1759–61.
This chart is the westernmost section of a segmented map that Cook and Samuel
Holland prepared during the Quebec campaign. Note the volume of navigational
detail depicting the strategic basin below the Citadel and Cook's soundings of the
Traverse downstream from the Île d'Orléans.

North American squadron left Halifax on April 22, 1760, rushing back to
Quebec to relieve a siege of the garrison by French reinforcements from
Montreal. Leaving two weeks earlier than the year before, this time they
met a vast amount of ice. Indeed, Cook reported that on the second night
at sea, *Northumberland* was "fast in the ice." In his report, the ship's captain,
Lord Alexander Colville, wrote of anchoring off Quebec "after a most
tedious and troublesome passage, being almost continually impeded, by
running amongst great Quantities of loose Ice, and confused by thick
fogs."[15] This event perfectly anticipated Cook's longer and more dangerous
forays along the edges of the Antarctic and Arctic ice packs.

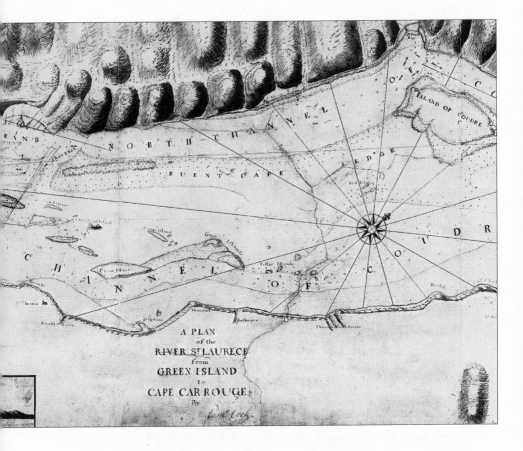

After relieving Quebec, the larger portion of the British fleet, but not including *Northumberland,* sailed upriver for the final mop-up action against the French in Montreal, effectively ending the war. Cook spent an uneventful summer anchored in the Quebec basin. When *Northumberland* sailed for Halifax in October 1760, Cook left the St. Lawrence for the last time. Halifax Harbour was his home for the next two years. He supervised the maintenance of the ship and continued to perfect his craft as a cartographer by working on his Nova Scotian charts. His diligence and fruitfulness was notable; indeed, he received a bonus payment of £50 at the direction of Lord Colville, who wrote that this sum was "in consideration of his indefatigable Industry in making himself Master of the Pilotage of the River Saint Lawrence."[16] "Indefatigable industry" proved to be Cook's emblematic characteristic.

Of his home port, Cook wrote: "The harbor of Halifax is without doubt one of the best in America sufficiently large to hold all the Navy of England with great safety. Both its in and outlet is very easy and open in the most severest frosts."[17] Cook's exposure to North Atlantic climate proved to be good preparation for the balance of his career because routine exposure to seasonal cold and ice helped forge his later resolve when making probes at the ends of the world. Among other things, he learned the great value of proper cold-weather gear for seamen and that precautions needed to be taken against weaker crewmembers being strong-armed out of their coats or the weak-minded selling them for alcohol.

Cook's quiet but productive phase in Halifax lasted until France concocted the last-ditch strategy of seizing the British port of St. John's on Newfoundland. This had less to do with reversing fortunes on the mainland than securing a bargaining chip in the final settlement with Britain. The French wanted to strengthen their hand in negotiations over access to the rich fishing grounds off the coasts of Newfoundland, and they took lightly defended St. John's in June 1762. Another future explorer, Jean-François de Galaup, comte de La Pérouse, who would follow Cook's wake into the Pacific in the 1780s, was a part of this command. When news of the attack reached Halifax, Colville reassembled a squadron from various stations in Nova Scotia. The counterattacking flotilla, which included Cook as sailing master aboard the flagship *Northumberland,* circled the opening to St. John's Harbour on September 12, 1762. The French ships were enclosed within it, and a British armed regiment was landed, but when a gale blew *Northumberland* and her sister ships out to sea, La Pérouse and his fellow officers sailed into the Atlantic, avoiding capture. The now doomed French garrison was quickly routed on September 18. With St. John's reversion to British control, *Northumberland* sailed in the next day, giving Cook, in Suthren's words, his first view of "the steep-sided, rocky harbour that he would come to know well over the next five years."[18]

The St. John's campaign served as Cook's introduction to the place that would lead to his postwar breakthrough – the Newfoundland survey. His immediate task was development of a chart of St. John's and adjoining harbours on the Avalon Peninsula, a task that customarily fell to masters of warships. These anchorages were deemed essential to potential defence needs should the French return. They were also considered central to the fishery. This work would be later incorporated into Cook's master map of Newfoundland. Instead of returning to Halifax, the *Northumberland* sailed for England on October 7, arriving home the next month. Cook had been away for more than four and a half years. Given the great length of Cook's Pacific voyages, his power of endurance was tested and proved early. His tour of duty to Canada during the Seven

Years' War was the longest of his career. Reaching England, Cook was paid, and he and the rest of *Northumberland*'s crew were dispersed. Shortly thereafter he married Elizabeth Batts.

⚓

Peacetime limited the Royal Navy's need for sailors, but Cook's emergent skill matched up well with one immediate postwar necessity – incorporating the vast extent of territory acquired from France into British cognizance through cartography and resource inventories. Decades before the British Hydrographic Office was established in 1795, surveys of coastal areas and interior regions were commissioned by the British Board of Trade, the Admiralty, and the War Office. Within this encompassing need for geographic comprehension, one area needed special attention. In the peace treaty signed in February 1763, French diplomats retained the right to fish off the Newfoundland shore. This zone ran from Cape Bonavista to the northern tip of the island and then around it on the Gulf of St. Lawrence side as far as Pointe Riche. To shelter her fishermen, France reserved the islands of St. Pierre and Miquelon southeast of Newfoundland (and maintains sovereign control to this day). The British were agreeable provided the French kept strictly within these limits. The problem for British enforcement was that there were no reliable maps of the infamously complex Newfoundland coastline. Cook was the solution. His new assignment was rooted in a letter written on December 30, 1762, by Lord Colville, his former captain and now admiral of the North American squadron. Reminding the Admiralty of Cook's previous work on the St. Lawrence and Nova Scotia coast and citing his own experience with "Mr Cook's Genius and Capacity," Colville thought him "well qualified ... for greater Undertakings of the same kind."[19]

With that recommendation, Cook was commissioned in March 1763 to survey Newfoundland, the world's sixteenth largest island. Now under the command of the new military governor, Captain Thomas Graves, Cook sailed aboard *Antelope* for St. John's on May 15. Graves wanted St. Pierre and Miquelon charted before they were turned over to France. On June 13, Cook and several assistants were dispatched on *Tweed*, captained by Charles Douglas, to conduct this insular survey, which concluded on July 31. This was fast-paced work performed while anxious French officials awaited its completion. It was finished so quickly, said Douglas, because of "the unwearied assiduity of Mr. Cooke."[20]

Tweed proceeded to St. John's, where Cook learned that Graves had secured an American-built schooner, renamed *Grenville* after the prime minister, for his exclusive use as a survey vessel. This was a savvy move. Newfoundland's complex latticework of headlands and harbours had heretofore thwarted precise measurement even though it had been much visited for centuries by the seamen of many nations who sailed to its rocky shores in the wake of the Vikings. The island's intricate coastline thus became for Cook the perfect laboratory in which to perfect the trademark characteristics of his later career: perseverance and exactitude. Cook's eventual delineation of Newfoundland's outline would result in one of his greatest cartographic accomplishments. Within a week of her purchase, *Grenville* was under way, with Cook serving as sailing master and commanding officer, toward the northern tip of Newfoundland. With his twenty-man team of sailors and cartographic assistants, Cook started the survey of the island proper near the famous Viking settlement site of L'Anse aux Meadows. From there, he proceeded across the Strait of Belle Isle to York Harbour and Chateau Bay on the Labrador Coast. It was the only visit to that shore and, at 52° N, the northernmost point reached during his multiyear survey.

By early October, *Grenville* was back in St. John's, where over the course of the next five weeks Cook and his draftsmen started finalizing the charts and sailing directions for the northern harbours and the French islands. Graves was impressed with the work. He told the Admiralty that Cook's "pains and attention are beyond my description." Graves added that he was sending Cook home for the winter so that he would "have the more time to finish the Surveys already taken." He was confident the Admiralty's board would readily perceive "how extreamly erroneous" the previous maps were, adding "I have no doubt in a Year or two more of seeing a perfect good chart of Newfoundland, and an exact survey of most ye good harbours in which there is not perhaps a part of the World that more abounds."[21]

In 1764, Graves was replaced as governor by Hugh Palliser, Cook's former captain aboard *Eagle* early in the Seven Years' War. In May, Cook carried back to Newfoundland and resumed command of *Grenville*. After drafting men from several other ships in St. John's for another surveying season, he stood out to sea on July 4. At the northern tip of Newfoundland, he started where he had left off the year before. The ship's log for July 14 is representative of Cook's laconic writing style and the systematic approach to surveying he had learned from Holland: "PM went into the Bay sacre, measured a Base Line and fix'd some Flaggs on Different Islands."[22] Cook was preparing to round the island to chart the long, straight coastline that faces the Strait of Belle Isle and leads to the Gulf of St. Lawrence when a powder horn blew up in his right hand. The injury that he suffered on August 6, 1764, was serious enough that *Grenville* hailed a French ship to find a surgeon capable of stitching the wound. In the short term, this accident limited Cook's ability to sketch and write, so the survey was temporarily suspended. The enduring significance of this incident is that it left Cook with a pronounced scar running between the thumb and forefinger to his wrist, a disfigurement that fifteen years later would be used to identify his remains.

Grenville anchored at Noddy Harbor until August 25 while Cook recuperated. Repairs were made, and spruce beer was brewed. Similar decoctions would often be prepared by Cook for his crews in the Pacific basin using the formula he learned here (just as Jacques Cartier had learned from Indigenous peoples during the winter of 1535–36). When the survey resumed, *Grenville* made it as far as St. Margaret Bay, a tenth of the way down the west coast of the island, by the end of September. With winter coming on, Cook closed down operations for that year, heading to St. John's and then on to England.

Grenville returned the following spring, but instead of resuming the survey where he had left off, Cook proceeded to Cape Race at the tip of the Avalon Peninsula to concentrate on the southern coast. The rationale was Palliser's concern about incursions by French fisherman into those waters from St. Pierre and Miquelon. Cook's effort during the summer of 1765 was prodigious, as Newfoundland's southeastern shoreline is the island's most intricate. Returning to St. John's, *Grenville* sailed with the fleet back to England in early November 1765.

Sailing for North America with his crew of seventeen on April 20, 1766, Cook directed *Grenville* toward Cape Race, which she rounded amidst "many Islds of ice," the then common term for icebergs.[23] During the course of his multiyear survey of Newfoundland, Cook learned the importance of knowing the local ice cycle. The fishing grounds off Newfoundland cleared of ice in April and May and started to refreeze in early October. Leaving the survey too late at the commencement of fall risked getting trapped in the ice, and a too early arrival after the spring sail from England could force a delay. Picking up at Bay d'Espoir, where he had left off the previous summer, Cook and his surveyors worked west on the southern coast.

During this season, Cook expanded his scope of endeavour, seeking, in Suthren's words, "unofficial membership in the scientific fraternity of the day" by observing an eclipse of the sun. Astronomy and botany were the most fashionable disciplines in Enlightenment science, and though

Cook would become strongly associated with both during the balance of his career, the former was integral to his daily life as a navigator. Cook's report on the eclipse was later read into the record of the Royal Society. Cook, though not present at the time, was described to members as "a good mathematician, and very expert in his business, having been appointed by the Lords Commissioners of the Admiralty to survey the sea coast of Newfoundland."[24]

Cook's 1766 survey ended at Cape Anguille, the headland that forms Newfoundland's southwestern tip. The vexatious southern coast had been mapped in exhaustive detail over the course of two summers. *Grenville* sailed for England on November 4 and was snug in berth at Deptford on the Thames by the last day of the month. At this point, only the long line of the western coast remained unmapped. Cook, perhaps overly eager to resume the Newfoundland project, left England on April 10, 1767, ten days earlier than the previous year (and the earliest of any trip). Accordingly, he encountered a great number of icebergs emerging from Davis Strait between Greenland and Labrador during the first week of May. He was in place off Cape Anguille by May 15, ready to close the gap. The most difficult challenges were the eponymous Bay of Islands and Bonne Bay and the saltwater access to Gros Morne, the grand canyon of Atlantic Canada. Cook's work, doggedly pursued for four years, was finished on September 24. By November 15, *Grenville* was back at Deptford.

Coincidentally, Joseph Banks, the soon-to-be-famous naturalist and Cook co-voyager, was in the waters of Newfoundland and Labrador that same year aboard *Niger*. Cook and the wealthy young botanist (and later patron of exploration as president of the Royal Society) would soon have their fates intertwined in a formidable fashion, but they indirectly crossed tracks, this time by accident, literally. Banks had left a birchbark canoe in St. John's to be transported back to England as deck cargo, which turned out to be on Cook's ship. However, *Grenville* struck a shoal off the coast of England, and though her planking had not stove, the topsail, yards, and crossarms crashed to the deck, ruining Banks's canoe. As the only significant navigational accident during the *Grenville*'s multiyear survey of Newfoundland, this incident was soon forgotten. Cook returned home to his burgeoning family. He finalized his charts and the next set of sailing directions, fully expecting to return to Newfoundland one last time to polish his previous work and close some minor gaps on Newfoundland's east coast. But when the April 1768 sailing season arrived, the Admiralty lords had other plans for him. *Grenville* departed with Michael Lane as sailing master instead, and it was he who would formally conclude the

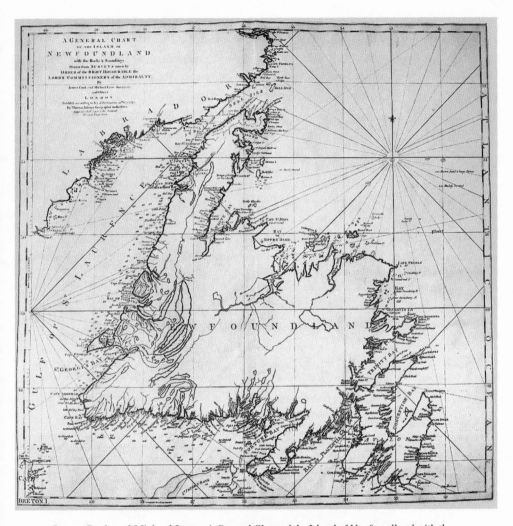

James Cook and Michael Lane, *A General Chart of the Island of Newfoundland with the Rocks and Soundings,* 1775. Cook surveyed the Newfoundland coast during summer from 1762 to 1767. This composite map of that effort, drawn from a series of harbour and coastline charts, was a stunning cartographic achievement that presaged Cook's ability to project an accurate image of the entire Pacific basin.

survey. With permission of the Admiralty, Cook's four charts of Newfoundland, one for each of his seasons, were published by year's end, and his sailing directions, meant as narrative accompaniment, were issued in 1769. He had perfected the techniques that Holland had taught, Simcoe had encouraged, and Colville and Graves had admired. A composite map showing a comprehensive view of the island was published in 1775, and it is this image that Vanessa Collingridge popularized in her 2007 television documentary series.[25]

The Newfoundland survey was a remarkable body of work for a man with agrarian roots, and it seeded the most extraordinary career in maritime history. Cook's requests for more sophisticated equipment and logistical support, which in turn produced greater exactitude, combined with his self-directed experiment monitoring a solar eclipse, show that he was becoming aware of his gifts and the privileged perspective they yielded. Having been introduced by Holland to the higher mathematical principles of the scientific survey, Cook began to grasp the potential of the Enlightenment's dyadic doctrines of specificity and universality. Once the grid of latitude and longitude had been established on a plane and the coordinates of any and all discernible places inscribed, a unitary vision of the world, geometrical and verifiable, was possible. This insight enabled Cook to become not merely the best explorer of his or perhaps any age but, borrowing an expression from Felipe Fernández-Armesto, an explorer of European exploration into distant seas.[26] Over the course of the next, and last, decade of his life, Cook's journals, and the published accounts that grew out of them, catalogued, and in some measure popularized, the voyages of Magellan, Tasman, and Bering, which preceded him into the Pacific.

This was crystalized for Cook in 1755 at Cape Town, near the end of his second voyage, where he had a coincidental meeting with fellow explorer Julien Marie Crozet. After comparing notes with him, Cook reflected: "Probably more authentick accounts may be got here after, but it will hardly be necessary to resume the Subject unless all the discoveries, both Ancient and Modern, are laid down in a Chart and then an explanatory Memoir will be necessary and such a Chart I intend to construct when I have time and the necessary materials."[27]

Indeed, after travelling more extensively over broad swaths of the Southern Hemisphere than any other explorer, who was better positioned to do this than Cook? This ambition also sheds light on Cook's motivation for taking on the quest for the Northwest Passage, a voyage that, if nothing else, would provide him with insight into hidden aspects of northern geography that matched what he had discerned in the south. In the end, when the Admiralty posthumously published the master chart of Cook's three circumnavigations, the result was the first modern map of the world. Such were the results of a chance encounter at Kennington Cove on the Nova Scotia coast. Cook was ordained by Holland into the priesthood of empirical rationality; his coordinates and the charts drawn from them created a way to record, share, replicate, and therefore verify the truthfulness of his travels. Cook became the chief priest of this new way of knowing, and a distinguished set of disciples – Vancouver, La Pérouse, Malaspina, Bellinghausen, and Wilkes – followed his tracks and attempted to emulate his style.

TWO

The Republic of Letters

James Cook was a good explorer, possibly the best, but his influence exceeded that of navigation and geography. Cook thought of himself first as a seaman, then as a navigator, and in the end as a captain of men – practical skills all. But he was also part of a seventeenth- and eighteenth-century intellectual revolution – the Enlightenment – that swept Europe with its array of inquiring and observant methodologies. Some of these disciplines, such as geology and meteorology, were in their prototypical stage in Cook's time. Others, including astronomy and botany, were well developed. Practitioners in these fields were known as naturalists, not scientists (a term that was not common until the middle of the nineteenth century). Cook grew in stature in direct proportion to his association with these currents of thought, as in the case of the solar eclipse he observed off the coast of Newfoundland.

⚓

A vital phase in Euro-American intellectual history, the Enlightenment was also known as the Age of Reason because it espoused rational thought. Its ferment propagated theoretical development in the natural sciences, the nascent social sciences (for example, anthropology), and liberal political philosophies advocating constitutional government in lieu of monarchies and the separation of church and state. The early iconic figures were the mathematician Isaac Newton and the political theorist John Locke and, closer in time to Cook, chemists Joseph Priestly and Henry Cavendish. Sweden's Carl Linnaeus, the pre-eminent botanist, was another towering figure by virtue of his revolutionary classification system. Philosophers of the moral universe included Adam Smith and the young Thomas Jefferson. The best-known North American thinker of that era was Benjamin Franklin, that rare man of learning who bridged the divide between experimental physics and politics.

The Enlightenment's empirical attitudes infused the search for the objective realities of planetary and cosmological processes, on the one hand, and ethical sentiments such as those found in the preamble to the American Declaration of Independence, on the other. One of its principal achievements was a school of thought on the nature of man. The central development here was Jean-Jacques Rousseau's *Discourse on Inequality*, published in 1755. His thesis held that the life of "primitive" peoples was both more joyful and virtuous than that of supposedly advanced civilizations such as Europe's. Rousseau's thinking was later expanded and amplified into the trope of the "noble savage," developed by Dennis Diderot after his reading of the Tahitian experiences of Cook and fellow Frenchman Louis-Antoine de Bougainville. Scholarly discourse about the uncorrupted indigene was, J.C. Beaglehole noted, "more diligently cultivated in France than in England," but the serial accounts of Pacific islanders related by Cook did more to sustain and popularize the theory than any publication emanating from Paris.[1]

Though elements of the Enlightenment were still an operative feature of discovery in the early nineteenth century (Jefferson's instructions to Meriwether Lewis in 1803 were modelled after Cook's), the Age of Reason's plain rhetorical style was gradually supplanted by a new cultural mode of thought and expression – romanticism, which valued personal sentiment and subjectivity over logic and intellect. For the explorers in the Romantic Age that succeeded Cook, the empirical cataloguing of new lands or people was replaced by narratives stressing heroic stances and intimate, emotional responses to nature. By the time John C. Frémont followed Lewis and Clark into the American West in the 1840s, the romance of exploration was the dominant paradigm.

Another Enlightenment hallmark was a fraternity between scientists that transcended borders. This sensibility was noteworthy, given the growth and maturation of imperial nation-states in that age, their rivalries, and the near constant conflict that resulted. But war occasionally took a holiday in the interest of furthering knowledge. The best-known instance came during Cook's third voyage when Great Britain was at war with rebellious colonies in America plus France. Both adversaries issued a form of diplomatic immunity for Cook and his ships, protecting them from search and seizure at sea. This professional courtesy fell within the tradition of international collegiality called the Republic of Letters. Cook himself used the term in his account of his second voyage.

This phrase reflected both the age's democratizing urge and the desire of its scholarly elite to freely share research results in order to diffuse knowledge. Writing elaborate letters to fellow scholars was still fashionable, but the printing press was increasingly a vital information technology.

Testing and argument, or more generally, discourse – the competitive and often collaborative engagement of minds – was the principal mode of expression. One vehicle that facilitated this dialogue was the learned journal, which through subscription also created a sense of community among scholars. Allied forms were pamphlets, books, and the libraries that held them; museums and botanical gardens, which collected specimens of human achievement or of the natural world; and private cabinets of curiosities.

One institution was central to this diffusion of knowledge – the learned society. These organizations did not so much compete with universities (another maturing node of critical thought) as complement them. Their great advantage was that, as member networks, they could transcend the limits of a college, a nation, or a faith tradition. And, often, several of these organizations worked across borders on projects of mutual interest. The premier institution of this kind in Great Britain was the Royal Society of London for Improving Natural Knowledge, founded in 1660. A century later, the Royal Society's scientific agenda enveloped James Cook's career.

⚓

A key project of the late Enlightenment (1750–80), and one imbricated with Cook, was the transit of Venus, projected to occur for the second time in eight years in June 1769. Planetary transits by Mercury and Venus are rare astronomical phenomena during which these celestial bodies pass directly between Earth and the sun, giving the appearance of crossing, or transiting, the solar face. (This is similar to what happens during a solar eclipse, when the moon obscures the sun, only the transiting image is much more visible; indeed, on occasion, it can be "total." By extension, on Mars, one could theoretically see the transit of Earth.) The transit of Venus, first observed in 1639, occurs in a four-part pattern in a cycle that repeats every 243 years. Twinned appearances are spaced by eight years between gaps more than a century long. (There have been only four Venus transits since 1769, most recently in 2012.)

In 1716, the astronomer Edmund Halley, in a presentation to the Royal Society, theorized that a transit of Venus could, through very complicated trigonometric formulas known as the parallax calculation, yield the sun's diameter, the distance of Earth to Venus and, most famously, Earth's distance from the sun. The experiment required precisely measuring the minor variability in the amount of time it took the transit to transpire, as observed simultaneously from disparate points on Earth. The best way to think of the solar parallax is as a triangulated survey, like the system Cook learned on Cape Breton Island, except on a cosmic scale. But for a planetary transit, rather than sighting a headland and rolling out the chain to create a baseline of a few dozen yards, the baseline between far-flung stations was derived from exact determinations of their respective latitude and longitude.

There had been a transit in 1761, and even at the height of the Seven Years' War, it was observed at over 120 stations around the world by astronomers from nine European nations. However, local weather was not conducive to taking measurements at many locations, and there were technical problems with instrumentation. When the data were collated, the experiment was deemed unsuccessful because of a lack of precision. Since the next transit was only eight years away, there was some urgency in the international scientific community to get the next one right. Otherwise, astronomers would have to wait until 1874.

In 1761, the leading promoter was France's J.N. Delisle. The British were relatively minor participants, certainly in proportion to their ability. (Two British stargazers playing a part were Charles Mason in Sumatra and Jeremiah Dixon at Cape Town. Both were later remembered for establishing the Pennsylvania-Maryland border, the line separating free and slave states.) At a meeting of the Royal Society in November 1767, planning was accelerated for the ensuing transit on June 3, 1769. The society's officers determined to contribute in a substantial fashion by sending more observers to more places than they had in 1761. Doing so would not only be commensurate with Britain's perceived status, the society's astronomers could also complement the efforts of other nations, all with the common aim of securing the most accurate measurements possible.

Some experimentation would take place close to home. Jefferson viewed the 1769 transit from the hilltop at Monticello. The Royal Society made a three-dimensional model of Earth and Venus revolving around the sun to introduce King George III to the astrophysical dynamics so he could track the transit in London, or at least part of it, since the entirety of the hours' long passage was not visible in western Europe. Accordingly, professional astronomers went to more distant parts. Abbé Jean-Baptiste Chappe d'Auteroche, a Catholic priest and member of the Royal Academy of Sciences in France who had gone to Siberia in 1761, chose, for 1769, to go to Cabo San Lucas at the tip of Baja California. The latter seemed like a better choice of destinations; however, he died within two months of the transit from an outbreak of yellow fever. His records did make it back to Europe, though, which enabled cartographers to fix the southerly extent of that peninsula. William Wales, who would sail on Cook's second voyage, was dispatched to the mouth of the Churchill River on Hudson Bay. William Bayly, an astronomer on both Cook's Antarctic expedition and the third voyage, was dispatched to the northern cape of Norway, where he was assisted by Dixon.

For the experiment to be successful, the sun needed to be well above the horizon and visible during the event. Because transits occur on the cusp of solar solstices (each pairing alternates between June and December), they are widely visible in one hemisphere's summer when the Earth is tilted toward the sun, but only a narrow cone of visibility prevails in the shorter daylight of the opposite hemisphere's winter. Since the 1769 transit was to occur at the onset of the northern summer, which had a long boreal day, there were many choices above the equator. Despite its more limited prospects, an extension of astronomical triangulation into the Southern Hemisphere was also desirable, especially a station in the wide expanse of the Pacific. But where to place an observatory during the austral winter was an open question. In February 1766, Thomas Hornsby prepared a table of sketchily known islands discerned by Iberian or Dutch explorers between 4° S and 21° S, the projected southern cone of visibility, including the Marquesas and Espiritu Santu, part of Vanuatu.

When Hornsby's paper was taken up by the Royal Society in November 1767, a committee of astronomers studied the question of how to complement the observations on Hudson Bay and Norway with one from the South Pacific. One of their number, Nevil Maskelyne, the astronomer royal, refined Hornsby's parameter by shaving off ten degrees of longitude at each end but extended it to 30° S. As delineated, this expanse formed an inverted trapezoid whose longer top line ran just south of the equator while the shorter parallel passed below the Tropic of Capricorn, enclosing within its perimeter modern French Polynesia, American Samoa, and Tonga. But the terrestrial contents of Maskelyne's zone were still largely unknown in the 1760s because no European vessel had ever explored the south-central Pacific above 20° S, so the problem as to where to send an astronomer within that sector lingered.

The solution came by accident and almost at the last minute. The end of the Seven Years' War brought with it a peace dividend that allowed Great Britain to divert naval resources into exploration and a new phase of empire building. Having bested France in North America, Britain's new strategy was loosening Spain's grip on the Pacific basin. However, France itself planned to re-establish an empire in the Southern Hemisphere, having lost most of its holdings in the North to Britain. This was the circumstance behind Louis-Antoine de Bougainville's plan to establish a chain of outposts south of the equator, starting with the Îles Malouines (also known as the Malvinas or Falkland Islands) and using displaced Acadians from Canada as a core population. His scheme included complementing the way station in the Malouines with another in the South Pacific, which would link France with markets in the East Indies and China. (On the basis of Bougainville's subsequent voyage, France was able to perfect its claim on Tahiti, even though the British visited that island first.) However, after the Spanish Crown raised objections with a fellow Bourbon monarch in Paris, Bougainville was directed to cede the Malvinas to Spain. As a consolation, he was offered the opportunity to extend the mission into a scientific round-the-world voyage, France's first.

Britain's goal in the Pacific was the discovery of new lands not previously sighted by the Iberians, and, by necessity, it wanted to beat France to the punch. Any discovery would be triply sweet – Britain's gain and Spain's and France's loss – particularly if the find was of continental proportions. Britain was allured by the prospect of discovering new colonies in the Pacific's temperate or semitropical zones that could feed mercantile empires like those instituted during the Columbian era. In Beaglehole's words, this golden southern continent, perhaps the great discovery of all time, "waited merely to be adequately known." He suggested that Britain's lackadaisical approach to the colonial rebellion in the run up to what became the American War of Independence was careless overconfidence based on the premise that the rich resources of North America could be readily replicated by an "untouched reservoir" in the form of this unknown southern continent.[2]

Within this context, two separate voyages were commissioned, starting in 1764 with that of Commodore John Byron, captain of the *Dolphin* (and grandfather of the poet). Byron's voyage had an ambitious discovery agenda but ended up being a spectacular failure. His cursory effort (it was the first circumnavigation to take less than two years) was compounded by his penchant for attracting inclement weather (he was known to his men as "Foulweather Jack"), but Byron's lack of success had more to do with his personal attributes than with external conditions. His instructions called for pressing into the Pacific and then toward the west coast of North America, visited previously by Francis Drake. He would then start looking for a passage through the continent back to England around 38° N; if the corridor was not found, he would return by way of the Cape of Good Hope.

Byron abandoned the plan to follow Drake's track off the coast of Chile because he deemed *Dolphin* insufficiently seaworthy. Sailing home, he managed to miss every major tropical island group, but in passing several atolls in the Tuamotu Archipelago, he recorded an absence of a sea swell from the southwest, which was taken to signify the existence of a large mass of land in that quarter. This was sufficient evidence to send *Dolphin* to those waters again. Captain Samuel Wallis left England in 1766. On this second and more important voyage, Wallis had a consort, *Swallow,* commanded by Philip Carteret. His was the first expedition with the stated purpose of looking for the unknown southern continent, Terra Australis Incognita, that Byron was thought to have detected. These vessels became separated while transecting the Magellan Strait, and while both ships managed to circumnavigate the globe, only Wallis's return to England in the spring of 1768 figures in our story.

With one exception, *Dolphin*'s voyage under Wallis was no more note-worthy than Byron's, for he too was an unexceptional explorer. However, as Beaglehole states, Wallis "had one piece of amazing good luck in discovery which would mark out his voyage forever, and was of the utmost significance, not for that voyage alone, nor alone for the great voyages that succeeded it, but for the whole history of the western mind." In June 1767, Wallis "came to an island such as dreams and enchantments are made of, real land though it was: an island of long beaches and lofty mountains, romantic in the pure ocean air, of noble trees and deep valleys, of bright falling waters." The climate of this island was delightful and almost magic-ally curative to the sickly seafarer, and it was inhabited by a captivating people, the women especially, who displayed, Wallis noted, "wanton gestures" whose meaning "could not possibly be mistaken." Wallis called it King George III Island, but we know it as Tahiti. Soon visited by Bougain-ville and Cook, Tahiti was Enlightenment Europe's most exciting and disruptive discovery. In *Civilisation and Its Discontents* (1930), Sigmund Freud argued that "the three great causes of unhappiness in the modern age were religion, psychoanalysis itself and the consequences of eighteenth-century exploration of the Pacific."[3]

The plan, prior to Wallis's return to England on May 20, 1768, was that the British astronomers destined for the South Pacific would establish their station somewhere within the vast expanse that Hornsby and Maskelyne had delineated and then conduct their experiment. But this enchanting isle of Tahiti, right in the centre of the astronomers' quadrilateral, immediately suggested itself as the most appropriate location. One of the Royal Society's favourite tactics was propagating voyages of discovery whose scale and expense usually required the in-kind support of government. A recent embezzlement by the organization's clerk had exacerbated matters. So, even before Wallis returned, the society was already beseeching the Admiralty for a ship that would sail to some spot in the South Pacific to observe the transit.

After Wallis's propitious return, an equally beguiled Admiralty exhibited increasing enthusiasm for another voyage to the South Pacific. Provocatively, Wallis's sailing master, George Robertson, spoke of an ephemeral sighting of "a very high land to the southward" of Tahiti. This tantalizing prospect, obscured in the moment by weather that was "so thick and hazey we could not see it ... for certain," was imagined as the northern promontory of the unknown southern continent.[4]

And so, virtually at the same time that Wallis returned, James Cook, an up-and-comer from the yeoman class with a remarkable tour of duty in North America on his resumé, walked onto the stage. Promoted from sailing master to lieutenant on May 25, 1768, Cook took command of the *Endeavour,* a former collier renamed by the navy to capture the kinetic spirit of the Enlightenment. Cook would sail her to Tahiti for the transit. Though he was entitled to the courtesy title of "captain" and is commonly referred to as "Captain Cook" in histories of all three voyages, he did not achieve the formal title until after his second expedition, having been promoted after the first to "commander."

Cook's was a meteoric rise, and in Great Britain's aristocratic culture it could only have happened in the Royal Navy, the only "meritocratic avenue of advancement." Cook had clearly established himself as a quality navigator, and he was already the best chart maker in the navy. He also had well-known mentors, such as Colville and Palliser. According to historian Martin Dugard, Cook became "the first man in Royal Navy history to rise from the bottom ... ranks to an officer's commission and command."[5]

⚓

Like the oldest child whose imagery dominates the family's photo album, Cook's first voyage dominates the scholarly and popular record of his career. Its novelty set the stage for the two voyages that followed, but the *Endeavour* voyage has been overemphasized, especially when we consider that its main goal – observing the transit – was not entirely successful. A durable interpretive axis, running from Great Britain to its antipodes in the southwest Pacific, is the principal cause of this exaggeration, starting with the celebrity of Joseph Banks in England. Banks, a nobleman who sailed on *Endeavour* as a supernumerary (a civilian with elevated status on board a navy ship), served as the principal naturalist. His particular interest

Joshua Reynolds, *Sir Joseph Banks*, 1772–73. Banks sat for this
portrait after the *Endeavour* voyage. His status as a writer-traveller is
symbolized by such accoutrements as a free-standing globe and
pen with inkwell.

was botany, a favoured activity in elite circles. Banks's fame took such hold
that it is still occasionally argued that he and his associate Daniel Solander
were the first scientists to join a voyage of discovery, notwithstanding the
fact that Bougainville beat Cook into the South Pacific with a botanist and
an astronomer on board.[6]

At the other end of the axis, scholars and cultural institutions in New
Zealand and Australia have traditionally seen Cook's first voyage to their
quarter of the world as a kind of origin myth. Some of the greatest re-
positories of Cook exploration material are housed there, and, of course,
the editor of Cook's journals, J.C. Beaglehole, hailed from New Zealand.

It is no coincidence that the only replica of a Cook vessel that can sail, *Endeavour,* was conceived as Australia's leading bicentennial project in 1988, even though the underlying commemoration was not Cook-related but rather the so-called first fleet that founded the convict colony at Botany Bay, which came in his wake.[7]

Sensibly, the balance of Cook's journals, drawings, and charts is held in archives and libraries close to the Greenwich meridian, and British scholars have played a leading role in Cook scholarship, from the appearance of the first biography to this day. Cook's definition of New Zealand's insularity plus that of the east coast of Australia were assuredly notable geographic accomplishments. Nothing on that scale had ever been done before with such accuracy. But when the Cook cartographic oeuvre is taken as a whole, the findings of the first voyage are less significant than those of the second or the third. Still, in 2001, when the BBC settled on the idea of a Cook documentary, it chose his Australian survey as the emblematic element. This choice, of course, was made easy by the existence of the *Endeavour* replica and the spectacular imagery that tall ships engender. The title of the production and the companion book was, simply, *The Ship,* and it is a perfect example of the antipodal axis in action.[8]

The enduring value of the *Endeavour* voyage was its contribution to what can only be called a cultural revolution. An account of the expedition was included in a multivolume compendium edited by John Hawkesworth. Titled *An Account of the Voyages Undertaken by the Order of His Present Majesty for Making Discoveries in the Southern Hemisphere,* Hawkesworth's publication created a sensation. The first volume was a recounting of the expeditions of Byron, Wallis, and Carteret, but Volumes 2 and 3 were exclusively devoted to what was unquestionably considered the far more important voyage of Cook. Indeed, Volume 2 actually appeared first because of topical interest. Hawkesworth was given this prestigious assignment by the Admiralty and paid £600 by the publishers, William Strahan and Thomas Cadell, for the copyright. This was the largest sum paid for a literary effort that century, a circumstance that was not lost on Cook and all who sailed with him on subsequent voyages. Hawkesworth was heavily reliant on Banks's more polished journal because it was "so interesting and copious," an observation he made publicly in the introduction, somewhat awkwardly since the account was nominally written in the name of the commander, Cook.[9] In a way, this only served to reflect the underlying reality. Cook clearly relied on Banks for the framing of many of his own shipboard reflections.

Cook, under a form of tutelage to Banks similar to the one he had with Samuel Holland, now extended his already considerable skill as a surveyor and astronomer to newer disciplines. As Beaglehole phrased it, by the end of the *Endeavour* voyage, Cook was "a man whose intelligence had widened considerably to an understanding of the importance of all knowledge."[10] But the pupil also became the teacher as Cook taught Banks the value of healthy skepticism when it came to the experimental nature of scientific discovery. Hawkesworth's *Discoveries* was printed in 1773 while Cook was on his second voyage. It popularized expeditionary accounts with the reading public of Europe and America to such an extent that it can be said to have created a literary subgenre. The novel did not overtake travel literature in popularity with the reading public until the 1850s.

Exploration narratives had several levels of appeal. In part, they were an extension of older fictional or quasi-fictional travel stories (such as Marco Polo's). Felipe Fernández-Armesto calls this vein "the idealization of adventure," replete with "chivalric affectations" that ran the gamut from melodramatic exclamations to faithful canine companions. (The best-known examples in North American history are William Clark's "O! The Joy," when first coming into view of the Pacific Ocean, or Lewis's dog, "Seaman.") The tradition of knight errantry, or the "quixotic impulse" – escapism, adventure, asceticism, social ambition, dogged determination, vanity – were all extended to the personality profile of the Enlightenment explorer. Even certain geographic terms, such as "California" and "Ultima Thule," had their origins in chivalric fiction or other medieval narratives.[11] Bougainville's *Voyage autour du Monde* (1771), followed by Hawkesworth's work two years later, added an innovative scientific overlay to this genre. The new value of verisimilitude – the story's apparent eyewitness truthfulness tied to a specific chronology, with copious detail describing, with stylistic plainness, the exotica of South Pacific island life (plant, animal, and human) – transformed the old form of travel literature into the modern nonfictional exploratory account. Every explorer in Cook's wake – La Pérouse, Malaspina, Vancouver, Mackenzie, Lewis – envisaged his place within this literary milieu.

⚓

Though, in retrospect, the choice of Cook as commander appears well warranted if not inspired, it was not a popular decision in some circles. Alexander Dalrymple, a frequently misunderstood character, was the Royal Society's candidate. He had joined the East India Company in 1752 at age sixteen as a London-based clerk. Before the year was out, he sailed on a company ship to Madras, India, where he was stationed for many years and rose rapidly through the ranks, much like Cook did in the Royal Navy. With a view toward improving his employer's prospects (and a promotion), Dalrymple studied and organized corporate records, which served as his introduction to the company's voyages of exploration and its competitors in Asian commerce.

In 1759, while Cook was in Quebec, Dalrymple sailed to Malacca in the Dutch East Indies. During the journey, he received training in navigation and seamanship. Once he arrived, he boarded the company's *Cuddalore*, which set out on a voyage to China to find an alternative route other than through the Straits of Malacca (which were in international waters but dominated by the Dutch) in times of war. Dalrymple was not responsible for practical navigation or management of *Cuddalore*'s crew; that was handled by Captain George Baker. But as supercargo and agent for the company's mission, he provided strategic direction. Though Baker remained a lifelong friend, the dynamic of co-command did not wear well. At least that is what Dalrymple implied in 1802, when he asserted that a divided authority with Cook on *Endeavour*, similar to his arrangement with Baker, had been proposed but rejected by him as unworkable. If in fact this was offered, surely the idea originated with some of Dalrymple's friends in the Royal Society, not the Admiralty.

Dalrymple is caricatured by Cook's historians as an armchair geographer, as if it was patently preposterous for anyone to have thought of him and Cook as coequal contenders for the Tahiti mission. Barbara Belyea affirms that most historians, counterposing Cook's career in Canada, ridicule Dalrymple's pretentions to be the head of the expedition.[12] But he was actually reasonably well qualified. Dalrymple's many detractors have been overawed by Cook's lifelong accomplishments, punctuated by the great navigator's occasional disdain for Dalrymple's geographic hypotheses. Overlooked in the standard assessment of Dalrymple's candidacy is the fact that, going back to Elizabethan times, voyage commanders were not necessarily skilled navy men. William Dampier, the most important English sailor between Walter Raleigh and Cook, was a merchantman and privateer who circumnavigated the globe three times without rank in the Royal Navy. Indeed, Cook's most famous contemporary in global voyaging, Bougainville, commanded his great voyage with a portfolio more like Dalrymple's on *Cuddalore* than Cook's on *Endeavour*. Dalrymple had more practical experience in maritime surveying and navigation than Bougainville, who was neither a scientist nor a sailor but an army officer.

While Cook was placing buoys in the St. Lawrence, Dalrymple directed *Cuddalore* to Macao. The next year, he surveyed the coast of Hainan and sailed to Da Nang, Vietnam, where he skirmished with a Portuguese ship. Returning to China, word reached Dalrymple and other company merchants that the French had posted a squadron in the Sunda Strait between Java and Sumatra. Coming amidst the Seven Years' War, the gambit was to lay in wait for British East Indiamen. The other supercargoes asked Dalrymple if it was possible for *Cuddalore* to guide their ships through unknown waters to avoid the French trap. Dalrymple was agreeable, having copies of Dutch charts of the Celebes, an archipelago east of Borneo. Weighing anchor in December 1760, *Cuddalore* led the British merchantmen through a strait south of Luzon, taking soundings as they progressed, thence down the Makassar Strait east of Borneo, eventually reaching the island of Sumbawa, east of Java and Bali. They were the first British ships ever to run that course. Most of the fleet sailed into the Indian Ocean and home to England, but Dalrymple took *Cuddalore* north through the Makassar Strait to Manila, where he surveyed the bay, much to the anxiety of Spanish authorities, who were always suspicious of English intervention. Dalrymple arrived back in Madras in January 1762, concluding three years of voyaging.

Dalrymple had conducted the first British reconnaissance of the East Indies, which pointed the way toward the expansion of commerce in a zone previously dominated by the Spanish, Portuguese, and Dutch. In practical terms, he gained valuable experience in navigation, seamanship, and merchant diplomacy and found a route to Canton that avoided enemy fortresses and fleets controlling the Malacca and Sunda Straits. At this point, Dalrymple's contributions to Britain's imperial ambitions were equal to Cook's in North America, if not greater. As historian Howard Fry observed, Dalrymple's command of *Cuddalore* "had been a remarkable achievement for a young man in his early twenties, who, at its start, had no other experience of the sea than the voyage out to India, as a passenger, six years before." On the basis of these and Dalrymple's other contributions to the East India Company and the British Empire over the course of his career, Fry concluded that "the legend of his supposed enmity with Captain Cook has done his reputation positive harm."[13]

Dalrymple returned to the East Indies in 1762 as supercargo aboard *London,* captained by James Rennell, an officer on loan from the Royal Navy. After making arrangements for unloading cargo in the Sulu Archipelago and packing goods for the Chinese market, Rennell observed that Dalrymple "measured Bases to determine the positions of the neighbouring Islands and in order to compleat his work he now brought with him two good Telescopes and a Time-keeper to determine the Longitude of Sooloo by eclipses of Jupiter's Satellites; and a Land Quadrant for the Latitudes: an apparatus which I believe few People would have provided (chiefly at their own expence) who were not certain of perceiving a pecuniary reward."

This is not the Dalrymple we are accustomed to seeing in histories of Cook, where he is commonly portrayed as a geographic dilettante. Dalrymple had never served in the Royal Navy, nor had he commanded shipboard regimen for the East India Company, but contrary to the impression left by some historians, Dalrymple was familiar with life at sea and long-distance voyaging, and he had practical exploring experience.[14] His chart of the East Indies is not as proficient as Cook's early efforts in Quebec and Nova Scotia, but it is a serviceable representation.

After six years in the East Indies, Dalrymple returned to England in July 1765. Now an experienced agent of British trade with significant seafaring experience, he settled into the upper echelon of Enlightenment life and maritime affairs. He was nominated to membership in the Royal Society for contributions to geographic understanding (by Benjamin Franklin, with whom he had a close association until 1775, when the Pennsylvanian returned to America, including coauthorship of a culturally presumptuous pamphlet touting the quixotic notion of stocking New Zealand with plants and animals in the wake of Cook's first voyage). He attended his first meeting in March 1766. Dalrymple was present when the society's astronomy committee discussed the transit in November 1767. He published a pamphlet that year, *Discoveries Made in the South Pacific,* which drew on a wide assortment of French, Spanish, and Dutch accounts, some unpublished. He was now a recognized expert on the history of exploration but shared a weakness that was common among French geographers: an abhorrence of blank spaces on a map. This was Dalrymple's Achilles heel, because it enabled Cook to prove him wrong with regularity and in time made him the subject of ridicule by historians, starting with Hawkesworth and running through and beyond Beaglehole.

At the time, Dalrymple was the most prominent person in Britain promoting the idea that a large undiscovered continent existed in the mid- to high latitudes of the Southern Hemisphere. It was generally presumed that this mysterious land was located either southeast of Africa where the Atlantic and Indian oceans converge or, given the wide expanse

of the Pacific, southwest of Cape Horn. Dalrymple pieced together ephemeral sightings ranging from the supposed island of Juan Fernandez in the
eastern Pacific to New Zealand in the west to support his "conjecture"
that between those extremities one could find a continent 5,323 miles
wide with a population of 50 million. This continent, he contended, was
"as rich and plentiful, as any Countries on the Face of the Globe, without
exception." Extrapolating tidbits from actual voyages and then combining
those "findings" with archaic theories of global physiognomy, Dalrymple
posited the "seeming necessity for a *Southern Continent*" equal to the size
of Eurasia in the north. Since there was an equal amount of land and
saltwater in the Northern Hemisphere then surely, he reasoned, the south
had matching qualities; otherwise, a massive imbalance would make the
world wobble on its axis. Terra Australis Incognita required only an explorer to find it.[15]

Dalrymple was reiterating the ancient principle of counterpoise, which
dated to the classical era but was enjoying an odd resurgence. It had been
reintroduced to the intellectual milieu of the Enlightenment by Charles
de Brosses's *Histoire des navigations aux terres australe,* published in 1756.
In essence, the precision of Newton's mechanical universe had fostered
an application on the terrestrial sphere. After all, the most accurate clocks
worked on the principle of a balancing mechanism: mass versus mass hung
on a wound spring. If the entire cosmos had been created with ordered
elegance, why not Earth? As Beaglehole phrased it, the southern continent
fit "rational form, the clearness of a balanced composition, an arithmetical
and happy harmony." Historian Felipe Fernández-Armesto explains that
a mostly oceanic world "seemed to defy every principle of order and symmetry, such as rational minds expected from a divine creator."[16] This influence was also evident in the cultural attitudes of Great Britain in the
Late Renaissance, which privileged balance and symmetry in architec-
ture, horticulture, and landscape design.

Within the councils of the Royal Society, Dalrymple was not merely
an advocate for observing the transit; he was the first to put forward the
idea of a multipurpose voyage to follow in *Dolphin*'s wake – that is, to
combine the transit mission with the broader purposes of geographic
discovery. For these reasons, and after a successful interview with the
astronomical committee, many members of the society saw him as the
most logical leader of the forthcoming expedition. Naively, the society
and Dalrymple thought that since they had first proposed the voyage, the
Admiralty's assent to providing a ship meant that one of their number

would be the principal astronomical observer and overall commander. When the Admiralty appointed Cook to fill the latter role, Dalrymple withdrew from the venture. He did not wish to serve under naval authority. Many in the society sympathized with his disappointment.

Cook's skills as a navigator trumped Dalrymple's commercial connections and reputation as a theoretical geographer. The Admiralty may have let itself be convinced by the influential Royal Society into providing a ship for the transit, but too much was at stake to trust the stewardship of one of the King's vessels to the command of anyone but a navy man. It was not Dalrymple's deficiencies as a mariner or geographer that doomed him; rather, it was a damaging combination of personal attributes – occasional outbursts of ill temper combined with fervent self-promotion. Dalrymple was too eager, and that made the Admiralty queasy. Cook, by contrast, seemed to be professionally capable yet was a modest man with a dispassionate demeanour.

Historians since Beaglehole have suggested that Dalrymple could not possibly have been a credible candidate to lead the transit voyage. The corollary – that each regarded the other as their archnemesis – is an exaggeration and particularly unfair to Dalrymple, because his 1767 pamphlet and professedly speculative charts about Terra Australis did much to create the context that made Cook's first two voyages possible. Cook and Dalrymple were occasionally critical of each other, but Cook was an equal-opportunity destroyer of cartographic myth during all three voyages, whether the legends were the handiwork of Dalrymple or of explorers from other countries who preceded him into the Pacific. Indeed, Cook routinely corrected his own work. This is what the empirical dialectic of the Republic of Letters required. It was not personal for Cook, just the business of discovery.

⚓

The bivalent nature of *Endeavour*'s voyage – one conceived by a community of natural philosophers but largely funded by an imperial government – is evident from the basic formulation of the venture. Cook had *three* sets of instructions: two from the Admiralty and one from the Royal Society. The first dictated the voyage's itinerary and timeline. Cook was directed to Tahiti via the route that all European explorers who preceded him had taken: sail to the tip of South America and then west into the Pacific. The Admiralty appended Astronomer Royal Maskelyne's "Limits for the Southern Observation of the Transit of Venus" in case Tahiti could not be found.[17] A second, confidential set of Admiralty instructions would be disclosed after the transit transpired. This was a gesture toward diplomatic sensibilities, intended to separate the collaborative nature of the international scientific experiment from the imperial mission, which Wallis's hint of more southerly lands portended. This second directive told Cook to extend the voyage by searching for a continent southwest of Tahiti.

Although "secret," some of what this second set of instructions contained leaked into public circulation in August 1768. A newspaper report alluded to a forthcoming search for new discoveries in the expanses of the Pacific. Much of what these directions conveyed could be inferred from Hawkesworth's account, but the actual document was not discovered until 1928. Cook was told that after the transit he was "to proceed to the southward in order to make discovery of the Continent" as far as the fortieth parallel (Tahiti is approximately 17° S), "unless you sooner fall in with it." Failing that, he was to head west "until you discover it, or fall in with the Eastern side of the Land discover'd by Tasman and now called New Zeland." The Admiralty also provided guidelines for recording navigational details and observing the natural resources found on any new lands.[18] With refinement and subsequent publication in the official accounts of his second and third voyages, these procedures became the standard practicum for explorers from many nations over the next quarter century. They took a particular hold on Jefferson for his guidance to Lewis.

A third set of instructions from the Royal Society framed the voyage's concurrent scientific agenda beyond the transit and provided counsel on how to comport with Indigenous peoples. In addition to the naval crew, *Endeavour* was to be staffed by a team of naturalists tasked with cataloguing novel plant life and making other observations about what the South Pacific held, including its people. The leader (the term is barely sufficient) of this group was Banks, a rich young nobleman, one of the major botanists of his day (Linnaeus was one of his correspondents), and a recent addition to the august ranks of the Royal Society. (In 1778, Banks would become president of the society, a position he held until his death in 1820.) A confidant of the Admiralty lords, Banks cajoled his way onto the expedition in June 1768, relatively late in the planning. The Admiralty agreed to the society's request to add Banks as salve for the organization's wounds in the wake of the Dalrymple affair and scientific cover for the imperial projection into previously uncharted waters. Many contemporary observers in the press and polite society thought Banks was the mainspring of the enterprise because of his nobility and Royal Society connections, plus he made a sizable £10,000 subvention. That was certainly the consensus after *Endeavour* returned, a view reinforced by Hawkesworth's commentary on the value of Banks's journal in comparison to Cook's.

Banks was only the most luminous member of the team of botanists, artist-illustrators, and assistants that sailed with Cook. The Swedish botanist Daniel Carl Solander, one of Linnaeus's protégés, was also notable as a frequent shoreside companion to Banks and Cook. Banks had two artists in his retinue, Sydney Parkinson and Alexander Buchan; a personal secretary, Herman Sporing; and four servants. (In a residual affect from exploration's chivalric roots, Banks also travelled with two canine companions.) The lead astronomer was Charles Green, heretofore an assistant at the royal observatory at Greenwich. As a civilian and the Royal Society's delegate for the observation of the transit, Green was the key figure in *Endeavour*'s central mission. Cook held an appointment from the society as second astronomer by virtue of his work on the eclipse in Newfoundland.

This was a distinguished group, which prompted Beaglehole to confuse matters by observing that the Royal Society team made the *Endeavour* "the first voyage of discovery equipped with a scientific staff." As noted earlier, that honour should go either to Bougainville, who in November 1766 set sail in a two-ship flotilla, or perhaps to Edmond Halley, who voyaged to the southern Atlantic to study terrestrial magnetism on the *Paramour* from 1698 to 1670. Halley did not have geographic discovery on his agenda, but his encounters with tabular icebergs introduced the phenomenon to British science, anticipating the definitive experience of Cook's second voyage. Bougainville was eclipsed because of the superiority of Cook's charts. The Frenchman added little to European understanding of the South Pacific other than Tahiti. Indeed, charting was such a relatively minor concern that Bougainville's cartographer was on the consort vessel, something that would have been unimaginable with Cook. Furthermore, Bougainville's scientists debarked on the Indian Ocean island of Mauritius, depriving him of their insights when he sat down in Paris to write his account.[19]

The Royal Society's instructions were offered "to the consideration of Captain Cooke, Mr. Bankes, Doctor Solander, and the other Gentlemen who go upon the Expedition on Board the *Endeavour*."[20] The guidance was carefully characterized as a series of "hints," probably in deference to Cook's status as a naval officer and his superior allegiance to the Admiralty's orders, as well as to the upper-class sensibilities of Banks, who probably felt he carried his instructions in his head. They were written by the Royal Society's chairman, James Douglas, the Earl of Morton, but they drew on a long tradition. Lawrence Rooke, one of the founders of the organization, composed an essay titled "Directions for Sea-Men Bound for Far Voyages," which appeared in the first volume of the society's journal, *Philosophical Transactions* (1666), and served as the fount from which Morton drew his guidelines. These hints dealt with three major themes that offer great insights into Enlightenment ideals (if not practice).

The *Endeavour*'s prime directive, to use Gene Roddenberry's adaptive phrase, dealt with relationships with Indigenous peoples. The ship's gentlemen were advised: "To exercise the utmost patience and forbearance with respect to the Natives of the several Lands where the Ship may touch. To check the petulance of the Sailors, and restrain the wanton use of Fire Arms. To have it still in view that sheding the blood of those people is crime of the highest nature: – They are human creatures, the work of the same omnipotent Author, equally under his care with the most polished Europeans; perhaps being less offensive, more entitled to his favor."[21]

The guidelines then shifted to the question of sovereignty over inhabited lands and the related body of international common law that legal historian Robert Miller calls the Doctrine of Discovery.[22] Indigenous peoples of newly encountered lands, the guidelines specified, "are the natural, and in the strictest sense of the word, the legal possessors of the several Regions they inhabit. No European Nation has a right to occupy any part of their country, or settle among them without their voluntary consent. Conquest over such people can give no just title; because they could never be the Agressors." Morton well knew of the likely course of events, so he advised: "They may naturally and justly attempt to repell intruders, whom they may apprehend are come to disturb them in the quiet possession of their country, whether that apprehension be well or ill founded. Therefore should they in a hostile manner oppose a landing, and kill some men in the attempt, even this would hardly justify firing among them, 'till every other gentle method had been tried." Morton added this aviso: "If during an inevitable skirmish some of the Natives should be slain; those who survive ... when brought under should be treated with distinguished humanity."[23] With his faults duly noted in the ship's log and his journal, this is the ethical guide under which Cook conducted Indigenous affairs during the course of three voyages in the Pacific.

Additional hints related to the recording of any new land's "animal, vegetable and mineral" forms. And any circumstance where gold was "met with" was to be examined "minutely." Precious stones, the instructions continued, were another "valuable part of Natural History, and are therefore a considerable object of enquiry." Almost as an afterthought, Morton added the desirability of securing vocabularies "of the names given by the Natives, to the several things and places which come under the Inspection of the Gentlemen."[24]

In sum, the Royal Society's instructions amplified Cook's instructions from the Admiralty, including the sensitivity he was to bring to encounters with Indigenous peoples. In this regard, few representatives of European society at that time would have been as diligent as Cook. Historian John Robson observes that recent criticism of Cook centres "on the deaths of Pacific people during contact with the British ships. Nobody was more saddened than Cook by such deaths, and it is a measure of Cook's leadership and respect for other people that so few deaths actually took place." In Robson's estimation, circumstances were "bound to lead to some friction, and on Cook's voyages there was, thanks largely to him, much less than on other European voyages." Cook could never quite surmount the problem that the restraint he showed during most of his beach encounters was not always reciprocated. As Lynne Withey puts it, Cook "interpreted theft, especially repeated theft, as a challenge to his authority, and indeed it often was."[25]

Given the ethnological emphasis of the instructions guiding the voyage, disquisitions on Indigenous peoples necessarily became one of the three central themes in the *Endeavour* journals of Cook and Banks. The second theme was key moments of exploratory drama, such as *Endeavour*'s near sinking on the Great Barrier Reef. Last, but not least, came the summative geographic or scientific analysis. This triptych dominated Cook's reporting for all three voyages, helping to substantiate and vitalize the Republic of Letters.

THREE

The South Pacific

Endeavour sailed from Plymouth on August 25, 1768. On board with James Cook were ninety-four other travellers, Joseph Banks's dogs, and a few barnyard animals, including a goat that became the mascot. The ship stopped for refreshment at Madeira and Rio de Janeiro. In Brazil, Cook became entangled in a diplomatic rumpus with the Portuguese governor, who, notwithstanding his nation's alliance with Britain, conducted his relations with the English visitors in an officious manner. From Rio, Cook set a course for the dangerous waters enveloping Cape Horn, and he sighted Tierra del Fuego on January 11, 1769.

The cape, at 56° S, was a gloomy place, even in austral midsummer. The cold was constant and accompanied by "frequent showers of Rain Hail and Snow."[1] (Cook's native Yorkshire and Sitka, Alaska, occupy a comparable northern latitude but enjoy pleasant summer weather.) Cook had chosen not to land at the Falkland Islands after leaving Rio, which annoyed Banks, but at Tierra del Fuego he was cajoled into looking for a harbour that would facilitate some botanizing. This led to a notorious incident when two of Banks's servants froze to death during what was supposed to be a brief jaunt. Said Banks: "We were caught in a snow storm in a climate we were utterly unacquainted with but which we had reason to beleive was as inhospitable as any in the world, not only from all the accounts we had heard or read but from the Quantity of snow which we saw falling, tho it was very little after midsummer."[2]

The long voyage from Cape Horn across the Pacific to Tahiti was uneventful. On this leg, Cook began implementing dietary controls as a part of his regular shipboard template, the start of his legendary success in battling scurvy, brought about (we now know) by vitamin C deficiency. Less than thirty years earlier, two-thirds of the 1,900 men who sailed in George Anson's six-ship squadron, which plundered Spanish shipping, had succumbed to this scourge. Cook's deserved reputation for preserving his men's health was mostly the result of his leadership skills and savvy understanding of group dynamics. When the crew rebelled at the strict regimen of sauerkraut, he cagily directed officers to be seen eating it. He wrote: "The Moment they see their Superiors set a Value upon it, it becomes the finest stuff in the World."[3] In truth, sauerkraut was of limited nutritional value. For this voyage and the two that followed, it was Cook's tireless commitment to shipboard cleanliness, fresh food (of whatever kind available, including, notoriously, walrus meat in the Arctic), and the constant replenishment of fresh water for the casks that sustained his crews to great contemporary acclaim. James Lind's 1753 treatise recommended lemon juice as an antiscorbutic, but the Admiralty did not mandate its use until 1795.

After entering the Pacific, Cook noticed the absence of strong currents, which signalled that no continent was near "because near land are generally found Currents."[4] Banks reflected forlornly: "When I look on the charts of these Seas and see our course, which has been Near a streight one at NW since we left Cape Horne, I cannot help wondering that we have not yet seen land." Although Banks disapproved of "theoretical writers" who described "these seas without having themselves been in them," he identified with them psychologically. Geographical speculators "generaly supposd that every foot of sea which they beleivd no ship had passd over to be land," but he himself held out hope for a southern continent until the circumnavigation of New Zealand was completed nearly a year later. Banks and Cook had seen enough of the expansive Pacific prior to arriving at Tahiti to realize that the principle of counterpoise, popularized by De Brosses and Dalrymple, was intellectually bankrupt. Banks determined the "number of square degrees of their land which we have already chang'd into water sufficiently disproves this, and ... we need not be anxious to give reasons how any one part of it counterbalances the rest."[5]

Tahiti came into view on April 11, 1769, well before the transit's target date in June. *Endeavour*'s early arrival allowed plenty of time for the Englishmen to conduct beachside bargaining for sex and supplies and observations of local customs. Banks took a special interest in Tahitian surfers and the "incredible swiftness" of their approach to shore.[6] But as to the transit experiment, the central mission for the voyage, it almost came to naught. A month beforehand, a quadrant was stolen, creating great concern. As Hawkesworth phrased it in Cook's voice for the official account: "Without this, we could not perform the service for which our voyage was principally undertaken." The instrument was recovered after considerable negotiation on Cook's part. The weather was not consistently clear either; some days were sunny, others not. Banks said this made the party "not a little anxious for success."[7] As a contingency plan, Cook sent back-up observers, including Banks, to the neighbouring island of Mo'orea.

The observation itself, if not a failure exactly, was not completely successful. The problem was determining with precision just when the transit began. Venus's atmosphere blurred its appearance at first contact with the sun's edge, making it a matter of judgment as to when it started. Cook and Charles Green, at the same place with identical equipment, differed over the timing of Venus's immersion by five seconds and over its exit by ten. Other astronomers around the world, all with equally effective optical instruments, noticed the same effect. Nevertheless, French astronomer Jérôme Lalande assessed the collated results and approximated the sun's distance from Earth to within 97 percent accuracy of its actual 92.9 million miles.

Cook and his officers were so focused on the transit that the command structure was stretched thin. Some seamen used the distraction to purloin the *Endeavour*'s stock of nails to trade for sexual favours. In effect, they robbed the bank. As Banks noted, this was an offence of a "serious nature as these nails if circulated by the people among the [Tahitians] will much lessen the value of Iron, our staple commodity," meaning the currency to buy food.[8] Managing the purchasing power of this ersatz monetary system was a vital function, and Cook handled this responsibility assiduously over the course of three voyages, even at the cost of rapport with the crew, famously so on the third voyage.

Geology was one of the Enlightenment's nascent sciences. Volcanism held a particular fascination because the catastrophic nature of its processes seemed contrary to the steady state or uniform nature of creation found in biblical accounts. Banks deduced the volcanic nature of Tahiti's origins and engaged in a last-ditch effort to reconcile observed reality with Dalrymple's counterpoise theory, which, at the surface level, he deemed "totaly demolishd by the Course our ship made from Cape Horn to this Island." The "nesscessary continent may have been sunk by Dreadfull earthquakes and Volcanos 2 or 300 fathoms under the sea, the tops of the highest mountains only still remaing above water in the shape of Islands."[9] As it turns out, Banks was on to something. Modern geographers consider New Zealand the top of a submerged continent, Zealandia (or, more colourfully, Tasmantis, a hybrid neologism derived from Abel Tasman, the Dutch explorer, and mythological Atlantis).

After the transit, Cook surveyed Tahiti's coastline in the Newfoundland style. He then made a cursory inspection of some neighbouring isles, the communal aspect of their co-location providing inspiration for his naming of the Society Islands. To facilitate the examination, *Endeavour* left Tahiti with a new supernumerary, Banks's newfound friend Tupaia, a priest, linguist, and island navigator extraordinaire. The ship passed to the north of Huahine and essentially circumnavigated Raiatea and Tahaa. Though Cook was aware of Mo'orea and Bora Bora, he did not chart them. Over the balance of the voyage through Polynesia, Tupaia served principally as an expert pilot but on occasion as translator and cultural intermediary. In this respect, Tupaia anticipated the role played by Indigenous "conductors" in the exploration of North America, whose emplacement with interlopers could prove peaceful intentions, Sacagawea most famously.

⚓

On August 10, Cook began the search for Terra Australis Incognita. When he reached 40° 22' S, bad weather drove him west. Cook did not fight these conditions. At this point in his career, concerned about how southern continent enthusiasts such as Dalrymple might interpret his conduct, Cook established a defence in his journal. He said he would have headed farther "southward if the winds had been moderate." But thinking he had "no prospect of meeting with land" in that zone anyway, the tempestuous weather led him to consider it "more advisable to stand to the Northward into better weather least we should receive such damages in our sails and rigging as might hinder the further prosecutions of the Voyage."[10]

Cook set a course toward New Zealand, the next best prospect for finding the southern continent and, for this voyage, his last hope. New Zealand had been sighted in 1642 by Tasman, who thought he had encountered the northern extremity of a continent. Onboard the *Endeavour,* land was descried by a youthful seaman, Nicholas Young, on October 6, 1769. As Hawkesworth phrased it: "This land became the subject of much eager conversation; but the general opinion seems to be that we had found *Terra australis incognita.*" This optimistic view was derived from Banks's journal, which referred to the "many conjectures" suggesting "that this is certainly the Continent we are in search of." Cook's journal is silent on the matter, but his unspoken skepticism speaks volumes.[11]

Cook stood in for an inlet later denominated as Poverty Bay. His stay there was a disaster, and notwithstanding the orthodox criticism that has grown up around Cook's deportment with Indigenous peoples during his final voyage, it was one of his greatest failures. Cook's small party went ashore. Four Māori circled behind them when they landed and attempted to seize the yawl. Shots were fired. One Māori fell dead; his compatriots fled. A chastened Cook returned the next day, this time accompanied by Tupaia and the ship's marines. He now faced a larger number of Māori, but to the surprise of everyone, they understood Tupaia. But the Māori did not yield to Tupaia's blandishments; they snatched the Englishmen's weapons instead. Cook directed gunfire at the perpetrators, resulting in another fatality, before disengaging to look for a source of fresh water and friendlier locals. After failing to intercept some Māori in canoes, a shot was sent over their heads. Rather than intimidating them into acquiescence, it sent the Māori into a rage. When they sent missiles his way, Cook felt "this obliged us to fire upon them and unfortunatly either two or three were kill'd, and one wounded." Four or possibly five Māori were dead, contravening the Royal Society's guidance on cultural encounters.[12]

There had been some violence earlier, but as historian Nicholas Thomas remarks: "No contact at Tahiti had gone as badly as this." The incident prompted considerable regret by Cook and Banks. Cook believed "most humane men who have not experienced things of this nature will cencure my conduct in fireing upon the people in this boat nor do I my self think that the reason I had for seizing upon her will att all justify me." Had he anticipated "the least resistance I would not have come near them, but as they did I was not to stand still and suffer either my self or those that were with me to be knocked on the head." For Banks, it was "the most disagreable day My life has yet seen, black be the mark for it and heaven send that such may never return to embitter future reflection."[13]

By late October, *Endeavour* had rounded the east cape on the north island. In an astronomical coincidence of the first order, the date for the transit of Mercury was approaching, and Cook recorded his desire "of being in some convenient place to observe" it.[14] Settling into a cove, which Cook named Mercury Bay, Cook and Green went ashore to view it on November 9. Lieutenant John Gore was left in command of the ship, which led to another noteworthy transgression of Royal Society protocol. Gore had more experience in the South Pacific than anyone on board, having sailed with both Byron and Wallis, but showed little wisdom and restraint in this instance. Gore and a young Māori man had agreed on an exchange of the young man's cape for a piece of cloth. When the Māori man rowed off without completing the transaction, Gore shot him dead. Learning of the episode when he returned to the ship, Cook was outraged at the loss of life over such a trifling. It was the most brazen murder perpetrated by any member of Cook's three crews and, it is worth noting, it occurred on the celebrated first voyage.

Cook rounded New Zealand's North Cape, and by mid-January 1770 he was in Queen Charlotte Sound, near the modern city of Wellington. From the adjoining heights, he discerned the bi-insular nature of New Zealand, but before exiting the strait, now named for him, he conducted a sovereignty rite. After a quick jaunt up the east coast of the north island to reconnect with the starting point of his survey at Poverty Bay, Cook reversed course and proceeded along the coast of the southern island. By mid-February, *Endeavour* had turned Banks Peninsula, near modern-day Christchurch. The pulse of many on board now quickened because it would soon be clear if the southern island was an extension of or adjacent to a continent. According to Banks, Gore believed he saw what "might be land, so he declard on the Quarter deck: on which the Cap[tain] who resolvd that nobody should say he left land behind unsought for orderd the ship to be steerd SE," that is, out to sea. Cook stated he was "very certain that it was only Clowds which dissipated as the Sun rose."[15]

No land came into view. Gore was insistent. An exasperated Cook wrote that there was no longer any hope of land "because we must have been more than double the distance from it at that time to what we were either last night or this morning." He was confident on this point because "the weather was exceeding clear and yet we could see no land either to the Eastward or Southward." Gore was adamant, so Cook maintained the southeasterly course for another day. At last, Cook bore southwest toward the coast to resume his survey of the southern island. Banks concludes his account of this episode by noting that everybody, except Gore, was now convinced that it was impossible to have seen land.[16] For Cook, the incident proved a valuable lesson. When Gore revived his penchant for exuberant optimism on the Northwest Passage voyage, Cook was quick to discount his enthusiasm.

Banks and Gore were the principals in what Banks had playfully called a contest between the "we Continents" versus "the no Continents." Beaglehole observed that Banks "was letting the continental theory go hard." As *Endeavour* proceeded south along the New Zealand coast, Banks repeatedly referred to the adjoining land as "the continent," as if saying so made it so. Every new vantage raised his hopes. The final denouement began on March 1–2, 1770, when a strong gale from the west blew *Endeavour* offshore. After Cook regained land on March 4, he was confident, by reading the sea more than the horizon, that the island's southern tip was near. The expectant Banks "had the pleasure to see more land to the Southward," and when that trend continued the next day, he wrote that some skeptics began to think "Continental measures will at last prevail." This notion faded over March 9–10, when *Endeavour* rounded New Zealand's southern cape. "To the regret of us Continent mongers," Banks recorded, "a great swell from SW" signalled the absence of any land in that quadrant. On March 10, Banks's entry was terse: "Blew fresh all day but carried us round the Point to the total demolition of our aerial fabrick calld continent."[17]

Cook was never optimistic about finding the southern continent; more commonly, he derided the very idea. In contrast to Banks's melodramatic prose, Cook's entry for March 10 is bereft of rhetorical embellishment: "I began now to think that this was the southernmost land and that we should be able to get round it by the west, for we have had a large hollow swell from the SW ever sence we had the last gale of wind from that quarter [March 1–2] which makes me think that there is no land in that direction." Cook completed his circumnavigation of New Zealand on March 27, 1770. But before leaving the islands he summarized what he had accomplished since Tahiti, asserting that he had "set a side the most if not all the arguments and proofs that have been advance'd by different Authors to prove that there must be a Southern Continent, I mean to the northward of 40°s for what may lay to the Southward of that Latitude I know not."[18]

⚓

Closing out the search for a southern continent during his first voyage, Cook laid the groundwork for the next:

> I have given my Opinion freely and without prejudic[e] not with any view to discourage any future attempts being made towards discovering the Southern Continent, on the Contrary, as this Voyage will evidently make it appear that there is left but a small space to the Northward of 40° where the grand Object can lay, I think it would be a great pitty that this thing which at times has been the object of many ages and Nations should not now be wholy clear'd up, which might be easily done in one Voyage without either much trouble or danger or fear of misscarrying as the Navigator would know where to go look for it.[19]

Here, Cook provides an important insight into the evolution of his strategic thinking and self-perception. The hundreds of small islands that Cook found during his three Pacific voyages, including those he refused to survey, were, to a degree, the functional outcome of serendipity. The Hawaiian Islands are the best example. On his second and third expeditions, Cook perceived himself as an explorer trying to solve geographic problems on a continental, not an insular, scale. The *philosophes* in the salons of London and Paris might have been enraptured by tropical islands; to Cook, they became merely warm-weather tactical resorts.

With the New Zealand survey complete, Cook "resolved to quit this country altogether and to bend my thoughts towards returning home by such a rout as might conduce most to the advantage of the service I am upon." He wanted to return by way of Cape Horn, and by doing so he could have transected the one remaining quadrant of the Pacific Ocean that might "prove the existence or non existence of a Southern Continent." However, to do that, *Endeavour* would have had to sail "in a high latitude in the very depth of winter." The ship was ill-prepared to run that risk because of the deteriorated condition of its sails and rigging. Sailing directly across the Indian Ocean to the Cape of Good Hope was ruled out because "no discovery of any moment could be hoped for in that rout." So Cook determined "to return by way of the East Indies" by striking the east coast of New Holland (Australia) then coursing toward its northeastern extremity.[20]

Cook's officers supported this decision, but Banks viewed the itinerary with "much regret." "That a Southern Continent realy exists," Banks wrote, "I firmly beleive." He confessed his reasoning was "weak; yet I have a prepossession in favour of the fact which I find it dificult to account for." What gnawed at Banks were the icebergs he saw near Cape Horn. Relying on the old theorem that ice "is formd by fresh water only," he assumed that land (with rivers) had to exist "to the Southward for the Coast of Terra del Fuego is by no means cold enough to produce such an Effect." Here, Banks broached, late in Cook's first voyage, the question of the origin and extent of oceanic glacial phenomena, which would dominate the second and third expeditions. Banks believed the icebergs sighted at Cape Horn must have been conveyed by the westerly winds and currents from a continent "well away to the Westward," in the "very high latitudes" of the Pacific Ocean.[21]

To find this elusive continent, Banks dictated a methodology that would be "the best and readyest way" of finding it, in effect outlining the course of Cook's second voyage. He advised leaving England in the spring, stopping at Cape Town for refreshment, then, for the same purpose, arriving at "any of the numerous harbours at the mouth of Cooks streights" by October. This was springtime in the Southern Hemisphere, what Banks deemed "the good season" for exploration. Aided by the westerlies, ships could range across "any Latitude," he continued. Should a follow-up voyage "not fall in with a Continent," the venture "might still be of service by exploring the Islands in the Pacific Ocean." Banks argued that this project ought to be "promoted by the Royal Society," but the expenses would have to be borne by the British government. He dismissed potential naysayers by weighing the "trifling" expense against the acclaim that the nation whose seamen executed the plan would secure. Beaglehole believed Banks's plan was in actuality Cook's thinking, and perhaps the itinerary was, but the patriotic vision belonged to Banks.[22]

⚓

The coast of Australia was sighted on April 19, 1770, south of present-day Sydney. Sailing to the north, some members of the crew proved they were not only tired and bored but cruel. A small gang cut off the ears of Richard Orton, the captain's clerk, while he lay asleep. Cook was disturbed by this incident, but because he was unable to identify the perpetrators, he also saw it as a threat to his authority and a betrayal of his good will, which it was. If this had happened at any point on the third voyage, historians would use it as evidence of a commander close to losing his crew to mutiny. But, coming as it did a little more than two weeks before Cook's heroics on Endeavour Reef, the episode has been written off as a first-voyage anomaly.

The most celebrated aspect of the Australian tour, if not Cook's entire career, has been his handling of *Endeavour*'s near wreck on the Great Barrier Reef on June 11, 1770. Cook, like everyone else on board, had never experienced a coral reef before. The crew was practised in looking for sandy shoals or stray submarine pinnacles, but coral rock (the calcareous skeleton of invertebrate polyps rising from the ocean floor, reaching for sunlight) was a new phenomenon. Coral reefs are particularly dangerous because they appear unannounced from great depths, leaving little room to manoeuvre if they are encountered suddenly. The Great Barrier Reef is an unusually complex labyrinth that covers 215,000 square miles of seacoast, and the full extent of it would not be mapped comprehensively for another two hundred years. (This reef, like sea ice, is now under duress because of climate change. Saltwater captures carbon dioxide from the atmosphere, resulting in a higher rate of acidification, which slowly dissolves the coral.)

The serial telling of *Endeavour*'s impalement on the reef, twelve miles off Australia's east coast, established Cook's legend as a commander who remained cool in a crisis. In Banks's account, he states that, early on, the "fear of Death" was visible on every countenance. Disaster was averted because "every man exerted his utmost for the preservation of the ship." Manning the ship's pumps was strenuous, but the key to avoiding catastrophe was the absence of panic. Contrary to what Banks had heard about sailors – that they will "plunder and refuse all command" "as soon as a ship is in a desperate situation" – on *Endeavour,* calm reigned because of "the cool and steady conduct of the officers." At no point during the day-long crisis were they other than "perfectly composd and unmovd by the circumstances howsoever dreadfull they might appear."[23]

Sydney Parkinson's journal likewise contains all the standard elements of the saga, including Cook's direction that ballast, wood, water, and cannon be thrown overboard to lighten the ship. But Parkinson introduces a new element to the story by relating that when the pumps were engaged, "every man on board assisted, the Captain, Mr. Banks, and all the officers, not excepted; relieving one another every quarter of an hour."[24] Cook does not mention doing this. Banks describes his own labour and fatigue but not Cook's hand at the pump. Accordingly, Hawkesworth's account trumpets only the nobleman's contribution. But *Endeavour*'s collision with the reef also shows how quickly Cook and crew were able to reconcile in the wake of Orton's mutilation. Cook did not live long enough to effect a similar rapprochement after the so-called sugar-cane beer mutiny of the third voyage, and for that reason the discord off the Hawaiian Islands has had a disproportionate profile in Cook lore.

After making repairs on land and relaunching the ship, Cook backed out of the reef's maze and threaded his way to open water farther from shore. This decision came at a cartographic cost. If Cook kept land in sight to maintain an integral chart of New Holland's coast, he ran the risk of impalement on another coral reef and, possibly, the loss of his ship. Or he could choose discretion and sail well away from the shore, guaranteeing the continuation of the voyage. Facing this conundrum, Cook composed one of his most self-reflective journal entries:

> The world will hardly admit of an excuse for a man leaving a Coast unexplored he has once discover'd, [because] if dangers are his excuse he is thn charged with Timorousness and want of Perseverance and at once pronounced the unfitest man in the world to be employ'd as a discoverer; if on the other hand he boldly incounters all the dangers and obstacles he meets and is unfortunate enough not to succeed he is thn charged with Temerity and want of conduct. The former of these aspersins cannot with Justice be laid to my charge and if I am fortunate enough to surmount all the dangers we may meet the latter will never be brought into question.[25]

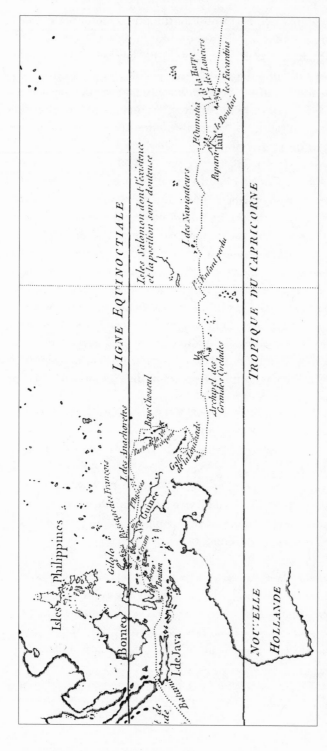

Louis-Antoine de Bougainville, *Chart Showing the Track around the World of the Boudeuse and Etoile under the Command of M. de Bougainville, 1766–1769* (detail, 1772). This map shows the west-central section of the Pacific covered by Bougainville in 1768. Cook's *Endeavour* track overlapped Bougainville's in Tahiti and then again in the vicinity of the Great Barrier Reef. Bougainville's image displays the hazy (pre-Cook) understanding of the relationship between Australia and New Guinea, including the possibility that they were the same mass of land. By navigating Torres Strait, Cook settled the matter once and for all.

Cook left the coast, resulting in a gap in his chart of the Queensland coast. Since there was nothing at issue relative to New Holland's interior, the matter was quickly passed over in the celebratory aftermath of the *Endeavour*'s voyage.

On his final voyage, a different set of circumstances – bad weather combined with strict adherence to his instructions – would take Cook far from the North American coast north of Nootka Sound. This also created a gap in Cook's cartography, but in this case, his thoroughness was questioned by some fur traders who sailed in his wake and then, much later, by historians of the Pacific Northwest who cast doubt on his third-voyage vigour.[26]

On August 21, 1770, Cook rounded Australia's northern promontory, what he named York Cape, an important benchmark in the course of the voyage. Cook's journal again takes on a reflective tone. After acknowledging the Dutch navigators who had first detected Australia's western shore, Cook retroactively claims the continent's east coast, "New South Wales," for Britain on the presumption that it "was never seen or viseted by any European before us."[27] This proved true, but Cook's proposition entailed some wishful thinking, for he would shortly learn in Batavia that Bougainville had just sailed in nearby waters, the probable cause for this inscription.

No Indigenous inhabitants engaged with the possession ceremony, contrary to Cook's instructions, but he was mindful of them:

From what I have said of the Natives of New-Holland they may appear to some to be the most wretched people upon Earth, but in reality they are far more happier than we Europeans; being wholy unacquainted not only with the superfluous but the necessary Conveniences so much sought after in Europe, they are happy in not knowing the use of them. They live in a Tranquility which is not disturb'd by the Inequality of Condition: The Earth and sea of their own accord furnishes them with all things necessary for life, they covet not Magnificent Houses, Household-stuff etc., they live in a warm and fine Climate and enjoy a very wholsome Air, so that they have very little need of Clothing and this they seem to be fully sencible of ...; this in my opinion argues that they think themselves provided with all the necessarys of Life and that they have no superfluities.[28]

With passages like this, Cook helped embed the idea of multiculturalism in Western civilization. As historian Felipe Fernández-Armesto phrases it, Enlightenment exploration frequently acquainted Europe with "morally superior" people who challenged "how society could be organized and life lived." Cook's exposition on Indigenous sensibilities was so radical and contrary to English cultural presumptions that Hawkesworth refused to include it in his account. Even Beaglehole, writing in the mid-twentieth century, excoriated Cook for deviating from the practical work of geography to write "nonsense about Australian society."[29]

Rounding York Cape and entering the Arafura Sea, Cook definitively proved that Australia was separated from New Guinea. Torres Strait – named after Spanish explorer Luís Vaez de Torres, who passed through it in the early seventeenth century – was so poorly understood in European geographic circles that Bougainville's master chart showed New Holland and New Guinea as a connected mass. Cook was now clear for a direct sail to Batavia (Jakarta), a Dutch port on Java where *Endeavour* could resupply for the concluding run to England.

⚓

Conventionally interpreted as empty of drama, several seemingly unrelated episodes on *Endeavour*'s transect of the Arafura Sea would prefigure better-known events during the concluding two months of Cook's final voyage. As soon as the ship entered the Arafura, Banks noticed a change in the crew's mood. Mere weeks after Cook had engineered a seemingly providential recovery on the Great Barrier Reef, Banks states: "The greatest part of them were now pretty far gone with the longing for home which the Physicians have gone so far as to esteem a disease under the name of Nostalgia; indeed I can find hardly any body in the ship clear of its effects but the Captn[,] Dr Solander and myself, indeed we three have pretty constant employment for our minds which I beleive to be the best if not the only remedy for it."[30]

Other than Beaglehole, who observed that it was one of the first recorded uses of the term "nostalgia," this comment has been little remarked on by historians. In effect, Banks was suggesting that, psychologically, the voyage was "over" for everyone except for himself, Solander, and Cook. Beaglehole concluded that *Endeavour*'s crew was "becoming bored with the voyage." They were well fed, brilliantly managed, and fit. After two years at sea, not a single sailor had died of sickness, an astonishing feat for that time. But what they wanted, Beaglehole averred, "was not the consolation of good health or reflections on the excellence of their commander's administration, but a known port, the sight of European faces," and most of all "a conventional voyage across known seas homeward."[31]

Once on the Arafura, Cook made a half-hearted attempt at delineating New Guinea's southern coastline. He wanted to "touch upon that Coast," but shoals made him head west "in hopes of meeting with fewer dangers and deeper water." This timidity is in contrast with Cook's determination to contest similarly shallow waters in Alaska's Norton Sound late in his Arctic expedition, when his skills as an explorer were supposedly exhausted.[32] Cook steered *Endeavour* to the northwest for a few more days in the hopes of hailing the New Guinea coast, all the while frustrated with the map in front of him, drawn by Robert de Vaugondy. After he "brought to," Cook went ashore "having a mind to land once in this Country before we quit it."[33]

This set the stage for an incident that underscored the truth of Banks's statement about the crew's lack of focus and growing feistiness. When Cook's party landed, the men thought the beach was uninhabited, but they were disabused of that notion after seeing footprints in the sand. Cook heard voices coming from the woods and concluded that it was not "safe to venture in for fear of an ambuscade." His instincts were rewarded, because after walking two hundred yards, the party was attacked. Cook retreated to the safety of the ship. As Hawkesworth later polished the tale, Cook was then "strongly urged by some of the officers to send a party of men ashore, and cut down the cocoa-nut trees for the sake of the fruit." This Cook "peremptorily refused, as equally unjust and Cruel." In his view: "The natives had attacked us merely for landing upon their coast, when we attempted to take nothing away, and it was therefore morally certain that they would have made a vigorous effort to defend their property if it had been invaded." Cook confided in his journal that had he been persuaded to land, men on both sides would have been killed, "and all this for 2 o[r] 300 green Cocoa-nutts which when we had got them would have done us little service."[34] Here, we see Cook's avid adoption of the Royal Society's hints for engagement with Indigenous peoples.

The next week, off the coast of Timor, Banks extended his psychological analysis, noting that the "wind came fair today and left our melancholy ones to search for some new occasion of sorrow." Cook was not happy either. His struggle to make sense of Vaugondy's maps continued. He groused about islands "wrong laid down, and this is not to be wonderd at when we consider that not only these Islands but the lands which bounds [sic] this sea have been discover'd and explor'd by different people and at different times." Cook said his predecessors were not faultless for this errant cartography, but the principal blame lay with "the Compilers and

Publishers who publish to the world the rude sketches of the Navigator as accurate surveys without telling what authority they have for so doing."[35] Cook's otherwise obscure critique of Vaugondy is important because it anticipated his engagement with dubious Russian cartography in the North Pacific, an episode historians have cited as evidence of the erratic explorer of the third voyage.[36] During all stages of his career, Cook was a severe critic of geographic malpractice whenever he encountered it.

What happened next validated Banks's theory. West of Timor, Cook discerned the promising small island of Savu. Steering toward it, he saw "Houses, Cocoa-nutt Trees and Flocks of Cattle grazing." This was a tempting scene, "hardly to be withstood by people in our Situation, especially such as were but in a very indifferent state of health and I may say mind too." Having refused to touch at Timor, Cook believed he "could not do less than to try to procure some refres[h]ments here as there appear'd to be plenty."[37] His textual shorthand seems to indicate a lot of grumbling aboard *Endeavour,* which is consistent with Banks's earlier critique.

These stories from the Arafura Sea – a Cook backwater if ever there was one – reveal that morale and cartographic challenges were issues during Cook's celebrated first voyage, not just the last one. They also indicate the onset of a portentous temporal threshold at play. By September 1770, *Endeavour* had been at sea for over two years. At this mark during all three expeditions, the mood and durability of Cook's crews began to wane, even though the conventional interpretation of the final voyage is that only Cook was worn out. Historians who see him losing grip on reality and unable to read his crew's emotional frame of mind when he failed to land in Hawaii in December 1778, two months before his death,[38] ignore his repeated refusals to alight during the transect of the Arafura Sea.

On October 1, 1770, Java Head came into view. Cook took possession of "the Officers, Petty officers and Seamens Log Books and Journals, at least all that I could find and enjoyn'd every one not to divulge where they had been."[39] This step, prescribed by naval tradition, meant that the discovery phase of the voyage had ended. When they reached Batavia on October 2, Cook and his shipmates were eager for news from England. Two Dutch East Indiamen were at anchor, and Cook dispatched Lieutenant Zachary Hicks to secure reports on recent events. Several items were of sufficient interest for Banks and Parkinson to record them in their journals, but the most salient was the incipient colonial rebellion. Parkinson refers to "fresh disturbances ... in America on account of taxes." Banks expands on the matter, allowing that "the Americans had refus'd to pay taxes of any kind," which had prompted "a large force being sent there." Cook makes no mention of this or any news from home. His journal emphasizes the navigational insights Hicks procured from Dutch captains regarding currents in the Indian Ocean.[40]

Ironically, just when Cook and his crew thought the voyage was effect-ively over, they faced their greatest tribulation. The crisis was not naviga-tional but pathogenic. Batavia was a notoriously unhealthy place because of its unsanitary canals and mosquito-infested swamps. Almost every crewmember was soon on the sick list because of pestilence: malaria, yellow fever, and dysentery. Tupaia, with his pristine immune system, was among the first to fall. Banks and Solander sought refuge in the country-side, a journey that took them away from the city's standing water. After reboarding, Solander saw mosquitoes breeding in the ship's scuttlebutt, but apart from recording their distasteful appearances, he drew no epi-demiologic conclusions. After refitting, *Endeavour* left port on December 26, but she became, as Beaglehole phrased it, "a hospital ship."[41] The Dutch commandant had congratulated Cook because so few of his men had died in Batavia, but the toll had just begun. Over the next six weeks, while sailing to Cape Town, twenty-three died. At one point, only ten men were available for deck duty. Among the dead were the naturalist Parkin-son and Green, the astronomer. Cook, who had done so much to pre-serve the health of his crew for more than two years at sea, was reduced to the role of writing obituaries in his journal as *Endeavour* sailed across the Indian Ocean. Nearly one-third of her complement died after the ship left Batavia.

It was in Cape Town that Cook, and Banks more particularly, learned that Bougainville had visited Tahiti with an unspecified "natural historian." Had Banks known that the scientist in question was Philibert Commerson, a celebrated botanist, he might have been even more alarmed than he was. Banks had joined *Endeavour* with the conceit that no one of his stature in the field of botany would have undertaken the risk he did. Only a light-weight, or a poseur, he thought, would dare it. In the moment, Banks exclaimed: "How necessessary then will it be for us to publish an account of our voyage as soon as possible after our arrival if we mean that our own countrey shall have the Honour of our Discoveries!" His concern was that if Bougainville made it into print first with an account of his voyage, the voyages of Wallis on *Dolphin* and Cook on *Endeavour* might be eclipsed. In the event, Bougainville published his narrative in 1771, two years before Hawkesworth's compendium appeared. As Banks feared, Bougainville ignored Wallis's voyage out of ignorance (when he was aware of British achievements, the French explorer cited them).[42]

Cook left Cape Town for England on April 16, 1771. A fortnight later, in a stroke of understatement, he wrote of having "cross'd the line of our first Meredean, viz. that of Greenwich having now circumnavigated the Globe in a west direction."[43] Cook was now among that elite fraternity who had captained a voyage "around the world," a phrase that resonated deeply in the travel literature of his time. On July 12, *Endeavour* anchored in the Thames. That evening, Cook, in faithful adherence to his instructions, went to the Admiralty to announce his return. Soon, all of London was ablaze with news of the voyage, and Banks played a starring role. Cook had his reward, though: within a month, he was promoted to commander in recognition of his accomplishments – navigational, cartographic, and managerial. In a letter to his old mentor John Walker in Whitby, Cook apologized for not writing sooner, citing Admiralty encumbrances not to discuss his voyage publicly. But he noted with pride "that the Voyage has fully Answered the expectation of my Superiors." He had the additional honour of a "Conference with the King ... who was pleased to express his Approbation of my Conduct in Terms that were extremely pleasing to me." Cook's next line was internally contradictory. He asserted that he had "made no very great Discoveries yet I have exploar'd more of the Great South Sea than all that have gone before me so much that little remains now to be done to have a thorough knowledge of that part of the Globe." With that glance toward future voyages, he also noted that the highest latitude he had ever reached with *Endeavour* was 60° 12' S, southwest of Cape Horn, and "here we had finer weather than in a Lower Latitude."[44] The bar was reset: for Cook, 60° S was the new 40° S.

PART TWO *A Frozen World*

FOUR

Toward the South Pole

Cook's first Pacific voyage so firmly established his navigational reputation that he was quickly tabbed as the captain of a two-ship flotilla and charged with conducting the definitive search for Terra Australis Incognita. Both craft were former colliers like *Endeavour. Resolution* was Cook's flagship while the consort *Adventure* was commanded by Tobias Furneaux.

Cook's second voyage is the only one that did not originate with the Royal Society. The Admiralty did all the planning. Hugh Palliser, head of the Navy Board, sent First Lord Sandwich a letter that reviewed *Endeavour*'s voyage and glanced ahead to the forthcoming expedition. Palliser noted that although discovery had not been the "first object" of Cook's transit voyage, "it enabled him to traverse far greater Space of Seas, before then unnavigated: to discover great Tracts of Country in high and low South latitudes, and even to explore and survey the extensive Coasts of those new discovered Countries; in short it was ... Cook's great Diligence, Perseverance and Resolution during the Voyage that enabled him to discover so much more, and at a greater Distance than any Discoverer performed before during One Voyage." As Peter Moore has recently observed, "perhaps the Admiralty's greatest discovery on the *Endeavour* voyage was Cook himself."[1]

Cook knew his star was rising. His meeting with King George III was proof. Yet in his letter to John Walker he professed to "have made no very great Discoveries." At least Cook was attempting to modulate his ego. Not Joseph Banks. Flush from his *Endeavour* success, Banks expected to be a part of the new venture. In a letter to Sandwich, Banks said he understood the purpose of the forthcoming mission was to "perfect the discoveries that had been begun in the last voyage." He was not shy about thinking of exploration in personal terms. In a December 1771 letter to the comte de Lauraguais, he wrote: "Next we shall sail upon a new undertaking of ye same kind in which we shall attempt the Souther[n] Polar Regions, O how Glorious would it be to set my heel upon ye Pole! and turn myself round 360 degrees in a second."[2]

Banks was either using a figure of speech or naively referring to an episode on *Endeavour*, when, west of Cape Horn, Cook briefly reached the sixtieth parallel in pleasant weather. Either way, he was tapping into two interrelated theories of ancient origin: saltwater did not freeze at a great distance from land, and seas near the poles were ice-free. If these age-old assumptions were correct, maybe Cook could sail all the way to 90° S and Banks could do his pirouette on deck. Of course, this was contingent on the legendary southern continent not blocking the way. An adroit observation by J.C. Beaglehole applies: "It is permitted to the geographer ... to have hypotheses; it is the task of the explorer to go and see."[3]

Cook, the explorer, would go and see, but Banks, the geographer, did not. He overplayed his hand by refitting *Resolution* to meet the tastes of an upper-class naturalist. A new deck, added to house his expansive retinue, made the ship top-heavy. When this was revealed after a sea trial, Cook recommended that these "improvements" be removed. The Admiralty concurred. When he saw the removal in progress, Banks stormed off the dock and then refused to proceed with the voyage except on his terms. Lieutenant Charles Clerke, who had befriended Banks on *Endeavour*, sent him a pleading letter, stipulating unequivocally that *Resolution* was "the most unsafe Ship, I ever saw or heard of." Nonetheless, if Banks wanted to "embark for the South Pole in a Ship, which a pilot ... will not undertake to carry down the River; all I can say, is, that you shall be most chearfully attended, so long as we can keep her above Water."[4]

Banks would not be cajoled. He conducted a vituperative correspondence with the Admiralty to no effect, and he also drafted a public letter criticizing the lords – which was never published – signed "Antarcticus." Thus, like Alexander Dalrymple who kept himself off *Endeavour*, *Resolution* sailed without Banks. The Admiralty's lack of support annoyed Banks but, as with Dalrymple, officialdom came down decisively on the side of naval prerogatives. With the passage of time, Banks realized he had acted presumptuously. He never blamed Cook, nor was there any long-standing animosity between the two men.[5]

Hurriedly responding to Banks's last-minute withdrawal, the father and son team of Johann and George Forster were boarded as naturalists. Johann Reinhold Forster was a year younger than Cook and a German polymath conversant in seventeen languages, ancient and current. He shared with Banks the status-symbol affectation, common among explorers, of having canine companionship, a spaniel, but when it came to shipboard temperament, he was the polar opposite of the congenial nobleman. Not that Forster was an inferior scientist, quite the contrary.

Both he and his son were far superior to Banks in one formidable aspect: they could reduce their studies to writing and get them published.

Formerly a clergyman in the German Reformed Church, Johann Forster's scientific specialization was zoology (with ornithology as a major subfield), but his interests were stunningly diverse, ranging from mineralogy to classic literature. He had arrived in London in 1766 from Danzig with his eldest son, George, hoping to gain entry to the world's premier circle of Enlightenment thought. The senior Forster had previously tried for an intellectual breakthrough with the St. Petersburg Academy, where he presented some of his zoological research. He did not receive an appointment there, but he did secure letters of introduction to scientific luminaries in Great Britain. Arriving in London nearly impoverished, Forster gradually eased his way into the Republic of Letters. Among his early acquaintances was Daniel Solander, a Swedish émigré and protégé of Linnaeus who would also assist Banks on *Endeavour.*

After delivering several papers on ancient inscriptions at meetings of the Society of Antiquarians in 1767, Forster was invited to become a fellow in a membership that included Solander, Banks, and Dalrymple. These contacts were Forster's entrée to the Royal Society, which he first attended that same year. Another new friend, and fellow immigrant, was Matthew Maty from Holland, the Royal Society's secretary who would later become an important figure in the denouement of Cook's third voyage. Before the year was out, Forster secured work as a tutor in modern languages and natural history at the Warrington Academy, an early forerunner of the modern liberal arts college. At Warrington, where Joseph Priestly also taught, Forster established himself as one of Great Britain's prominent naturalists by publishing a treatise in mineralogy that revolutionized study of the subject.

Forster's intellect, determination, and networking skills put him on a path to secure a coveted place on Cook's second voyage merely six years after immigrating to England. His trademark lack of tact cost him the position at Warrington at the end of the spring 1769 semester, about the time Cook was appointed *Endeavour*'s commander. Forster shifted to supporting himself, son George, and the balance of his family who had since moved to England by teaching languages in the local grammar school. He also translated the botanical works of Linnean travellers to North America. These books were well received, raising Forster's stock in Britain's scientific circle. But his most important connection, and the man to whom Forster hitched his fate, was Daines Barrington, a lawyer, fellow antiquarian with an interest in science, and vice-president of the Royal

Society. Propitiously, Barrington was also the foremost advocate in British circles for seeking the Northwest Passage, which would play a role in the evolution of Cook's second voyage account and the origins of the third.

After Cook returned to England in July 1771, word soon spread that the Admiralty had plans for a return voyage to the South Seas. That summer, Forster was going to press with his English translation of Bougainville's *Voyage autour du monde,* which he hoped might transform the *Endeavour*'s Banks-Solander duo into a *Resolution* triumvirate. In February 1772, Forster was elected a fellow in the Royal Society (having been nominated by Banks, Barrington, Solander, and Maty), but Banks settled on James Lind, a physician known for his antiscorbutic study and skilled in astronomy besides, to accompany him on Cook's next voyage. However, three months later, Banks was out of the picture. Forster was recommended as a replacement by the Royal Society and lobbied for by Barrington, who effectively served as his agent with the Admiralty.

Barrington approached Forster a mere six weeks before *Resolution* sailed. Forster agreed to join the expedition if he received the same salary Banks had been slated to receive. He also insisted that his now eighteen-year-old son George be signed on as an assistant naturalist and artist. (In the event, George would have a much bigger role in the history of the voyage than his title suggests.) Filling out the team of naturalists was Anders Sparrman, a young Swede and protégé of Linnaeus who enrolled at Cape Town. Another scientist aboard *Resolution* was the astronomer William Wales, who had viewed the transit on Hudson Bay. His counterpart on *Adventure* was William Bayly, an assistant at the Royal Observatory in Greenwich who had tracked Venus at Norway's northern extremity.

A parliamentary snafu in the wake of Banks's withdrawal forced the Admiralty to petition to King George III for an allotment from his royal funding to pay for Forster's salary and equipage. Forster would later conflate this circumstance into the presumption of a royal appointment, which he believed insulated him from other authorities, including Cook. Compounding matters, on the basis of a promise that Barrington had deployed to sweeten the offer, Forster undertook the voyage expecting he would write and profit from the voyage's official account. After the expedition, Cook and Forster became entangled in a contentious dispute because Cook presumed the same prerogative. Forster's journal never appeared in that form under his own name. Instead, as a means of circumventing Admiralty restrictions controlling unauthorized accounts, it provided the foundation for a book his son published on their return.

In Bernard Smith's estimation, George Forster's *A Voyage Round the World* (1777) "was the best written account to issue from all three of Cook's voyages." This is the more remarkable because it was rushed into print to beat the appearance of Cook's official publication. From the time of its appearance, informed observers thought the book was based on the father's journal, if not cowritten by him. (Wales, *Resolution*'s astronomer, who defended Cook against postvoyage disparagements from the Forster family, was the first to publicly suspect its synthetic origins.) Forster's *Voyage* was considered a paragon of Enlightenment sensibility with its fondness for reason, nature, and humanity. Although intermittently critical of Cook's decisions, Forster's account is a reliable complement to Cook's version of events because it does not stint on praise for Cook either. In the introduction to his book, George calls him "our skilful and famous navigator."[6]

Prevented from writing a connected travel account, Johann would use his journal to inform a compendium of essays on scientific topics. As a rule, the botanical work during the voyage fell to George and Sparrman, while Johann concentrated on zoology, geology, and ethnology. Johann and Cook engaged in a continuous scientific discourse over the reigning theory about the origins of icebergs and polar ice packs. Michael Hoare, editor of Forster's journal, concludes, "there must have been many ways scientifically in which Cook learned from Forster and vice-versa," one salient instance being "the theoretical discussion upon the origins of Antarctic ice, whether it was terrestrial or marine-formed."[7]

When Cook left on his second voyage no one, including Cook himself, could have anticipated that encounters with frozen masses would dominate the expedition. Historians Nicholas Thomas and Oliver Berghof cogently observe that received opinion on each of Cook's voyages focuses on aspects "extraneous to its chief purpose." *Endeavour* was launched to observe the transit, but accounts of the voyage are dominated by descriptions of Cook's delineation of New Zealand and the east coast of Australia. The historiography of the third voyage focuses on Cook's death in Hawaii, not his pivotal role in the evolution of the cartographic image of the Northwest Passage, a foundational element in the history of North America. For the second voyage, Cook's mission was to search the Southern Hemisphere's high latitudes for a continental mass. Cook would not find it, but he and Forster encountered astounding volumes of ice in different forms. They reflected on polar hydrology routinely and occasionally experimented with it. This is the great scientific legacy of Cook's second voyage but the scholarly discourse, dominated by an anthropological outlook, is fixated on cultural encounters in the tropics.[8]

⚓

Cook's second voyage was unlike the first in several important respects, including the complications brought about by sailing with a sister ship. Tobias Furneaux, *Adventure*'s commander, had served under Wallis as an inconsequential lieutenant. He was a serviceable seaman, but as Beaglehole observed, "not really an explorer" in temperament because he lacked imagination and curiosity.[9] Furneaux would lose contact with Cook twice, the second time permanently, and a portion of his crew was cannibalized in New Zealand, one of the most infamous episodes in South Pacific history. If Furneaux was not much of an explorer, there was one in the making on Cook's ship, midshipman George Vancouver. A mere fifteen-year-old when *Resolution* left England, Vancouver attended to his duties seriously. He learned not only how to reef the sails and steer the ship but also, more critically, how to read seas and shores, both with his eyes and through chart making. In maritime lore, Vancouver is reputed to have climbed out on *Resolution*'s bowsprit to claim the honour of being the farthest south.

As for the geographic scope of the second expedition, the southern continent was the entire focus, not an adjunct to the primary mission, as it had been on *Endeavour*. In the Admiralty's account, Cook describes his strategy as "keeping in as high a latitude as I could, and prosecuting my discoveries as near to the South Pole as possible, so long as the condition of the ships, the health of their crews, and the state of their provisions, would admit of."[10] The tropics were to be visited only for staging and refreshment. He reduced the exploratory agenda to "whether the unexplored part of the Southern Hemisphere be only an immense mass of water, or contain another continent, as speculative geography seemed to suggest." This, he continued, "was a question which had long engaged the attention, not only of learned men, but of most of the maritime powers of Europe." The objective was "to put an end to all diversity of opinion about a matter so curious and important."[11]

Resolution and *Adventure* left England on July 13, 1772. Unlike *Endeavour*, which had sailed into the Pacific against the wind and currents by taking a course west of Cape Horn, the ships cruised to Cape Town before heading southeast into the Indian Ocean and on to circumnavigate the Southern Hemisphere. To aid this daunting navigational effort, Cook and Furneaux had recourse to seasoned astronomers in Wales and Bayly, men schooled in the laborious lunar-distance method of calculating longitude. They also had two of the newest marine chronometers. These timepieces kept constant time for Greenwich by which the ships' longitude could be derived almost at a glance.

When Cook framed the expedition in his general introduction to the second voyage account, polar ice was not posited as either a strategic factor or a tactical impediment; ironically so, since this substance would prompt his legendary epigram. Beaglehole notes that "no one had yet made what could be properly called an antarctic voyage." Cold, stormy weather and dangerously high seas were to be expected, and given Cook's procurement of ice anchors and hatchets, some unspecified encounters with ice. But, Beaglehole added, "what picture otherwise he had formed in his mind we ourselves have no means of knowing."[12] Cook, unlike Banks, never speculated on how far south he might go.

On *Endeavour*, Cook had briefly encountered cold, icy conditions rounding Tierra del Fuego and a hint of similar weather running from Tahiti to New Zealand. So he had every reason to expect frigid conditions exploring the higher southern latitudes. Accordingly, as he recorded in the official account: "Some additional clothing, adapted to a cold climate, was put on board; to be given to the seamen whenever it was thought necessary."[13] Thus prepared, over the course of three successive southern "summers" – December through February of 1772–73, 1773–74, and 1774–75 – Cook swept the icy latitudes of the Indian, Pacific, and Atlantic oceans, respectively. Encounters with fog, snow, and ice became routine. During the austral winters, when mission-centric exploration was impossible, Cook returned to the salubrious climate of Polynesia.

⚓

After leaving Cape Town on November 22, 1772, the ships bore southeast into the Indian Ocean on the longest of Cook's three Antarctic ice-edge cruises. Anticipating the rigours ahead, George Forster stated: "Every person whose duty exposed him to the severity of southern climates, from the lieutenant to the sailor, was provided with a jacket and a pair of trowsers of the thickest woolen stuff called *fearnought,* or strong flannel, which kept out the wet for a long time." The lack of tailoring made the clothing too short for some sailors, but Forster believed these garments "helped the crew considerably." Though his men were equipped with heavy if ill-fitting woollens, Cook noted on December 6, at 48° S, that "the air so cold that every one complains." Temperatures were constantly at or slightly above freezing. Four days later, near 51° S, Cook saw an "Island of Ice to the Westward," his first sighting of an iceberg, which would soon be commonplace. The ships were buffeted by snow and sleet, and Cook related the detection of another iceberg, "which the Adventure took for land and made the signal accordingly."[14] It was a false alarm, one of many.

By December 12, the ships were near 53° S, and the size and frequency of icebergs increased. George Forster was stunned that "in the midst of December, which answers to our June ... we had already passed several pieces of ice." This he attributed to the "want of land in the southern hemisphere ... since the sea, as a transparent fluid, absorbs the beams of the sun, instead of reflecting them." His not-quite-correct hypothesis (land is warm because it absorbs and reradiates, rather than reflects, solar radiation) was emblematic of the long chain of discourse inaugurated that week in Cook's and Johann Forster's journals regarding the circumstances that had produced the ice floating around them. Forster found it "difficult to conceive where and i[n] what manner these huge Masses could be formed." His son George recorded that when the ships breasted an iceberg, "the wind blowing from thence" caused the temperature to drop four degrees, which "very perceptibly affected our bodies." He deduced that "the large masses of ice greatly contributed to refrigerate the general temperature of the air in these inhospitable seas." The net effect was that "the great degree of cold in these icy regions entirely precluded the idea of a summer, which we had expected at this time of the year."[15]

Young Forster also reported that icebergs "were hourly seen in all directions around the sloops, so that they were now become as familiar to us as the clouds and the sea." His father declared: "It requires really a steady, settled mind to view with composure the Quantities of Ice surrounding a Ship in about 54° Degrees of Latitude." In his published narrative, Cook likewise marvelled at the number and size of these icy islands, but he was also increasingly aware of the risk they presented. Some, he noted, were "two miles in circuit, and sixty feet high ... yet, such was the force and height of the waves, that the sea broke quite over them. This exhibited a view which for a few moments was pleasing to the eye; but when we reflected on the danger, the mind was filled with horror. For were a ship to get against the weather-side of one of these islands when the sea runs high, she would be dashed to pieces in a moment." To manage this uncertainty, we learn from George Forster, Cook divided the crew "into three watches, not into two, as is usual on warships." This meant the men "were less exposed to the changes of the weather, and it gave them time to dry their clothes when they got wet."[16]

On December 14, Cook was stopped at 54° 55' S by a new phenomenon, "an immence field of Ice to which we could see no end." To keep from getting stuck in this low ice at the ships' waterline, Cook pulled back, only to be carried into the midst of the icebergs. This he considered the lesser of two evils: "Dangerous as it is sailing a mongest the floating Rocks in a thick Fog and unknown Sea, yet it is preferable to being entangled with Field Ice under the same circumstances." The danger was getting stuck fast without considering the damage that could be inflicted if the ships were detained for any length of time. In the southern summer, Cook expected that the ice would be breaking up, just as it had near Greenland during a comparable season. For this insight, Cook drew on the experiences of shipmates who had witnessed North Atlantic ice phenomena first-hand, and he made the first of several references to their experiences in his journal. But there was no evidence of this dissolution in the Indian Ocean's high latitudes. There were some "hills" (i.e., icebergs) embedded in the ice, like those floating freely in the sea, "and some on board thought they saw land also over the ice." At first Cook concurred but changed his opinion "upon more narrowly examining these ice hills, and the various appearances they made when seen through the haze."[17]

Fugitive sightings of land – such as John Ross's 1818 "discovery" of the Croker Mountains seemingly astride Lancaster Sound in the Canadian Arctic – were a common, at times infamous, phenomenon in polar exploration. These mirages, a shimmering effect created by air turbulence when there are significant temperature differentials perpendicular to the Earth, are akin to atmospheric astigmatism – that is, a stronger curving of light on the vertical scale than the horizontal. The resulting illusion is called *fata morgana*. Light reflecting off ice filtered through several layers of increasingly warmer air (a temperature inversion) creates the illusion of a greyish rampart in the distance, which is often mistaken for land.[18]

Cook's ships sailed east along the edge of the ice between the fifty-fourth and fifty-fifth parallels for ninety miles. (This is well north of the seasonal mean for ice in modern times.) There was an occasional "bay" that Cook "looked into," but none provided a southerly opening. The constancy of the packed ice led him "to suppose that this Ice either joins to or that there is land behind it." This supposition mirrored Johann Forster's uncertainty about where and how to get to land "through the great impenetrable Fortification of Ice."[19]

⚓

Cook's oblique comments were the first of many reflections made during the second voyage on how sea ice was formed. His views, like Johann Forster's, would evolve after subsequent observations to such an extent that both men later disavowed much of what was thought to be known when the voyage commenced. In his polished, published narrative of this leg of the journey, Cook laid out the orthodox understanding and what it portended: "It is a general opinion, that the ice I have been speaking of, is formed in bays and rivers," that is, in juxtaposition to terrestrial influences. The belief was that sea ice materialized either in shallow saltwater close to shore (defined by Johann Forster as "bays, Straits, Sounds, harbours, Inlets and other such narrows near some land of considerable Extent"), called landfast, or it was the outflow of formerly frozen freshwater rivers. In the depths of an expansive sea, it was thought that saltwater could not freeze. As Forster phrased it, "it is well ascertained" that masses of oceanic ice "consist of fresh water, and must therefore come from some Land or other." In the moment, Cook supposed that "land was not far distant." Having already sailed 90 miles "along the edge of the ice, without finding a passage to the south," he sailed east another 120 miles hoping "to get to the southward, and, if I met with no land, or other impediment, to get behind the ice, and put the matter out of all manner of dispute."[20]

At this juncture – anticipating that glacial phenomena would become routine – Cook formally developed a lexicon to meet his narrative requirements. His typology was an amalgam of his own thinking; insights gleaned from two men on board "who had been in the Greenland trade," that is, the North Atlantic whale fishery; and Johann Forster's shared thoughts on what he had "read and heard" about the subject. In his polar nomenclature, Cook contrasted free floating "ice islands," known as tabular icebergs in modern scientific parlance, with "packed ice," or what Forster at one point called "table ice," which consisted of one contiguous and relatively flat mass that was "not above 3 or 4 feet or perhaps something more above the water, but extends in a continued body, often for miles together." (This is the kind of ice that can hold ships fast; most famously, Ernest Shackleton's near disaster in Antarctica's Weddell Sea and the USS *Jeannette*'s similar fate in the Arctic in the 1880s, which inspired the northern experimental expeditions of Fridtjof Nansen and Roald Amundsen in the late nineteenth and early twentieth centuries.) Cook differentiated pack ice from another low-rise form, which he denominated "field ice ... from its immence extent," consisting "of many pieces of various sizes both in thickness and Surface, from 30 or 40 feet square" to small chunks two or three feet wide. In contrast to the undivided pack, the key feature of this form, from a navigational perspective, is the occasional opening that avails between pieces of rubble.[21]

Cook did not think it was advisable to sail through broken-up field ice because it "would be found too hard for a Ships side that it is not properly armed against it; how long it may have or will lay here is a point not easily determined; such Ice is found in the Greenland Seas all the summer long and I think it cannot be colder there in summer that it is here." Whether the Southern Hemisphere might be colder than the North would become an oft-visited question by Cook and Johann Forster. Even though the temperature was regularly below freezing in "the middle of Summer," Cook was not yet ready to concede that point.[22]

⚓

On December 20, 1772, Cook's visibility was limited by sleet and snow. It was now obvious the austral summer was not going to relent from its role as winter in disguise, so the tailors lengthened the sleeves "of the Seamans Jackets" and made "Caps to shelter them from the Severity of the Weather." *Resolution* and *Adventure* spent Christmas at 57° 50' S, barely east of Cape Town's meridian. "The weather was fair and cloudy," Cook wrote, and "the air sharp and cold, attended with a hard frost." With near freezing temperatures in "the middle of summer with us," he doubted "if the day was colder in any part of England." In deference to naval custom, there was no exploring that day. He wrote: "Seeing that the People were inclinable to celebrate Christmas Day in their own way, I brought the Sloops under a very snug sail least I should be surprised with a gale wind with a drunken crew." But Cook did not stop studying the flat pieces of ice floating about, which appeared "from four to six or eight inches thick." He repeated that its appearance was "of that sort which is generally formed in bays or rivers."[23]

Over the last week of 1772, Cook sailed southwest and into the Atlantic. He was searching for "Cape Circumcision," a land somewhat miraculously sighted by the French explorer Lozier Bouvet in 1739 but, it turns out, one haphazardly charted: it was placed eight degrees too far east and slightly north of the true mark. The explorer's name for the island was derived from the liturgical calendar's former Feast of the Circumcision, January 1. Now known as Bouvet Island, it is the most isolated terrain in the world (at 54° 25' S and 3° 22' E). Cook's hypothesis was that Cape Circumcision was a projection of a larger land mass whose snow-covered terrain was the generator of all the ice he had been encountering. By December 30, Cook had reached west of Cape Town's meridian, though, at 60° S, his ships were five degrees closer to the pole than where he had turned away from the ice in the Indian Ocean in mid-December. This suggested that the pack had either moved or melted back. Cook saw penguins, considered "a sure indication" of land, so not encountering terra firma was puzzling. He surmised that these birds could roost on icebergs but concluded, correctly, that they needed land on which to raise their young. Without resolving the matter, Cook committed to taking continued "notice of these birds whenever we see them" and deferred judgment on what their appearance signified.[24]

On New Year's Eve, Cook broke off his westward run into the Atlantic. Spying an immense field of ice to the north that extended "farther than the eye could reach," he wanted to avoid an icy vise. Hauling southeast, *Resolution* and *Adventure* briefly carried south of the sixtieth parallel, the first of six times Cook would transect that line. He began to think that Bouvet had never sighted land and pronounced that what the French explorer had seen "was nothing but Mountains of Ice surrounded by field Ice." His own initial experience in these latitudes had shown how easily a voyager might be deceived. Cook also began to question his own supposition that the ice he had first encountered was "join'd to land to the Southward." With a modicum of experience at the ice edge now in hand, Cook expressed his first doubts about the prevailing theory on sea ice formation or, as he put it, the "general recieved opinion that Ice is formed near land."[25]

To address this quandary, Cook and Forster conducted an experiment to determine if saltwater froze. Lowering a thermometer into water at various depths, they found temperatures below that were required to freeze freshwater. Forster theorized that saltwater needed a colder temperature to freeze than freshwater and, alternatively, that the motion created by wind chop, tides, and currents prevented seawater from freezing. The latter theory was also Cook's emergent thinking. (After the voyage, Forster studied the scientific literature of ice formation and came to a firm conclusion, but suffice it to say here that we now know that freshwater begins to crystallize at 32 °F, whereas seawater will not freeze until the temperature approaches 28 °F, with minor variations depending on salinity.)

Cook did not yet understand how the ice floating about him was created, but he appreciated its utility. He deemed melting iceberg fragments for freshwater "the most expeditious way of Watering I ever met with." William Wales stated that the ships were "much distressd for want of Water" and considered melted ice of "more real value than Gold!" When Wales had traversed Hudson Strait on his transit mission, he had seen men capturing water as runoff from the sides of icebergs, a more dangerous manoeuvre than Cook's method of capturing a chunk, hoisting

William Hodges, *Ship's Watering by Taking on Ice in 61° S, January 4, 1773*. After leaving Cape Town in November 1772, Cook's ships regularly encountered icebergs on a track that took them above the Antarctic Circle. In addition to their mass, number, and aesthetic properties, icebergs were a convenient source of fresh water during Cook's three-month transect of the Indian Ocean's high latitudes and thereafter. In this drawing, Hodges depicts the harvesting of a chunk of ice to be hoisted aboard for chipping into smaller pieces and melting.

it aboard, chopping it into pieces, and then stuffing it into casks, where it melted. Johann Forster reported that when these barrels were taken down into the hold, the temperature dropped fifteen degrees. But this procedure raised eyebrows for another reason. George Forster recalled: "Several persons on board, unacquainted with natural philosophy," thought the ice "would burst the casks in which it was packed, not considering that its volume must be greater in its frozen than in its melted state, since it floated on the surface." To "undeceive them," Cook put some ice in a pot, "where it gradually dissolved, and in that state took up considerably less space than before." This "ocular demonstration" was better than the "clearest arguments." Just how this ice, from the saltwater of the Indian Ocean, yielded, as Cook said, "water perfectly sweet and fresh," remained unexplained. Forster also noted that the ice water was sweet, but as to "whether this happens in the very act of congelation, that only the fresh water particles freeze and the saline ones not, I cannot say." He would revisit this problem after the expedition in his scientific treatise.[26]

⚓

Cook was sorry to have spent as much time as he did on Bouvet's "imaginary lands." This was a replay of his frustration with Robert de Vaugondy's cartography on *Endeavour* and a foreshadowing of his examination of Russian maps on the third voyage, which historians see as an anomaly and explain as evidence of a diminished explorer.[27] Giving up, for now, on Bouvet's elusive cape, Cook veered southeast on a track that, on January 17, 1773, took him beyond the Antarctic Circle on the fortieth meridian (south of Madagascar). It was the first time any European had crossed this polar ring and would prove to be one of three such occasions during the voyage. Cook was aware of the historic nature of the moment. He wrote that in clear, 34 °F weather, "about a 1/4 past 11 o'Clock we cross'd the Antarctic Circle for at Noon we were by observation four Miles and a half South of it and are undoubtedly the first and only Ship that ever cross'd that line."[28] What is interesting about Cook's formulation is that he backtracked chronologically to determine the moment when *Resolution* had crossed the polar boundary. *Adventure* passed that mark at about the same time, but he makes no mention of the consort.

Cook was hoping for a long sail in open water toward the pole, but he soon reached what proved to be his Farthest South (67° 15') for the Indian Ocean ice-edge foray, a limit set by the reappearance of the Antarctic ice pack. Rarely is Cook's vigour as a navigator more clearly evident than in his journal entry for the next day, January 18:

> From the mast head I could see nothing to the Southward but Ice ... without the least appearances of any partition, this immense Feild was composed of different kinds of Ice, such as high Hills or Islands, smaller pieces packed close together and ... of such extend that I could see no end to it ... I did not think it was consistant with the safty of the Sloops or any ways prudent for me to persevere in going farther to the South as the summer was already half spent and it would have taken up some time to have got round this Ice, even supposing this to have been practicable.[29]

Cook hauled northeast hoping to preserve the potential for another deep run to the south before summer ended. Though he did not know it, when Cook turned away from the ice edge, he was only seventy-five miles from Antarctica.

Cook's ships proceeded to the fiftieth parallel on the sixtieth meridian east of Greenwich, one-third of the way between Africa and Australia. On

January 30, he observed: "This is the first and only day we have seen no Ice sence we first discovered it," referring to the second week of December. This prompted another of his serial reflections about the ice's origins:

> I know it will be asked from whence this huge body of Ice comes, not one of which I am obliged to answer, at least I shall not do it now, but leave the further discussion of this subject, till some other oppertunity, as it can hardly be doubted but that I shall have occasion to resume it again, perhaps more than once in the course of the Voyage.

He did. As the voyage progressed, discussion about Terra Australis Incognita vanished, replaced by a running dialogue on polar hydrology.[30]

⚓

Cook now sought Kerguelen Island, named after French navigator Yves-Joseph de Kerguelen-Trémarec, who had discovered it the year before. Kerguelen, like Bouvet, had extrapolated from a single ephemeral sighting of land to extravagantly claim that he had discovered a headland of the southern continent, La France Australe. Cook had learned of Kerguelen's find in Cape Town, a report that guided his route away from the ice edge to the northeast. But the Indian Ocean's long heavy swell convinced him that no continent lay in this vicinity, so after cruising the fiftieth parallel for a week, he cut bait, not wishing to engage his time and the crew's patience in a rerun of the search for Cape Circumcision. Concluding that if Kerguelen's sighting was real, it could only be an island, Cook moored his search for it. He would later find this rocky outcrop on his third expedition 480 miles east of its posted location.

At this juncture, *Resolution* and *Adventure* became separated in fog. According to a preset plan, Cook and Furneaux conducted a maritime minuet, spending three days looking and signalling for each other via cannon and fire without success. By Cook's previous order, in anticipation of just such an eventuality, Furneaux headed directly to the rendezvous in New Zealand, where they would meet again. Cook was confident Furneaux would reach Queen Charlotte Sound comfortably. In his journal, he showed no concern about sailing *Resolution* unaccompanied. Not so George Forster and, if he is to be believed, everyone else:

> We were obliged to proceed alone on a dismal course to the southward, and to expose ourselves once more to the dangers of that frozen climate, without the hope of being saved by our fellow-voyagers, in case of losing our own

vessel. Our parting with the Adventure, was almost universally regretted among our crew, and none of them ever looked around the ocean without expressing some concern on seeing our ship alone on this vast and unexplored expanse."[31]

Cook may have been serene, but it would be wrong to conclude that he was pleased by this turn of events or the permanent separation from *Adventure* that happened later in the voyage. No one has ever blamed Cook for these incidents, which occurred when scholars presume he was at the peak of his seafaring skills, but he never lost contact with his consort in the Arctic fog during his search for the Northwest Passage.

Adventure may have disappeared from view, but Cook was not done with his high-latitude survey of the Indian Ocean. By February 22, not quite two weeks after the separation, *Resolution* was again approaching the sixtieth parallel on a southeasterly glide in sleet and snow. Halfway between Africa and Australia, Cook stopped near a small iceberg to secure water, but it turned turtle just as the boats were lowered, pitching *Resolution* with a huge wave. Disaster was barely averted. Ironically, just when a consort vessel might have been desirable, Cook did not have one. With his ship "surrounded on every side with huge pieces of Ice equally as dangerous as so many rocks" and the days getting shorter, Cook was increasingly concerned about navigating around "mountains of ice which in the night would have passed unseen." Accordingly, he abandoned his plan for "crossing the Antarctick Circle once more." Standing north in a hard gale made for "great distruction among the Islands of Ice."[32]

This sequence prompted Cook into a rare engagement with the sublime, Edmund Burke's construct for understanding emotional reactions to exotic or foreboding landscapes. As found in the published account, Cook reflected:

The large pieces which break from the ice islands, are much more dangerous than the islands themselves. The latter are so high out of the water, that we can generally see them, unless the weather be very thick and dark, before we are very near them. Whereas the others cannot be seen in the night, until they are under the ship's bows. These dangers were, however, now become so familiar to us, that the apprehensions they caused were never of long duration; and were, in some measure, compensated both by the seasonable supplies of fresh water these ice islands afforded us, (without which we must have been greatly distressed,) and also by their very romantic appearance, greatly heightened by the foaming and dashing of the waves into the curious holes and caverns which are formed in many of them; the whole exhibiting a view which at once filled the mind with admiration and horror, and can only be described by the hand of an able painter.[33]

Cook's description of the event intertwined Enlightenment observation with a proto-Romantic sensibility that seemingly inspired Meriwether Lewis's discourse on the majestic nature of Missouri's Great Falls a generation later.

George Forster said Cook's decision to turn away from the ice was a popular one. Echoing the captain, he noted that the "favourable season for making discoveries towards the frozen zone" was closing. Daily temperatures "became more sharp, and uncomfortable, and presaged a dreadful winter in these seas; and, lastly, the nights lengthened apace, and made our navigation more dangerous than it had hitherto been. It was therefore very natural, that our people, exhausted by fatigues and the want of wholesome food, should wish for a place of refreshment, and rejoice to leave a part of the world, where they could not expect to meet with it." His father concurred: "We were happy enough to meet with no land to the South; which might have seduced us to spend a cold season somewhere on it, and to experience the rigors of an Antarctic winter."[34] During all three expeditions, the crew's taste for voyaging went stale long before Cook's did.

With the austral winter coming on, Cook ran the sixtieth parallel eastward across the width of Australia but well south of it, ending the second voyage's first of three seasons of polar exploration. It was, in Beaglehole's estimation, a "remarkable achievement of seamanship" because Cook, as a novice, conducted it "without accident in one of the stormiest oceans of the world, hampered by fog, and constantly among icebergs." Others followed in Cook's wake, but "none again was faced with the long-continued crisis of first experiment."[35] Cook also honed his narration skills, especially when describing icy phenomena. He had previous exposure to glacial substances, probably in the Baltic and certainly in maritime Canada, and he knew the dangers of navigating around them. In his initial encounters with Antarctic ice, Cook was relatively indifferent, but over time he was increasingly fascinated by it. By voyage's end, ice dominated his thinking.

⚓

Cook was ambivalent about his course exiting the Indian Ocean. On March 14, 1773, at 58° 22' S on the meridian parallel to the east coast of

Australia, he lamented that unexpectedly pleasant weather "made me wish I had been a few degrees of Latitude farther South." He was tempted "to incline a little with our course that way," but the weather turned, convincing him he was in the proper latitude. Cook reasoned that there was no further point to enduring the Southern Ocean's "intense cold" when the large swell from the southwest led him to conclude that there was "no land behind us in that direction." Cook bore away to the northeast "with a resolution of making the best of my way to *New Holland* or *New Zealand.*"[36]

In the event, Cook chose New Zealand. On the way, though, he passed on an opportunity to explore Van Diemen's Land (modern Tasmania), whose status as an island or extension of New South Wales was still in doubt. Cook was content with this decision on two counts. First, the winds were not favourable; they continually bent to the east. Second, he believed that Furneaux would have already cleared up that point on his separate track to Queen Charlotte Sound. This was an overestimation of his colleague's zeal. Once rejoined with *Adventure* and debriefed, Cook would never expect ambition on Furneaux's part again.

⚓

On March 26, 1773, Cook reached Dusky Sound at the tip of New Zealand's southern island. The first order was freshening the ship's larder because a handful of men seemed on the verge of scurvy. Green vegetables were hard to come by, so Cook brewed spruce beer. The recipe was modelled after a method he had seen employed in Newfoundland and consisted of conifer branches, shrubbery, molasses, and thickened wort. Looking ahead to a key episode that would inform historians' criticism of Cook's deportment near the end of his life, he cut off the supply of grog.[37] This beverage, one part rum and three parts water, was a Royal Navy tradition advised because of the typically foul condition of water stored in casks. On occasion, such as the need to husband the supply of spirits, a "small" or weak beer would be substituted for grog, which was Cook's reasoning here. According to Beaglehole, "the majority of the crew took to it very well," but off the Big Island of Hawaii during the third voyage a similar step precipitated considerable discord.[38]

The forge was landed to repair metal work, and the sailmakers began repairing canvas. The naturalists were busy botanizing, sportsmen hunted fowl, and in time even a scenic jaunt or two was arranged. The falls at Cascade Cove elicited the attention of the Forsters and the painter William Hodges, and they prompted William Wales to inscribe several stanzas from

James Thomson's lyrical paean to nature, *The Seasons,* in his journal. (Thomson was popular with both Enlightenment-era explorers and proto-Romantics such as Meriwether Lewis, who was introduced to the poet through reading the literature of Cook.[39]) After six weeks of recuperation, Cook sailed out of Dusky Sound aiming for Furneaux in the waters separating New Zealand's largest islands, the designated rendezvous. In mid-May, *Adventure* caught sight of *Resolution* and welcomed her with the flash and bang of cannon to Ship Cove, an arm of Queen Charlotte Sound.

This was the first time Cook had visited a Polynesian homeland twice. The problematic dynamic of such encounters informed his evolving outlook on the cultural ramifications brought about by voyages such as his. Seeing Māori men trafficking younger women in the sex trade, Cook observed: "Such are the concequences of a commerce with Europeans and what is still more to our Shame [as] civilized Christians, we debauch their Morals already too prone to vice and we interduce among them wants and perhaps diseases which they never before knew and which serves only to disturb that happy tranquillity they and their fore Fathers had injoy'd. If any one denies the truth of this assertion let him tell me what the Natives of the whole extent of America have gained by the commerce they have had with Europeans."[40]

This is one of Cook's most perspicacious observations. Nevertheless, it provides his modern critics with ammunition, notwithstanding the fact that he identified with the interests and fate of Indigenous peoples to a greater degree than he has been given credit for and never really believed that his "discovery" of them was a good thing. As Vanessa Collingridge points out, Cook may have become more short-tempered with his own crew, but "the second voyage had been marked by a growing tenderness ... towards the people he encountered on his journey."[41]

George Forster was also mindful of the encounter's problematic nature. His narrative contains a similar, if not more damning, indictment. To him, the loss of life was an "unhappy" if "unavoidable consequence of all our voyages of discovery," but he felt this "heavy injury" was "trifling when compared to the irretrievable harm entailed upon them by corrupting their morals. If these evils were in some measure compensated by the introduction of some real benefit in these countries, or by the abolition of some other immoral customs among their inhabitants, we might at least comfort ourselves, that what they lost on one hand, they gained on the other; but I fear that hitherto our intercourse has been wholly disadvantageous to the nations of the South Seas."[42]

⚓

Cook was a restless explorer. It is one thing to provide refreshment for yourself and the crew, another to while away time until the optimum moment for resuming the search for the southern continent. So he settled on a course that called for sailing east from New Zealand toward a band of oceanic expanse between the forty-first and forty-sixth parallels. This range was below *Endeavour*'s westbound track but not so far south as to tempt fate with the Antarctic winter. Reaching east of Tahiti's meridian, Cook would then sail north and, "providing no land was discovered," bisect *Endeavour*'s route. Once that space was closed, he would cruise west to Tahiti. This inside loop of the South Pacific would be completed by returning to New Zealand.[43]

As Cook stated in the voyage's account, this plan was intended to result in the mid-latitude exploration of "all the unknown parts of the sea between the meridian of New Zealand and Cape Horn." Informed by his recent experience sailing south of Australia, Cook explained the risks and potential benefits to be derived from his itinerary:

> Some may think it an extraordinary step in me to proceed on discoveries as far south [as] 46° degrees of latitude, in the very depth of winter. But though it must be owned, that winter is by no means favourable for discoveries, it nevertheless appeared to me necessary that something should be done in it, in order to lessen the work I was upon; lest I should not be able to finish the discovery of the southern part of the South Pacific Ocean the ensuing summer. Besides, if I should discover any land in my route to the east, I should be ready to begin, with the summer, to explore it. Setting aside all these considerations, I had little to fear; having two good ships well provided; and healthy crews.[44]

It took less than two months to complete this more compact of the two wintertime circuits Cook conducted through Polynesia in July and August 1773. He ran the forty-fifth parallel nearly to the meridian of Pitcairn Island before turning north and passing by (but not stopping at) that soon-to-be-famous refuge of the *Bounty*'s mutineers, thence proceeding to Tahiti and New Zealand. At the eastern end of this leg, and after eliminating one more sector of the South Pacific that could hold a continental mass, Cook concluded that "circumstances seem to point out to us that there is none." Still, this was "too important a point" to be left to conjecture; only "facts" could settle the matter, and collecting them could only be done by "viseting the remaining unexplored parts of this Sea which will be the work of the remaining part of this Voyage."[45]

Here is Cook the empiricist at his best, but his curiosity about South Pacific islands, apart from their ability to provide refreshment, was beginning to wane. Having already passed on surveying Tasmania, Cook forsook following up on Bougainville's discovery of islands near Pitcairn. He asserted that "no discovery of importance can be made, some few islands" excepted, while he remained in the tropics.[46] This was one of two clues Cook left in the record of his second voyage that indicated he was losing interest in becoming an insular cataloguer.

The fabled island of Tahiti came back into view on August 16. Cook's description of this moment was reserved, but George Forster was captivated by the scene: "It was one of those beautiful mornings which the poets of all nations have attempted to describe"; a "faint breeze ... wafted a delicious perfume from the land." Forster had finally secured for himself the already classic South Seas experience that he thought he had signed up for. But the romance soon wore off, replaced by the hard realities of the encounter. He reflected on "the introduction of foreign luxuries" that were likely to hasten what he presciently termed "that fatal period," wondering whether it was "better for the discoverers, and the discovered, that the South Sea had still remained unknown to Europe and its restless inhabitants."[47] With this passage, Forster introduced the concept of the fatal impact that later became the dominant trope in postcolonial interpretations of Cook's voyages.[48]

Similar themes emerged in Cook's journals and others, foremost among them descriptions of Polynesian thievery. On Huahine, Anders Sparrman made the mistake of going on a botanical search by himself. Polynesians beat and stripped him of all his possessions, including most of his clothes, but allowed him to return to the ship. Cook went to Oree, the Headman, to complain and seek a return of the stolen materials. What happened next bore a remarkable resemblance to what would occur at Kealakekua Bay when a Hawaiian Chief refused to board Cook's launch.

The cultural dynamic is best introduced by George Forster: "The old chief immediately resolved to assist captain Cook in the search after the thieves, but his noble resolution filled all his relations with terror. Upwards of fifty people of both sexes began to weep when he stepped into the boat; some with the most pathetic and moving gestures tried to dissuade him; and others held him back and embraced him; but he was not to be prevailed upon and went off with us." The launch travelled along the shore until it reached a point where Oree sent his men inland to retrieve some of Sparrman's clothing but not all his equipment. At this point, Cook determined on returning to *Resolution*. To his surprise, Oree, without a specific invitation, insisted on accompanying him back to the ship "in spite of the oppossission he met with from those about him." Afterwards, Oree was carried ashore, "where some hundreds waited to receive him" with embraces and "tears of joy in their eyes." This incident concluded to just about everyone's satisfaction except Johann Forster's. He was disappointed that Cook did not "think it convenient to take any other step for to bring the Thieves to any punishment." Michael Hoare, editor of Johann's journal, reminds us that Forster had "paid and equipped Sparmann" and that the episode became a foundational element in his estrangement from the captain.[49]

⚓

Cook completed his quick wintertime swing through Polynesia with a short stay at Tonga followed by his return to New Zealand, the northern tip of which came into view on October 21, 1773. He intended to secure refreshment before heading into the Pacific's high latitudes for his second ice-edge foray. Johann Forster anticipated botanizing during the austral spring when many plants would be in flower. These agendas were interrupted by a storm that blew hard from the northwest, a gale of sufficient strength that *Resolution*'s topgallant mast was carried away. Cook had no chance of getting the ships to the safety of Cook Strait, so he decided to ride out the storm by reefing all the ships' sails, allowing the wind to carry them away from land. At this point, southeast of Cape Palliser, *Resolution* and *Adventure* were separated again, this time permanently. When the storm abated, Cook sailed into Queen Charlotte Sound. He settled into Ship Cove on November 3 fully expecting to find Furneaux already there.

But *Adventure*, leaking badly and almost unmanageable in the storm, found refuge in Tolaga Bay, halfway up the east coast of the north island, on November 9. By this time, Cook had been waiting at Queen Charlotte Sound for six days. After repair and refreshment, Furneaux was back at sea on November 16, just as Cook began fretting about his consort's fate. He was "totally at a loss to conceive what is become of her." Cook surmised that *Adventure* had sought refuge in a nearby harbour "to compleat her wood and Water." This conjecture held for a while, but after twelve days Cook considered it improbable. Johann Forster stated that everyone on *Resolution* now "began to be alarmed." Cook spent the balance of his three weeks in Ship Cove refitting and resupplying *Resolution* while awaiting *Adventure*. This proved too brief for Forster, who failed to reach his botanizing goals in part because springtime weather was tardy; midday temperatures were in the low fifties. He did not know if this was "owing to a cold winter, or whether this is the annual constant state of the Climate."[50]

Strong winds kept beating *Adventure* off the coast, further frustrating Furneaux's attempts to reach the rendezvous. He finally reached Ship Cove in the strait on November 30, but Cook, restless with the austral springtime upon him, had weighed anchor and sailed away on November 25. Furneaux found a message in a bottle from Cook that included a rough outline of the commander's itinerary, including a projected stop at Easter Island and another return to New Zealand via Tahiti, but it did not formally stipulate a new rendezvous. Furneaux finally cleared Cook Strait on December 23. As Beaglehole phrased it, "a more imaginative man" might have headed south knowing that was Cook's intended direction, or he could have waited for him at one of the islands listed in the note. Instead, Furneaux headed for Cape Town and then home. He reached England on July 14, 1774, two years after departing and a year before Cook did.[51]

Furneaux's lagging interest in exploration was heavily influenced by an incident that occurred near Ship Cove before he left. For refreshment, he sent ten men to nearby Grass Cove to look for greens. When they did not return, Lieutenant James Burney was dispatched to find them and, as Beaglehole phrases it, came on "startling evidence of slaughter and cannibalism." This incident not only demoralized Furneaux, it centralized cannibalism in the literature on Cook.[52]

After leaving Ship Cove, Cook briefly looked for Furneaux, but with "all the officers being unanimous of opinion that the Adventure could neither be stranded on the Coast or be in any Ports in this Country," he determined to spend "no more time in serch of her, but to proceed directly to the Southward." Cook did not consider the absence of a consort an impediment to "fully exploring the Southern parts of the Pacific Ocean." His intention was to spend the forthcoming season exploring, and if he did not find "a Continent or isle between this and Cape Horn in which we can Winter perhaps I may spend the Winter within the Tropicks or else proceed round Cape Horn to Faulkland Islands."[53] This plan varied from the one outlined in the note he left Furneaux. Cook was keeping his options open.

Cook would end up spending the ensuing winter in the tropics, somewhat to the frustration of those, such as Johann Forster, who by then were wishing for a quicker conclusion to the voyage. As Cook's journals made clear, only the southernmost reaches of the Pacific now appealed to him as an explorer. Tropical islands were now mere stations for refreshment as the seasons turned. Even the South Atlantic was now more intriguing to him. The crew was upbeat, even without *Adventure*. Applying a phrase that became a common literary trope for the explorers who followed him, Cook reported that "not a man was dejected or thought the dangers we had yet to go through were in the least increased by being alone, but as cheerfully proceeded to the South or wherever I thought proper to lead them as if she or even more Ships had been in our Company."[54]

FIVE

The Limit of Ambition

In December 1773, James Cook began conducting his second extensive and gruelling exploration of the high southern latitudes, sailing a discursive route across the Pacific Ocean. This historic segment of the voyage started auspiciously. Southeast of New Zealand, Cook, the Forsters, and other colleagues celebrated one of Enlightenment-era exploration's most symbolic moments: reaching the Antipode of Greenwich, 180 degrees east and west of London and as far south of the equator as the British capital is north.

At first, they discovered that the Southern Ocean was strangely bereft of icebergs. For December 10, George Forster recorded reaching "59° south, without having met with any ice" but that on the same date the year before "we fell in with it ... between the 50th and 51st deg. of south latitude." Drawing on his father's journal, George found it difficult to account for the difference and speculated that "a severe winter preceding our first course from the Cape of Good Hope" might have been the reason, a view consistent with talk in Cape Town that it had been "sharper there than usual." Johann, referencing Bouvet's putative discovery, posited that "perhaps the vicinity of Land" in that sector accounted for their seeing ice "so early" and in "so low a Latitude."[1]

Resolution finally encountered an iceberg two days later at 62° 10' S, 11.5 degrees farther south than the first sighting, made after leaving Cape Town the year before. Three foggy and snowy days later, *Resolution* found low sea ice at 66° S. In his published account, Cook described it as an immense formation of ice, "close packed together; in other places, there appeared partitions in the field, and a clear sea beyond it." He did not think it was safe to venture forward because "the wind would not permit us to return the same way that we must go in." Indeed, the "loose ice" surrounding his ship was "rather more dangerous than the great islands," an observation he had also made the previous summer. Veering northeast, Cook got clear, "but not before we had received several hard knocks from the larger pieces, which, with all our care, we could not avoid." One danger succeeded another for "the weather remained foggy, and many large islands lay in our way." This combination of difficulties, "together with the improbability of finding land farther south, and the impossibility of exploring it, on account of ice, if we should find any, determined me to get more to the north." A westerly gale with snow followed *Resolution*, coating the ship's rigging and sails with ice. This was a particularly hazardous situation for Cook because *Resolution* was "hourly meeting with some of the large islands, which, in these high latitudes, render navigation so very dangerous."[2]

According to Cook, the icebergs "were very near proving fatal to us," some coming within a length or two of the ship. Citing an old proverb, he professed that "a miss is as good as a mile." When the weather cleared, bringing a favourable northwest wind, he bent to the east. Cook was eager to steer for the Antarctic Circle again but was thwarted when conditions changed to "thick fog, snow, sleet, and rain, which constitutes the very worst of weather." The rigging "was so loaded with ice that we had enough to do to get our topsails down." Frigid air seeped into the ship's accommodations. On December 16, Johann Forster's cabin was so cold (37 °F) it was as if he were "quartered in the open Air."[3]

Resolution crossed the Antarctic line for the second time on December 20, 1773. Cook noted this matter of factly. The icebergs, however, continued to impress him. They were "very high and rugged, forming at their tops, many peaks; whereas the most of those we had seen before, were flat at top, and not so high." These new ones were three hundred feet tall and two or three miles in circumference, "with perpendicular cliffs or sides, astonishing to behold." Johann Forster was also captivated by "the enormous size of these icy masses" and marvelled at "the great number of them." He noted 186 icebergs "in sight from the mast head," none smaller than "the hull of a ship." This observation prompted him to revisit a theme first raised on the Indian Ocean ice swing. Forster puzzled over "the degree of warmth and cold under the same corresponding degree of Latitude" in the opposing hemispheres. "Some parts" of the Northern Hemisphere, he commented, "are inhabited and even cultivated and are very fertile," and in summer they sometimes see "an intense heat." However, in the high latitudes of the Southern Hemisphere, "we have passed two Seasons called Summer ... and have hitherto found no Land at all." Instead: "Whenever we attempted to dive into the Antarctic-Circle we found all the Sea covered with solid masses of Ice, very probably extending to the very pole." Forster would return to the question of whether the

William Hodges, *Ice Island:* Resolution *Passing a Tabular Iceberg*, 1774. After leaving New Zealand in November 1773 for the Pacific's icy latitudes, Cook spent the austral summer by crossing the polar circle twice, with icebergs almost a constant companion. Having been separated from *Adventure* in New Zealand, Hodges took the artistic licence of putting *Resolution* into the distant scene, although he was actually aboard her at the time.

Southern Hemisphere was colder than its northern counterpart in his postvoyage treatise. But in the moment, he observed that "these truely inhospitable climates can hardly be expressed by words. The Sun is seldom seen. All is fogg and mist around us: hardly any birds are observed," though they saw whales occasionally.[4]

On December 22, *Resolution* reached 67° 31' at a point halfway between New Zealand and Cape Horn. Cook announced that this was "the highest we had yet been in." The next day, the ship "fell in with such a vast quantity of field, or loose ice, as covered the whole Sea from South to East and was so thick and close as to obstruct our passage." With moderating winds, Cook set out two boats to hoist in chunks of ice. This was "cold work," a factor that limited the amount that could be hauled aboard. A strong gale "attended with snow and sleet" terminated the effort. Freezing weather made "the ropes like wires, and the sails like boards or plates of metal."[5]

Like the previous southern "summer," winter-like conditions prevailed. Writing on Christmas Eve, 1773, Cook stated:

The cold so intense as hardly to be endured, the whole Sea in a manner covered with ice, a hard gale and a thick fog: under all these unfavourable circumstances it was natural for me to think of returning more to the North, seeing there was no probability of finding land here nor a possibility of get[ting] farther to the South and to have proceeded to the East in this Latitude would not have been prudent as well on account of the ice as the vast space of Sea we must have left lying to the north unexplored, a space of 24° of Latitude in which a large track of land might be, this point could only be determined by makeing a stretch to the North.[6]

Cook was reflecting on the practicability of employing the strategy used the previous year of moving into an untracked quadrant while also seeking temporary respite from the ice and cold. The intention was not to return to the tropics or even the subtropical latitudes of the low thirties, but to travel sufficiently north to get a reprieve from polar conditions while crossing previously unexplored stretches of sea, thereby ruling out the existence of a large land mass. As executed, Cook's course resembled the Greek letter omega (Ω), whose apex was just north of 50° S. In Martin Dugard's colourful characterization, "Cook chose to sail the world as if he were wandering in a park on a sunny day," in contrast to the straight-line mode of Magellan, Tasman, and Drake.[7]

Heading north, *Resolution* transited a winter-like wonderland. Icebergs were constantly in view. Reflecting an increasingly sour outlook, Johann Forster said with foreboding that the scene reminded him of "the wrecks of a destroyed world." Cook had positioned his ship such that "she drifted along with the ice, and by taking advantage of every light air of wind, was kept from falling aboard any of these floating isles." Cook stated: "Here it was we spent Christmas day, much in the same manner as we did the preceding one. We were fortunate in having continued day-light, and clear weather, for had it been as foggy as on some of the preceding days, nothing less than a miracle could have saved us from being dashed to pieces." As Johann Forster phrased it, "everyone of them threatens us with impending ruin." When accounting for the ship's isolated situation, compounded by the customary frivolity of the holiday and its "parcel of drunken Sailors hollowing and hurraing about us, and peeling our Ears continually with Oaths and Execrations, curses and Dam's," he found the situation "no distant relation to the Image of hell, drawn by the poets." If not "for the pinching cold, we would really think it were still more similar."[8]

As he steered north for two weeks after Christmas, Cook filled his journal with routine observations on the winds, bearings, birds, marine life, the "piercing cold" temperatures, the distances travelled, and the ever-present icebergs and occasional snow shower. He makes no mention of the shipboard dynamic. George Forster paints a distinctly different picture: "A general languor and sickly look ... manifested itself in almost every person's face, which threatened us with more dangerous consquences. Captain Cook himself was likewise pale and lean, entirely lost his appetite, and laboured under a perpetual costiveness." Twelve crew members, including George's father, were confined to bed with "rheumatic pains," which probably informed the gloomy outlook of his narrative. Indeed, in his own journal, Johann said his damp, dark cabin "looks more like a subterraneous mansion for the dead than a habitation for the living." Michael Hoare argues that Johann had entered "his deepest period of depression on the voyage."[9] But these circumstances had no appreciable effect on Cook's resolve to conduct the voyage; indeed, his most legendary moment as an explorer would soon follow. He seems to have never tired of exploration, even when sick.

By New Year's Day, 1774, *Resolution* was north of 60° S, where the master map of Cook's voyages stated there was "no ice to be seen." On January 4, the noontime temperature was above 40 °F for the first time since December 7, a balmy 46 °F. Two days later, about six hundred miles below Tahiti, Cook determined it was time to head south again: "it is not probable there can be any land" between 50° S and the tropical isles, "and it is less probable there can be any to the west from the high vast billows we now have from that quarter." Cook flattened the parabolic trajectory of his course away from the ice and reached a northerly limit of 47° 51' S on January 11 before bending back into the icy latitudes.[10]

⚓

The Forsters were coming to the same conclusion as Cook regarding a southern land mass. George stated: "It is sufficient for us, to have proved that no large land or continent exists in the South Sea within the temperate zone." As Johann phrased it in his journal: "A great continent cannot be there, for there are our two tracks that won't admit of it." George added that if a southern continent "exists at all, we have at least confined it within the antarctic circle." He made another point pregnant with meaning for the future of Cook's career, or rather historians' interpretation of it. George stated that it was "obvious that to search a sea of such extent as the South Sea, in order to be certain of the existence, or non-existence of a small island, would require many voyages in numberless different tracks and cannot be effected in a single expedition."[11] This insight was probably derived from a conversation he and Johann had with Cook, because both the captain's and his father's journal address the terrestrial holdings of the Southern Ocean at about the same time. In any event, George's comment reflected the great navigator's forthcoming discovery agenda, for the remainder of the second voyage and the next, which was becoming ever less focused on the insular particularity of the Pacific basin.

George's conclusion about the absence of a southern continent, notwithstanding its conformance with the captain's views, was a narrative mask because, unlike Cook, he and his father were content to simply assume that verdict and head home. The brief return to the fiftieth parallel had been greeted warmly by all aboard, but the thought of making another run toward the pole was universally detested, if George is to be believed. As *Resolution* proceeded south, he asserted: "These cold climates began now to hang heavily on our crew, especially as it banished all hope of returning home this year, which had hitherto supported their spirits. At first a painful despondence, owing to the dreary prospect of another year's cruize to the South, seemed painted in every countenance." The crew "resigned themselves to their fate" by degree until their outlook took on the cast of "a kind of sullen indifference." The worst part was "ignorance of our future destination, which, without any apparent reason, was constantly kept a secret to every person in the ship." Johann concluded that since it was apparent "that we are to stay out one year more," the "general opinion" was that "we are to winter at some Portugueze settlement in South America." But Cook was offering no such hints at this time.[12]

Johann Forster considered Cook's decision to head south a form of harassment. All "reasonable and good natured" people, he averred, would be "well contented with what we have hitherto done, provided we spare ourselves for to bring the news of our Discoveries home." Forster believed he had no choice but to submit to individuals such as the unnamed Cook, people "who are hardened to all feelings, and will give no ear to the dictates of humanity and reason." Mocking Cook's sense of "*good conduct*" and "*perseverance*," which dictated leaving "nothing to *chance*" (journal text that George did not convey to print), Johann ironically highlighted the qualities that made Cook a good explorer. In a stunningly prescient glance into futurity, he stated: "These people should be constantly employed by Government upon such Schemes: as for instance the N.W. or NE. Passage; there they will find a career to give their genius full Scope."[13]

All this grumbling took place in the middle of a voyage that is conventionally recognized by historians as Cook's greatest, and this was only the start of the complaining. George Forster's concurrent reflections on Cook's decision to head south are telling because he is attempting to recapture the mood of Cook's first voyage even though he was not party to it. In a kind of oral history based on unnamed informants who were *Endeavour* veterans, leavened by a heavy dose of wishful thinking, young Forster stated:

> This voyage was not to be compared to any preceding one, for the multitude of hardships and distresses which attended it. Our predecessors in the South Sea had always navigated within the tropic, or at least the best parts of the temperate zone; they had almost constantly enjoyed mild easy weather, and sailed in sight of lands, which were never so wretchedly destitute as not to afford them refreshments from time to time. Such a voyage would have been a party of pleasure to us; continually entertained with new and often agreeable objects, our minds would have been at ease, our conversation cheerful, our bodies healthy, and our whole situation desirable and happy. Ours was just the reverse of this; our southern cruizes were uniform and tedious in the highest degree; the ice, the fogs, the storms and ruffled surface of the sea formed a disagreeable scene, which was seldom cheered by the reviving beams of the sun; the climate was rigorous and our food detestable. In short, we rather vegetated than lived; we withered, and became indifferent to all that animates the soul at other times. We sacrificed our health, our feelings, our enjoyments, to the honour of pursuing a track unattempted before.[14]

Few passages in the entire body of Cook literature offer as much insight into the internal dynamics of Cook's three voyages as George Forster provides here.

The jibe directed at Cook's desire to run new tracks across the high southern latitudes is only the most obvious contrast between the captain's outlook and Forster's own. A voyage of discovery with Cook, according to this grievance, was supposed to be refreshing, entertaining, and good for one's health besides. It is easy to appreciate how that would be attractive, but the Forsters should have anticipated that exploration in less salubrious climes would involve all the opposite characteristics. Cook's fidelity to the mission of high-latitude discovery, notwithstanding the Forsters' criticisms, only puts the great navigator in a more positive light. Cook was, simply, remarkably diligent and durable, far more than most, perhaps all.

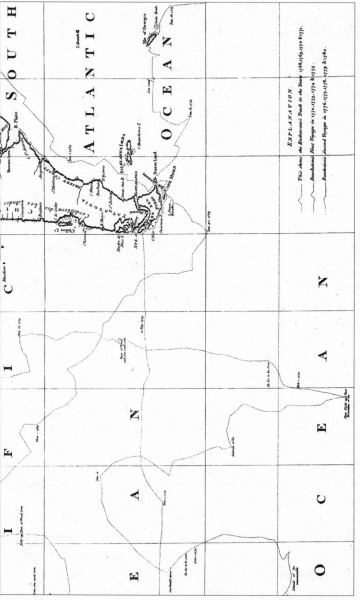

Resolution's Antarctic Probe, January 1774: *A General Chart Exhibiting the Discoveries of Captn. James Cook, in This and His Two Preceding Voyages; with the Tracks of the Ships under his Command, by Lieut. [Henry] Roberts of His Majesty's Royal Navy* (detail, 1784). This inset from the larger work completed after Cook's final voyage shows *Resolution*'s needle-nose-like penetration toward Antarctica in January 1774. It was on this leg that Cook reached his Farthest South. Cook deviated from his second voyage "in and out" of the ice method during his third voyage, where he conducted an extensive lateral manoeuvre at the edge of the Arctic ice pack for many degrees of longitude.

⚓

Notwithstanding George Forster's pining for the kind of pleasant voyage Banks had been on, Cook's next leg was his highest reach in either hemisphere, on any voyage. With rheumatic Johann sequestered in his cabin, obsessively recording how cold the interior temperature was and anticipating "the same scene as in the *Endeavour* at *Batavia*," Cook set the stage for his most famous moment. After his respite in the fifties, Cook sailed south in a swift, needle-nose-like run toward the South Pole. By January 18, 1774, he was above 60° S in temperatures hovering around 40 °F. Now southwest of the tip of South America, he had entered the Pacific's last prospect for a sizable land mass. Icebergs came back into view at 62° 34' S. Cook said one was "as large as any we had seen." It was around two hundred feet high and crowned by a cupola that reminded him of St. Paul's Cathedral in London. Thoughts of home did not distract from his strategic discernment. A great swell from the west, he wrote, "continues a probable certainty there is no land between us and the Meridian of 133 ½° which we were under when last in this Latitude." Cook was referring to his break to the north from the ice edge after Christmas hundreds of miles to the west. Presaging the conditions that would soon facilitate Cook's Farthest South, when the ship crossed the sixty-fifth parallel, temperatures moderated to a point Cook deemed "not unpleasant; and not a bit of ice in view. This we thought a little extraordinary, as it was but a month before ... we were in a manner blocked up with large islands of ice in this very latitude."[15]

This curious circumstance, Cook noted, "causes various opinions and conjectures," but he did not elaborate. Setting a high-latitude record was probably discussed. *Resolution* transited the Antarctic Circle for the third and last time on January 26, 1774. What was thought to be "an appearance of land to the East and SE" quickened pulses. Cook "haul'd up for it and presently after it disapeared in the haze." This ephemeral sighting prompted a depth sounding, but they "found no ground with a line of 130 fathom." As the ship stretched southeast in the direction of a potential landfall, Cook's mood was tinged with disappointment when he wrote "our supposed land was vanished into clowds." Johann Forster noted: "We have never before been so far South, and God knows how far we shall still go on." Indeed, echoing the prevoyage bravado of Joseph Banks, he avowed that "if Ice or Land does not stop us, we are in a fair way to go to the pole and take a trip round the world in five minutes."[16]

Cook once again encountered smaller ice floes on the afternoon of January 28, some of which were hoisted aboard for fresh water. He was soon in "so thick a fog, that we could not see two hundred yards round us; and as we knew not the extent of the loose ice, I durst not steer to the south till we had clear weather." His patience paid off. The next morning, "the Sky cleared up," and with a northeast wind *Resolution* "bore away SSE passing several large Ice islands." In pleasant weather, pushed by "a gentle gale," Cook plunged deeper, crossing 70° S for the first and only time. Optimism reigned. Charles Clerke found it a "very extraordinary incident in this part of the World – its the first day I can by any means denominate pleasant that I ever met with either within, or near, the Antarctick Circle." George Forster stated that "the mildest sunshine we had ever experienced in the frigid zone" was raising spirits and expectations. His father had "ventured upon deck for the first time after a month's confinement." These changes, he continued, meant "we now entertained hopes of penetrating to the south as far as other navigators have done towards the north pole," a reference to European voyages to Svalbard and the northerly extents of Baffin Bay, near 80° N.[17]

This was Nature's tease. The long austral mid-summer morning of January 30, 1774 (at *Resolution*'s latitude, sunrise was at 1:43 a.m.), proved to be a benchmark for the great navigator's career. Cook wrote:

A little after 4 AM we precieved the Clowds to the South near the horizon to be of an unusual Snow white brightness which denounced our approach to field ice, soon after it was seen from the Mast-head and at 8 o'Clock we were close to the edge of it which extended East and West in a streight line far beyond our sight; as appear'd by the brightness of the horizon; in the Situation we were now in just the Southern half of the horizon was enlightned by the Reflected rays of the Ice to a considerable height. The Clowds near the horizon were of a perfect Snow whiteness and were difficult to be distinguished from the Ice hills whose lofty summits reached the Clowds.[18]

This phenomenon is called ice blink, which is caused when sunlight reflects off snow or ice into the atmosphere. Earlier in his account, George Forster explained: "We were certain of meeting with ice in any quarter where we perceived a strong reflexion of white on the skirts of the sky near the horizon." Johann described this sensation further in his postvoyage treatise: "Whenever we approached large tracts of solid ice, we observed, on the horizon, a white reflexion from the snow and ice, which the Greenlandmen call *the blink of the ice.*" When it appeared, "we were sure to be within a few leagues of the ice."[19]

Although the term "blink" does not appear in Cook's journal for the South Pole expedition (he would, however, use it when he encountered the phenomenon in the Arctic Ocean while searching for the Northwest Passage), it established the limits of exploration in the search for the unknown southern continent. Cook wrote:

> The outer or No[r]thern edge of this immence Ice field was compose[d] of loose or broken ice so close packed together that nothing could enter it; about a Mile in began the firm ice, in one compact solid boddy and seemed to increase in height as you traced it to the South; In this field we counted Ninety Seven Ice Hills or Mountains, many of them vastly large. Such Ice Mountains as these are never seen in Greenland, so that we cannot draw a comparison between the Greenland Ice and this now before us: Was it not for the Greenland Ships fishing yearly among such Ice (the hills excepted) I should not have hisitated one moment in declaring it as my opinion that the Ice we now see extended in a solid body quite to the Pole, and that it is here, i.e. to the South of this parallel, where the many Ice Islands we find floating about in the Sea are first form'd, and afterwards broke off by gales of wind and other causes.[20]

Once again, Cook was grappling with glaciological theory; in this case, he was suggesting that icebergs originated amidst field or pack ice, a theory he would reject before the voyage was over.

⚓

Cook was reaching the apogee of his descriptive powers. He was also approaching what most historians cite as the summit of his career (ironically, given the dominance of the palm-tree paradigm). When he stretched to 71° 10' S, Cook offered this initial assessment in his journal: "I will not say it was impossible anywhere to get among this Ice, but I will assert that the bare attempting of it would be a very dangerous enterprise and what I believe no man in my situation would have thought of." Then he added one of the most memorable lines in the history of exploration: "I whose ambition leads me not only farther than any other man has been before me, but as far as I think it possible for man to go, was not sorry at meeting with this interruption, as it in some measure relieved us from the dangers and hardships, inseparable with the Navigation of the Southern Polar regions." Since he could not "proceed one Inch farther South," Cook turned north.[21] His southern record would not be broken until 1823, when James Weddell reached 185 miles farther into the South Atlantic.

Since the Pacific, because of its breadth, was considered the best prospect for finding Terra Australis Incognita, and having run the length of its southern perimeter, Cook believed this was the pivotal moment of the voyage. And so he slightly recrafted the original journal entry into a more polished reflection in anticipation of the published account. He replaced "among the ice" with "farther to the South." The inherent risk became "a dangerous and rash enterprise." He then added a line that sought the authority that comes from consensus:

> It was indeed my opinion as well as the opinion of most on board, that this Ice extended quite to the Pole or perhaps joins to some land, to which it had been fixed from the creation and that it here, that is to the South of this Parallel, where all the Ice we find scatered up and down to the North are first form'd and afterwards broke off by gales of Wind or other cause and brought to the North by the Currents which we have always found to set in that direction in the high Latitudes.[22]

Cook reiterated the landfast origination theory at a moment when Johann Forster was moving away from it.

According to George, his father believed "that all the south pole, to the distance of 20 degrees, more or less, is covered with solid ice, of which only the extremities are annually broken off by storms, consumed by the action of the sun, and regenerated in winter." This conformed to Cook's view, but, coming to a more radical point, George held that his father's opinion was "less exceptionable, since there seems to be no absolute necessity for the existence of land towards the formation of ice, and because we have little reason to suppose that there is any land of considerable extent in the frigid zone."[23] The first half of this proposition is correct as it relates to pack and field ice but not to icebergs, which originate from ice that accumulates on land. As detailed later, Cook never quite came to believe that seawater could freeze far from shore, but he was right (where the Forsters were wrong) about an extensive tract of land at the South Pole. He cited the appearance of penguins and other birds as evidence of land nearby.

For Cook to bring his published narrative into strict accordance with the Enlightenment's skeptical sensibilities, he stated that the ice was no longer "fixed from the creation" but dated to "the earliest time," and the phrase "if there was any land behind this Ice" appeared in print as "there must be some to the south behind this ice." And as for his famous line, with a slight change in tense it read: "I, who had ambition not only to go farther than any one had been before, but as far as it was possible for man to go, was not sorry at meeting this interruption."

The preceding excursion into Cook's emblematic declaration provides insight into the dynamics of how exploratory narratives were constructed. The text we read in published accounts is not always an authentic relic from the moment of discovery but rather a honed reflection that draws on literary convention. Indeed, Nicholas Thomas argues convincingly that Cook's famous expression echoes a line from Alexander Dalrymple's history of South Pacific voyages, published in 1770–71, that challenges would-be discoverers: "What has *he* done which no one else ever *did* before, or *can* do after him?"[24]

Beaglehole found Cook's avowal of ambition, instead of his usual understatement, noteworthy, but considered his welcoming of an "interruption" more significant. Comparing this moment with the Great Barrier Reef heroics, Beaglehole noted that here it was Cook, not the sea, dictating events, and he found the factor of human agency probative: "The decision when to turn back is one which, at some time or other, faces every explorer. There is always the desire to go on still a little farther – a desire which may, and often does, lead to disaster. It requires courage to turn back in time to face ... those who will perhaps say that you were prudent to save your own skin." Beaglehole concluded that making decisions "at the right time ... is the mark of the great explorer." In the moment, George Forster offered a concurring opinion. Having expressed misgivings about this run to the south before it commenced, he noted pensively that the ship made it to "less than 19 deg. from the pole; but as it was impossible to proceed farther, we put the ship about, well satisfied with our perilous expedition, and almost persuaded that no navigator will care to come after, and much less attempt to pass beyond us."[25]

On February 1, 1774, only two degrees north of his turn-back point, Cook was startled to discern that "the Sea was pretty clear of ice." Sensing a possible shortage of drinking water going forward, floating chunks of ice were hauled aboard that yielded "Tons of Water." Sailing on "with a gentle breeze," Cook observed light air and an occasional snow shower or "island of ice." There was "nothing else worthy of note." Anticlimactically, Cook recrossed the Antarctic Circle northbound in "pleasant weather," setting the stage for his next two, and conclusive, sweeps of the Pacific Ocean, first west and then eastward.[26]

⚓

With two summers of exploration behind him, Cook came to some preliminary conclusions about the southern continent and its relationship to the ice he repeatedly encountered. As he phrased it in his journal, he was proceeding "directly to the North as there was no probability of finding Land in these high latitudes, at least not on this side [of] Cape Horn and I thought it equally as improbable any should be found on the other side." This last observation was a barb aimed in the direction of Alexander Dalrymple, whose speculative charts implied such a prospect in the South Atlantic. But in any event, Cook continued, Terra Australis Incognita would have to wait for another season because proceeding immediately to the Falkland Islands, or even to Cape Town for purposes of staging, would waste half a year "without being able in that time to make any discovery."[27]

Cook was laying the foundation for extending the voyage into another year, during which he could investigage the South Atlantic, the last potential home of Terra Australis. He had proved that the Pacific did not contain the classic version of the unknown continent, unless it was "far to the south." But there was still some work in the Pacific's temperate latitudes that could fill the expedition's time until the South Atlantic could be investigated. But since that prospect was necessarily ten months away, what to do in the meantime? In a nuanced line of text, Cook observed: "There remained room for very large islands in places wholly unexamined; and many of those which were formerly discovered, are but imperfectly explored, and their situations as imperfectly known," meaning inexact latitude and longitude. The best example in the latter category was Easter Island, "whose situation was known with so little certainty, that the attempts lately made to find it had miscarried." Another was Espiritu Santo, discovered by Pedro Fernandez de Quiros in the East Indies and thought by Cook to be a "land as being large, or lying in the neighbourhood of large lands; and as this was a point which M. de Bougainville had neither confirmed nor refuted, I thought it was worth clearing up."[28] Nothing raised Cook's spirits more than testing the efforts of fellow explorers.

Cook did not think he could "do better than to spend the insuing Winter within the Tropicks." His plan entailed another grand loop of the South Pacific. After sailing west to New Zealand, he would "steer to the South and so back to the East between the Latitudes of 50° and 60° intending if possible to be the length of Cape Horne in November next, when we should have the best part of the Summer before us to explore the Southern part of the Atlantick Ocean." Cook said his crew "heartily concur'd" with the proposed route. The seamen were transported with glee, rejoicing "at the Prospect of its being prolonged a nother year and soon enjoying the benefits of a milder Climate."[29]

George Forster's narrative confirmed Cook's assessment of improved morale aboard *Resolution,* especially his and Johann's, in no small measure because the captain had finally clued everyone into his strategy: "We were now told that we should spend the winter season, which was coming on apace, among the tropical islands of the Pacific Ocean, in the same manner as we had passed that immediately preceding. The prospect of making new discoveries, and of enjoying the excellent refreshments which those islands afford, entirely revived our hopes, and made us look on our continuance on the western side of Cape Horn with some degree of satisfaction."[30]

SIX

Temporizing in the Tropics

When *Resolution* transected 40° S on the way to Easter Island in February 1774, James Cook noted a "sensible change in the weather." On February 20, the temperature reached 66 °F. This he deemed "the only summer's day we had had since we left New Zealand" the previous November. Cook was looking for the obscure island of Juan Fernandez, a place the Spanish had touted as the southern continent's headland. Unlike his search for Cape Circumcision, Cook did not trouble himself much with this elusive isle. Sailing by its supposed location, he remarked that he was sure that, even presuming it existed, Juan Fernandez "can be nothing but a small island; there being hardly room for a large land, as will fully appear by the tracks of Captain Wallis, Bougainville, of the Endeavour, and this of the Resolution."[1] Juan Fernandez exists, but it is thirty degrees of longitude closer to Chile than previously charted. More importantly, Cook's inconspicuous comment confirms his increasingly dismissive attitude toward cataloguing the Pacific's insular holdings. Large tracts of land still mattered, as did correcting previous explorers, but he only sought islands for occasional refreshment.

⚓

On March 11, Easter Island came into view. George Forster wrote: "The joy which this fortunate event spread on every countenance is scarcely to be described. We had been an hundred and three days out of sight of land" (not that anyone was counting). His list of tribulations was long, starting with "rigorous weather to the south, the fatigues of continual attendance [to duty] during storms, or amidst dangerous masses of ice, the sudden changes of climate, and the long continuance of a noxious diet." These trials took a toll on the crew, including the captain. Cook was gravely ill, and was nourished back to health through the sacrifice of Johann Forster's dog, whose meat was rendered into stock for a broth, the only food Cook kept down. Cook described his condition as "bilious cholic," a fill-in-the-blank explanation that reflected the era's immature understanding of internal medicine.[2]

A host of speculations about Cook's illness have been posited. Frank McLynn offered several speculative diagnoses ranging from scurvy to roundworm and theorized that if the captain had a parasitic infestation, such a condition could have created a "deep-seated and long-lived" sickness whose "corrosive and pernicious effect ... might even account for Cook's odd behaviour on his third voyage." Sir James Watt observed that, "by concentrating on vitamin C ... historians may have ignored for too long the serious effects on decision-making resulting from vitamin B deficiencies." Both of these explanations fit comfortably within the Cook-as-diminished-explorer trope. Johann Forster – who, unlike modern historians, had no interpretive objective associated with his diagnosis – thought it was something as simple as a gall bladder inflammation.[3]

Ever laconic, Cook did not elaborate on his illness other than to credit his recovery to nourishment "from food which would have made most people in Europe sick." George Forster rendered a fuller, if somewhat sanitized, explanation. At first, Cook tried concealing his intestinal blockage "from every person in the ship, at the same time endeavouring to get the better of it by taking hardly any sustenance." This self-treatment "in the space of a few days confined him to his bed." An emetic induced "a violent vomiting." A "dreadful hiccough" followed. It lasted "for upwards of twenty-four hours" and hit "with such astonishing violence that his life was entirely despaired of." Cook lay "above a week in the most imminent danger." We learn from Johann's journal that repeated enemas finally cleared Cook's obstruction. Cook's problems had been made worse, Johann argued, because Cook ignored the symptoms at their onset, choosing to attend to deck duty instead.[4] No surprise there. (During Cook's supposedly subpar third voyage, he was never as sick or below his usual standard of vigorous leadership as found here in his legendary second voyage.)

The stay at Easter Island (now Rapa Nui) was short. The famous statues fascinated everyone, but this was no Tahiti. The principal shortcoming was water quality. A few casks were filled, but Cook noted that the water "was not much better than if it had been taken up out of the Sea." He did not linger because there was still "a long run to make." Before departing, Cook made the salient observation that the island's inhabitants were the same "People as the New Zealanders" and other intermediate places he had visited. This conclusion was based on "the affinity of the Language,

Colour and some of thier customs." He was beginning to lay out the fundamentals of one of his two enduring anthropological findings: the breadth of the Polynesian dispersion. Cook, who as a navigator fully appreciated the distances covered, offered a testimonial to the ancient explorers who preceded him: "It is extraordinary that the same Nation should have spread themselves over all the isles in this Vast Ocean from New Zealand to this Island which is almost a fourth part of the circumference of the Globe, many of them at this time have no other knowledge of each other than what is recorded in antiquated tradition and have by length of time become as it were different Nations each having adopted some peculiar custom or habit."[5]

Leaving Easter Island, *Resolution* set a course for the Marquesas, first charted by Álvarado Mendaña in 1595. Running down their latitude, Cook reached them on April 7, 1774, his first visit. The Tahiti standard loomed large again. Cook considered the Marquesas an unlikely source for wood, water, and general provisioning, and did not think they were a good site for repairing the ship. He resolved to sail farther west "for some place that would supply our wants better, for it must be supposed that after having been 19 Weeks at Sea (for I cannot call the two or three days spent at Easter Island any thing else) ... we must want some refreshments."[6]

Cook next headed toward the southwest and Tahiti, which came into view on April 21. This famed island offered, as had now become customary, a friendly reception and provender. Cook committed to a long stay "to begin with the repairs of the Ship." *Resolution* was caulked and her rigging overhauled, all of which "the high Southern Latitudes had made highly necessary." During this interlude, the deteriorated state of provisions stored on board was discovered. Almost two tonnes of mouldy bread biscuits were thrown overboard. At first mystified because the hard tack had been "pack'd in good casks and Stowed in a dry part of the hold," Cook concluded that "the Ice we so frequently tooke in when to the Southward" had dampened the hold.[7]

Cook sailed from Tahiti on June 5, five years and two days after the transit. On leaving, he observed: "When I first came to these islands, I had some thought of visiting ... [Bora Bora]. But as I had now got on board a plentiful supply of all manner of refreshments, and the route I had in view allowing me no time to spare, I laid this design aside, and directed my course to the west; taking our final leave of these happy isles."[8] This was the second time Cook had passed on surveying Bora Bora, having also done so on *Endeavour*'s voyage.

Running westward, *Resolution* encountered the island of Niue. The landing attempted on June 21 did not go well. Coasting a mile offshore, Cook saw "some people" running along the beach. This seemed propitious, so he hoisted out two small boats, which were manned by marines. Cook was accompanied by the Forsters, Anders Sparrman, and William Hodges. The Niueans disappeared into the woods, which initially did not alarm Cook. The Forsters were botanizing "under the protection of the Party under Arms" when some Niueans appeared "in the Skirts of the woods not a Stones throw from us." This was not a metaphor for Cook, as "one of the two men who were advanced before the rest threw a Stone which Struck Mr Sparrman on the Arm." Sparrman fired his musket ("without order" Cook made a point to add), dispersing the crowd. The boats were "imbarqued" and relanded at another beach, which held four canoes but apparently no inhabitants. Stepping ashore, Cook laid some "trifles" in the canoes "to induce the Natives to believe we intended them no harm." After admiring the canoes for "a few minutes," they were attacked. As Cook phrased it, there was no point in "endeavouring to bring them to a parly."[9]

George Forster's narrative carries the story forward:

> The captain endeavoured to discharge his musket, but it missed fire. He desired us to fire in our own defence, and the same thing happened to us all. The natives threw two spears: captain Cook narrowly escaped one of them by stooping; the other slid along my thigh ... We tried to fire again, and at last my piece, loaded with small shot, went off, and Mr. Hodges fired a ball, which did no execution. At the same time a regular firing began behind us from our party, who having observed our retreat, had likewise viewed another troop of the natives coming down by a different path to cut us off. The effect of the small shot fortunately stopped the natives from rushing upon us, and gave us time to retreat.

The similarities between this Niuean tableau and events that would occur at Kealakekua in February 1779, when Cook was killed, are startling. In his journal, Cook noted that the spear Forster described had "pass'd close over my Shoulder."[10]

Cook's westward course took him next to Tonga and what he called the Friendly Islands. He had visited this chain during his tropical swing the previous October for "so short a time as four or five days"; this time, Cook again spent only a week transecting this elongatged archipelago.

This irritated Johann Forster to no end. He complained that many islands "within reach" were not visited, and the few times *Resolution* anchored, the stays were of short duration. Even then, the pattern was to "go late ashore, come early off again." This limited opportunities "where something might be gotten." In truth, on the second voyage, Cook never explored any part of the tropics exhaustively.[11]

For example, on July 17, Cook reached "Australia Del Espiritu Santu of Quiros or what M[onsieur] D. Bougainville calls the Great Cyclades." Cook named this large archipelago the New Hebrides (now Vanuatu), part of an ethnographic zone modern anthropologists define as Melanesia. In his scan of Ambrym Island, Cook noted its general features but acknowledged that "we did not see the whole of it." He later admitted that the cursory nature of his "survey" of this and neighbouring islands "may be faulty so far as it regards the line of the coast." Cook explained in the Admiralty account that "the word Survey is not here to be understood in its literal sense. Surveying a place, according to my idea, is taking a geometric plan of it, in which every place is to have its true situation, which cannot be done in a work of this nature."[12] This is a stark, little-remarked-on admission. If, in fact, Cook ever did get tired of the work of exploration, there is more evidence to suggest that his enthusiasm waned on this swing through the tropics rather than during the third voyage.

⚓

July 1774 marked *Resolution*'s second anniversary at sea, duly noted by Johann Forster, who groused that he had "probably gone through the greater part of those disagreeable circumstances, which we must undergo." Drawing on his father's text, George's account criticized Cook's plan to clear up minor matters of tropical geography related to earlier explorers, principally Bougainville. What Cook revelled in, George deemed indulgent:

> Thus, after spending two years in visiting the discoveries of former voyagers, in rectifying their mistakes, and in combating vulgar errors, we began the third, by investigating a group of islands which the French navigator, pressed by necessity, and ill fitted out, had left with precipitation. It was reserved for this last year to teem in new discoveries, and to make amends for the two first ... We had room to make a variety of observations on men and manners, which, though they ought to be the first objects of travellers, have still been postponed, even by those who have aimed at being looked upon by the world as the most enlightened.[13]

This last line was an oblique criticism of Cook but touched on a larger point. During each of Cook's voyages, the two-year anniversary typically marked the outbreak of disciplinary problems on the beach and disaffection on board. It happened during the *Endeavour* voyage when Cook entered the Arafura Sea, at this juncture on the second, and on the same timetable during the third. George's narrative also underscores that Cook's goal was not an exhaustive survey of the Pacific's numberless islands.

George Forster complained that since natural history was always "the secondary object in this voyage," the result was "little more than a new track on the chart of the southern hemisphere."[14] Playing second fiddle was a naturalist's occupational hazard, and he was not the first to be disappointed. Georg Steller had but a handful of shore excursions during Bering's 1741 voyage to North America, during which he was able to describe a few plants and two animals – a fox and, more notably, *Cyanocitta stelleri*, Steller's jay. Naturalists were quick to blame the overweening egos of ships' captains in such instances, but the root of their disappointment was the tendency to see navigational decisions in zero-sum terms; a gain in geography was perceived as a loss to their prestige and fame as scientists.

Proceeding through the New Hebrides, *Resolution* touched at Eromanga in early August 1774. Cook put ashore "in the face of a great Multitude with nothing but a green branch in my hand." He stepped out of the launch to supervise the procurement of fresh water "and a few Cocoa nutts." The "Chief" signed to Cook to haul the boat ashore, but Cook "gave him to understand" that he intended to head back to the ship and would return later. This was not communicated well. When Cook stepped back into his cutter, the Eromangans "now attempted by force to accomplish what they could not obtain by more gentler means, the gang-board having been put out for me to come in [and] some seized hold of it while others snatched hold of the Oars."[15]

This was another of Cook's potential close encounters of the fatal kind during the second voyage. In his telling, "upon my pointing a musquet at them they in some measure desisted, but return'd again in an instant seemingly ditermined to haul the boat up upon the Shore." The scene that unfolded also eerily foreshadowed Cook's demise at Kealakekua Bay. As he put it: "My Musquet at this critical Moment refused to perform its part and made it absolutely necessary for me to give orders to fire as they now began to Shoot their Arrows and throw darts and Stones at us." Actually, Cook's musket misfired several times, as did those of the others in the shore detachment. The few guns that worked finally dispersed the crowd

Jean François Rigaud, *Portrait of Dr. Johann Reinhold Forster and His Son George Forster, c. 1780.* The father and son team is shown handling a bird specimen and inscribing a field note. The Forsters admired Cook's navigational ability but often quarrelled with him over the limitations his pace of exploration had on studying nature, and later over authorial prerogatives.

(a lesson Cook learned too well). At least two Eromangans were killed, and one of Cook's crew was wounded. When Cook returned to the ship, he recorded that he had been "prevailed upon to fire a four pound Shott at them to let them see the effect of our great guns." The ball fell short "but frightned them so much that not one afterwards appeared."[16] Since neither he nor anyone in his command had been in acute danger, this was a gratuitously aggressive act, and worse, one that it appears he was cajoled by others to execute.

Cook would suffer much shipboard criticism for such behaviour during his third voyage.[17] George Forster's interpretation of this event during the second has received less attention. "For my own part," Forster wrote, "I cannot entirely persuade myself that these people had any hostile intentions in detaining our boat. The levelling of a musket at them, or rather at their chief, provoked them to attack our crew."[18] If Cook displayed poor judgment at Kealakekua, and by all accounts he did, it was not *sui generis,* as the episode at Eromanga reveals. For that reason, the Hawaiian incident should not be seen as a peculiar emblem for his deportment and capabilities during the third voyage.

Notwithstanding his occasional excesses, one theme that Cook carried over from the *Endeavour* voyage into his descriptions of the second was his misgivings about the encounters he was precipitating. Writing about the residents of Tanna in August 1774, he stated:

> We found these people Civil and good Natured when not prompted by jealousy to a contrary conduct, a conduct one cannot blame them for when one considers the light in which they must look upon us in, its impossible for them to know our real design, we enter their Ports without their daring to make opposition, we attempt to land in a peaceable manner, if this succeeds its well, if not we land nevertheless and mentain the footing we thus got by the Superiority of our fire arms, in what other light can they at first look upon us as invaders of their Country.[19]

This passage is reminiscent of his reflection on Indigenous Australians, and is in keeping with the Royal Society's hints proffered prior to the *Endeavour* voyage. It may have been self-recrimination for his conduct at Eromanga ten days earlier.

Five days later, Cook returned to this theme. The islanders were again assembled on the landing at Tanna when "unfortunately one of them was Shott by one of our Centinals, I who was present and on the Spot saw not the least cause for the commiting of such an outrage and was astonished beyond Measure at the inhumanity of the act, the rascal who perpetuated this crime pretended that one of the Natives laid his arrow across his bow and held it in the Attitude of Shooting so that he apprehen[d]ed himself in danger." Cook was so angry at this sentry's lack of discipline that the man was taken back to the ship and flogged. Some of his officers objected to the penalty, and an argument ensued. The rank and file favoured leniency because they believed the sentry had only been defending himself.

Cook prevailed, of course. According to midshipman John Elliott, though the captain was "a Most Brave, Just, Humane, and good man," in this episode he had "lost sight, of both justice, and Humanity." The islanders, not unexpectedly, evaluated Cook's response to the outrage differently. A missionary in the 1840s discovered that the Tannese still remembered the incident and were impressed with how Cook handled it.[20]

This story would be much better known if it had happened on the third voyage, where it would fit within the orthodox depiction of an out-of-touch commander at odds with his men. In this instance, fatigue brought on by a long voyage and an outbreak of what Joseph Banks had previously described as a nostalgic sentiment for home were having a negative effect on the beachside dynamic. These incidents induced Cook to think about becoming an active explorer again with a specific discovery agenda, for he was now merely touring the South Seas bringing trouble wherever he went. At the end of his Vanuatu tour, with the austral spring on the temporal horizon, Cook stated in his log: "I had no more business there, besides the Season of the year made it necessary I should think of returning to the South." In his journal, he expanded on his intentions: "I had yet some time left to explore any lands I might meet with between this and New Zeland, where I intended to touch to refresh my people and recruit our stock of wood and Water, for another Southern Cruse."[21] The casual nature of Cook's characterization of sailing into the high latitudes is noteworthy.

Cook's intuition quickly bore fruit. On September 4, 1774, on the track to his Antarctic staging station in New Zealand, he encountered and named New Caledonia. When gales and squalls pushed *Resolution* offshore, he "gave over all thought of returning to the land we had left." Expanding on this decision, Cook explained: "[Considering] the vast ocean we had to explore to the south; the state and condition of the ship, already in want of some necessary stores; that summer was approaching fast, and that any considerable accident might detain us in this sea another year; I did not think it advisable to regain the land." Then he added: "Thus I was obliged, as it were by necessity, *for the first time,* to leave a coast I had discovered, before it was fully explored."[22]

Cook's defensiveness was an exaggeration, since he had passed on insular exploration before. Nevertheless, New Caledonia was precisely the kind of "large island" that Cook had said he wanted to discover in the tropics, and yet he did not map it fully when he had the chance. The whole southwest coast of the island went unseen. Beaglehole was forced to annotate his map of Cook's course around New Caledonia by denoting "approx. limits of coastline charted."[23]

Cook's hurried pace sent the Forsters over the edge. Johann began to think his work was being purposely hindered, if not by Cook then certainly by others whose "base and mean, dirty principles" were "beneath any Man of Sense." George conceded that extenuating circumstances – shifting winds – had negatively affected their prospects for examining the Isle of Pines off the tip of New Caledonia, but he was not willing to let Cook off the hook either. He explained that New Caledonia, the largest island "hitherto discovered in the South Seas between the tropics, remain[ed] entirely unexplored on its south side. The direction and outline of its northern coast, was sketched out during the short time we could afford to spend on this valuable discovery; but its animals, vegetables, and minerals still remain untouched, and offer an ample field to the naturalist." The indictment continued: "Thus it still remains for future navigators, to continue our discoveries in the South Seas, and to take more time in investigating their productions."[24] In other words, well before the second voyage was over, and with significant discoveries in the Atlantic's icy latitudes yet to come, Cook routinely passed on opportunities to detect or fully delineate tropical isles. This pattern would continue into the third voyage.

Off the Isle of Pines, the noontime temperature on September 27 was 69 °F. Cook noted that this was "lower than it has been since the 27th of last February," or half a year. The tropical tour was coming to a close. The southern "summer was at hand," and since the high latitudes yet to be explored "could only be done in summer," he directed the ship to New Zealand for stores and repairs. Practical requirements were increasingly on Cook's mind. Off New Caledonia, he had prospected timber for spars, masts, and yards, but he did not pursue it. On October 10, another new island came into view. Seemingly uninhabited, Cook regarded it as "a kin to New Zealand" in terms of plant life, especially the "Spruce Pines which grow here in vast abundance and to a vast size." The famous Norfolk Island pines thus entered the cognizance of European sailors. The trees were thick and straight, and Cook's carpenter told him "the wood is exactly of the same nature as the Quebeck Pines." Cook concluded he had found a place "where masts for the largest Ships may be had." As he cleared the coast of "Norfolk Isle," he braced himself by recording a design "to touch at Queen Charlottes Sound in New Zealand, there to refresh my people and put the Ship in a condition to cross this great ocean in a high Latitude once more."[25]

⚓

Arriving safely in Ship Cove in Queen Charlotte Sound on October 19, 1774, Cook moored *Resolution* and refitted the ship to his specifications in anticipation of the upcoming leg. William Wales set up the astronomical observatory to recheck the accuracy of the chronometer used for establishing longitude. Cook had been blasé about losing *Adventure,* but he was now growing "very uneasy" about rumours that had begun trickling in from Māori sources that suggested some of the ship's crew had been killed, perhaps cannibalized. The reports were vague and in their extremity so alarming that Cook, as he stated in his journal, "began to think our people had Misunderstood them and that the story refered to some of their own people."[26] This proved to be wishful thinking, as he would learn in Cape Town.

The weather in late October 1774 was variably spring-like – windy, rainy, or pleasant. The generally pliant conditions facilitated the great body of repair work that Cook had deemed necessary. By November 3, the renovations were completed "except Caulking which goes on slowly, as having only two Caulkers and a great deal to do ... before we can put to sea." With time on his hands, Cook went on an excursion to trace the shore of Queen Charlotte Sound in one of *Resolution*'s boats. His small party, including the Forsters, rowed into a previously unexplored harbour not far from Grass Cove, where some of Furneaux's men had been killed. Cook's description of the ensuing events on the banks of Tory Channel is desultory: "A large settlement of the Natives who received us with great courtesy; our stay with them was short." On landing, he demurred on the invitation for an extended visit and returned to *Resolution* with "some fish we had got from the Natives."[27]

George Forster gave an entirely different cast to the event. As the launch came into the bay, he said he noticed its "shores were every where lined with natives." Cook's party soon "reimbarked," which was "adviseable, as many of the natives who arrived last, brought their arms, and the whole croud now amounted to two hundred and upwards, a much greater number than we had suspected the sound to contain, or had ever seen assembled." Cook was apprised that the fish had not been paid for, so he "took the last nail which was left, and calling to the native, threw it on the beach at his feet." The Māori man,

offended, or thinking himself attacked, picked up a stone, and threw it into the boat with great force, but luckily without hitting any one of us. We now called to him again, and pointed to the nail which we had thrown towards him. As soon as he had seen, and picked it up, he laughed at his own petulance, and seemed highly pleased with our conduct towards him. This circumstance, with a little rashness on our part, might have become very fatal to us, or might at least have involved us in a dangerous quarrel. If we had resented the affront of being pelted with a stone, the whole body would have joined in the cause of their countryman, and we must have fallen an easy prey to their numbers, being at the distance of five or six leagues from the ship, without any hopes of assistance.[28]

In this episode, Cook comes across as either fearless or oblivious. At a minimum, it was yet another occasion when events could have taken the course that probabilities finally dictated at Kealakekua Bay.

Outside of this incident, the stay in New Zealand was predictably refreshing. George Forster wrote: "If we came in ever so pale and emaciated, the good cheer which we enjoyed during our stay, soon rekindled a glow of health on our cheeks, and we returned to the south, like our ship, to all outward appearance, as clean and sound as ever, though in reality somewhat impaired by the many hard rubs of the voyage."[29] Hard rubs or not, Cook was ready to make one last run at the question of whether the Pacific Ocean held the southern continent. The plan was to transect the previous year's loop, which had alternated between the Tropic of Capricorn and the Antarctic Circle. Having already sailed through the zones adjoining the fortieth parallel, the route eastward this time focused on the lower fifties – a straight shot in the direction of Cape Horn.

⚓

The sail to Tierra del Fuego was so quick and anticlimactic that George Forster could cover it in three paragraphs. Exiting Cook Strait on November 10, 1774, *Resolution* steered southeast with "all sails Set," aiming for "the Latitude of 54° or 55°." Cook's intention "was to cross this vast Ocean nearly in these Parallels," passing over "those parts which were left unexplored last summer," meaning the discursive, omega-shaped track south of Tahiti. Within a week, *Resolution* had reached 52° S, where it encountered "a great Swell from the West." Steady winds and "clear pleasant weather" brought Cook to the targeted fifty-fifth parallel by November 25, so he bent *Resolution*'s southeasterly course to one more directly eastward.[30]

Cook had never filled a map, or more accurately vacated one, as quickly has he did on this leg. Johann Forster, for whom the voyage could not end soon enough, considered the expedited crossing "one of the most extraordinary Passages considering the Great Distance." Splitting the tracks of his earlier swings through the South Pacific on this voyage and on *Endeavour,* Cook vanquished another set of previously unexplored latitudes. On November 27, he covered 183 miles in one day, a Cook-voyage record accomplished at a speed that Forster considered almost preposterous. That same day, only halfway across the Pacific, Cook wrote: "I now gave up all hopes of finding any more land in this ocean, and came to a resolution to steer directly for the west entrance of the Straits of [Magellan]."[31]

Cook's pace was so rapid that on December 17, 1774, merely thirty-seven days after leaving New Zealand, the tip of South America came into view. This occasion prompted one of the most significant but overlooked reflections Cook ever inscribed. He began by crediting himself for "the first run that had been made directly a Cross this ocean in a high Sothern Latitude." Cook determined that he had never conducted "a passage any where of such length, or even much shorter, where so few intresting circumstance[s] occurred." He recorded a few particulars on the weather. It had been neither "stormy nor cold, before we arrived in the latitude of 50°," though the temperature fell gradually from 60 to 50 °F when the ship reached the fifty-fifth parallel. With that prosaic introduction, Cook then proffered a single line of text that prefigured his outlook when sailing below the equinoctial line during his third expedition: "I have now done with the SOUTHERN PACIFIC OCEAN, and flatter my self that no one will think that I have left it unexplor'd, or that more could be done in one voyage towards obtaining that end than has been done in this."[32]

This statement, almost verbatim, appeared in the Admiralty account, which affirms the vehemence behind it. By the time the report was in final draft form, Cook had already agreed to search for the North Pacific gateway to the Northwest Passage, and since he was returning to the Pacific basin, he could have excised this passage. The fact that he did not has a simple explanation: Cook had become bored with the South Pacific and its insular holdings and wanted his readers to know it.

SEVEN

Cook and Forster, on Ice

For James Cook, with the South Pacific behind him both literally and figuratively, there was only one more place – the icy latitudes of the Atlantic – that could hold Terra Australis Incognita. Johann Forster had already capitulated on this point, thinking no further reconnaissance was necessary. But Cook was determined to look; until the last recourse was exhausted, he would be open to the possibility that discovery might await. The first projected course from Cape Horn to Cape Town "was SE with a view of discovering that extensive coast which Mr. Dalrymple lies down on his Chart." The supposed feud and mutual jealousy between Alexander Dalrymple and Cook is one of the hoariest traditions in Cook scholarship and much exaggerated. But the Enlightenment's empirical method did necessitate verification by replicable experiment or observation, and so Cook set out "to explore the Southern part of this [the Atlantic] ocean."[1]

Dalrymple had placed this supposed coast at close to 60° S, but in early January 1775, when *Resolution* was a few days' sail east of Cape Horn, Cook "saw neither land nor signs of any." This made him "the more doubtfull of its existence." Forster, reflecting his own sour and fatigued outlook, avowed that many crewmembers dreaded falling in with land because "this might retard our early arrival at the Cape." Unlike his earlier forays in the Indian and Pacific oceans, Cook did not sail to the Atlantic's ice edge. This decision was informed by the loss of a main topgallant mast and his sense of the crew's growing impatience. After hauling from 58° S, *Resolution* ran the fifty-fifth parallel eastward. Cook noted "a Swell from ESE," which obviated the prospect of an "extensive tract of land laid in that direction," but he sensed that ice was near because the temperature was much colder than it had been since the ship left New Zealand. But instead of a passel of icebergs or packed ice, Cook came upon land "wholly covered with snow." Cook had discovered a notoriously remote and inhospitable place, and he eventually named it the Isle of Georgia after his sovereign. We know it as South Georgia, made famous in the early twentieth century by Antarctic explorer Ernest Shackleton.[2]

Resolution could not approach the island because of tumultuous seas. The temperature dipped to 35 °F, and sails were reefed until the storm abated. Three days later, a mountainous, snow-clad island came more clearly into view at 54° S. Approaching from the southwest, Cook went through his logic model for evaluating this late prospect for Terra Australis Incognita. A swell from the south "made it probable no land was near us in that direction: but the Cold air which we felt and the vast quantity of Snow on the land in sight induced us to think that it was extensive and I chose to begin with exploring the Northern Coast."[3]

South Georgia's "huge masses of snow or ice" dominated Cook's perception of this place, and his experience here had a profound effect on his thinking about glaciation and sea-ice formation theory. As he phrased it in his published account, the island "seems to abound with bays and harbours, the N.E. coast especially; but the vast quantity of ice must render them in-accessible the greatest part of the year." Cook found it "remarkable that we did not see a river, or stream of fresh water, on the whole coast." It seemed "highly probable that there are no perennial springs in the country; and that the interior parts, as being much elevated, never enjoy heat enough to melt the snow in such quantities as to produce a river, or stream, of water." One inlet terminated in "a huge Mass of Snow and ice of vast extent," presenting "a perpendicular clift of considerable height, just like the side or face of an ice isle." This was an epiphany. Pieces were breaking off "and floating out to sea." Dramatically, a huge chunk calved off the face of the cliff while *Resolution* was in the bay, making "a noise like cannon." J.C. Beaglehole said of this moment that Cook "was looking at glaciers, for which he had no word, and the birth of an iceberg."[4]

In *Farther Than Any Man*, Martin Dugard observes that, unlike other polar explorers who followed in his wake, "Cook never developed an affection for the ice." That is a fair assertion, for Cook was not a Romantic. But the formidability of the ice's various appearances – as icebergs, field ice, or a solid pack – held his attention, as evidenced by his many reflections on glaciology during his South Atlantic swing and earlier in the Indian and Pacific oceans. By way of contrast, a tired Johann Forster, nearly the same age as Cook, had by this time lost interest in making extended climatological observations altogether. He was now more interested in moralizing about the environment than studying it: "If a Capt, some Officers and a Crew were convicted of some heinous crimes, they ought to be sent by way of punishment to these inhospitable cursed Regions, for to explore and survey them. The very thought to live here a year fills the whole Soul with horror and despair!" Cook's description of Possession

Possession Bay in the Island of South Georgia 1777. Engraving by S. Smith, after a drawing by William Hodges, January 17, 1775. Cook's understanding of the origin of the icebergs circling the South Pole was crystallized by his encounters with the glaciers on South Georgia, including the one depicted here.

Bay (northwest of the harbour Shackleton glissaded into) was, on the other hand, classically sublime. The island's "lofty summits ... were lost in the Clouds," but more spectacularly, "the Vallies laid buried in everlasting Snow. Not a tree or shrub was to be seen, no not even big enough to make a tooth-pick."[5]

Coasting South Georgia's north shore, Cook marvelled at the "vast tracks of frozen snow or ice not yet broke loose." Proceeding to the southeast, he glimpsed a tantalizing "new land" shaped like "a Sugar Loafe," but this terrain did not live up to its promise because the next day it proved to be a small detached isle. Reaching South Georgia's southeastern extremity, Cook gained a view of the side of the island facing the pole. From this he discerned its relatively limited extent. (South Georgia runs 116 miles on a northwest-southeast tangent and is up to 20 miles wide.) Cook took a back sight toward the opposite point, which demonstrated "that this land which we had taken to be *part of a great Continent* was no more than an Island of 70 leagues in circuit." Cook named the southeasterly tip Cape Disappointment.[6] The emotive content of this name is instructive if not emblematic of the voyage as a whole, and especially South Georgia's role within it, suggesting a hint of regret that has not been remarked upon sufficiently in studies of Cook's career.

At this point, Cook's journal shifts from reflections on geography to glaciology:

Who would have thought that an Island of no greater extent than this is, situated between the Latitude of 54° and 55°, should in the very height of Summer be in a manner wholy covered many fathoms deep with frozen Snow, but more especially the SW Coast, the very sides and craggy summits of the lofty Mountains were cased with snow and ice, but the quantity which lay in the Vallies is incredible, before all of them the Coast was terminated by a wall of Ice of considerable height. It can hardly be doubted but that a great deal of ice is formed here in the Winter which in the Spring is broke off and dispersed over the Sea: but this isle cannot produce the ten thousand part of what we have seen, either there must be more land or else ice is formed without it.

Both surmises proved true, but given what he had said elsewhere, Cook placed more faith in the former than the latter.[7]

Cook summarized his sentiments about this two-day sequence, which had started with the sighting of a "new land" shaped like a sugar loaf and culminated with his rounding of Cape Disappointment, as follows: "These reflections led me to think that the land we had seen the preceding day might belong to an extensive tract and I still had hopes of discovering a continent. I must Confess the disappointment I now met with did not affect me much, for to judge of the bulk by the sample it would not be worth the discovery."[8] This last sentence is a rare, perhaps solitary, example of Cook rationalizing his lack of success. Notwithstanding his frequent scorn for the idea of a southern continent, it is clear from this passage that Cook held out hope, until the very end of the voyage's exploration phase, that he would discover, at a minimum, the northernmost extension of a southern continent. The pleasure Cook might have taken from finding a high-latitude continent was countervailed by his bias, held in common with his contemporaries, that a frozen land was worthless. In any event, no place name provides more insight into Cook's outlook as an explorer near the end of his epic second voyage than South Georgia's "Cape Disappointment."

Lieutenant Charles Clerke's journal underscores this hypothesis. When anchored in Possession Sound, he reported that "the Captain and Botanical Gentry went on shore to take possession of this new Country (Southern Continent I hope)." Anticipating Cook's reflections, Clerke confessed: "I did flatter myself from the distant soundings and the high

Hills about it, we had got hold of the Southern Continent, but alas these pleasing dreams are reduc'd to a small isle." As historian Lynne Withey points out, Cook was "a man of restrained emotion," but his Cape Disappointment nomenclature and reflections on its import were a rare "expression of depression or discouragement." In addition, Cook did not withhold his disappointment from the reading public; a polished version of the paragraph appeared in the Admiralty's account.[9]

⚓

The South Georgia encounter, especially the collateral realization about the amount of land required to spawn the Southern Ocean's icebergs, warmed Cook's enthusiasm for the still ghostly prospect of a southern (if mostly polar) continent. He quit South Georgia on January 20, 1775, to conduct the deepest plunge of his third and last season of high-latitude voyaging in the Southern Hemisphere. *Resolution* veered southeast, back into the five-degree-wide corridor in the high fifties that Dalrymple had posited as a possible home for the unknown land. Cook was under no illusions about what such a land would portend if he found it. Surely it was "doomed by Nature to Frigidity" like South Georgia. Given the lateness of the season and the wavering endurance of his shipmates, he "did not doubt but that I should find more land than I should have time to explore."[10]

(Here it is important to keep in mind that Cook's exploratory strategy in the South Atlantic was completely reliant on sketchy geography. We have been conditioned by accounts from Beaglehole to Collingridge to root for Cook as he demolishes Dalrymple one more time. However, Cook is eviscerated by these same historians for relying on equally vague Russian cartographic projections of Alaska to guide his course in the North Pacific.[11] Thus, on the second voyage – where Cook is perceived as a successful explorer – he is characterized as being thorough when he pursues the leads laid before him by Dalrymple and Bouvet. On the third, where his death intimates that the voyage was destined for failure, the very same pattern of testing geographic theories is used as evidence to show that Cook is either worn out or foolish. Conduct which on one voyage is a virtue becomes in the other a weakness. What historians admire in one expedition becomes grounds for impeachment in another.)

A week after leaving South Georgia, in a deep fog and expecting to hit the ice line soon, Cook reached what he reckoned was the sixtieth parallel. He did not intend to go farther "unless I met with some certain signs of soon meeting with land, for it would not have been prudent in me to have spent my time in penetrating to the South when it was, at least as probable, that a large tract of land might be found near Cape Circumcision." Cook was wrong to infer that Cape Circumcision was part of a larger continent, just as Bouvet was wrong to imply it. But the geographic implications Bouvet propagated had a strange grip on Cook's imagination. In a context related to his denomination of Cape Disappointment, his fascination reflects an emerging desire for what might be termed positive discovery within the zone of exploration specified by his instructions. Accomplishments savoured on the *Endeavour* voyage, or even the first two swings along the edge of ice that crossed the polar circle, no longer did. Cook wrote: "I was now tired of these high Southern Latitudes, where nothing was to be found but ice and thick fogs. We had now a long hollow swell from the West, a strong indication that there was no land in that direction. I think I may now venture to assert that that extensive coast, laid down in Mr Dalrymple's Chart of the Ocean between Africa and America ... does not exist."[12]

Beaglehole once termed Cook "the executioner of misbegotten hypotheses."[13] This is true, but Cook himself was not fully satisfied. In the South Atlantic, he expressed genuine frustration about looking for something that he could not find, now for the third year running. Cook's forthright admission of fatigue (something he never expressed during the third voyage) was felt by others, as George Forster's recollection indicates. The cold, snow, and fog combined with the "great masses of ice" surrounding the ship "stopped our farther progress [south], greatly to the satisfaction of all the crew, who were at present thoroughly tired of this dreadful climate, and exhausted by perpetual watching and attendance, which the frequency and sudden appearance of dangers required."[14]

Cook may have been tiring of the Antarctic region, resembling his earlier expression that he was "done" with the southern Pacific, but it was not quite done with him. In foggy conditions so severe that he could not see the length of the ship, in a sea strewn with both loose blocks of ice and mammoth icebergs, Cook steered *Resolution* northeast and away from the sixtieth parallel for the last time, heading toward what he acknowledged was the last prospect for Terra Australis Incognita: Bouvet's Cape Circumcision.

But on January 31, 1775, Cook recorded the discovery of what he later and somewhat elusively called Sandwich Land (after his sponsor, Lord Sandwich). In the haze, it was not clear whether the various promontories

Joseph Gilbert, *View of Bristol Island, South Sandwich Islands,* January 31, 1775. Gilbert, *Resolution*'s sailing master, graphically captured the turbulent seas of the Southern Ocean and the forbidding nature of Freezeland Peak and the adjoining coastline of Bristol Island in the South Sandwich Archipelago. Cook was intrigued by the various headlands he cryptically referred to as "Sandwich Land," thinking they may have formed the outer limit of Terra Australis Incognita. Cook considered this image so vital to his understanding of the expedition's circumnavigation of the South Pole that a version of it was included in the lower left corner of his landmark second-voyage chart, "Part of the Southern Hemisphere Showing Resolution's Track through the Pacific and Southern Ocean."

that came in and out of view to the north were a chain of islands or head-lands projecting from a larger mass. Was it conceivable that the long-wished-for continental discovery might occur on the last leg of a three-year circumnavigation of the South Pole? Later explorers would prove this was indeed an insular chain, called the South Sandwich Islands to distinguish them from Cook's third-voyage discovery of the Sandwich Islands – the Hawaiian archipelago. George Forster stated that Cook "did not venture to lose any time in the investigation of this coast, where he was exposed to imminent danger from the violence of westerly winds."[15]

This serendipity prompted one of the more evocative place names in Cook history. He called an outcropping of ice-covered rock at 59° 13' Southern Thule because "it is the most southern land that has ever yet been discovered." (This record was not broken until 1819, when the South Shetland Islands were discovered.) Like South Georgia, this island also showed "a surface of vast height, and is every where covered with snow." For Cook, the "great Westerly swell which set right upon the shore" made this "the most horrible Coast in the World." Johann Forster, who had been stunned by South Georgia's forbidding qualities, said that, with these headlands, "nature meant to convince us of her power of producing something still more wretched." The terrain was "absolutely covered with ice and snow (some detached rocks excepted) and in all probability incapable of producing a single plant." George maintained that it was Johann, on the basis of his study in the classics, who first suggested the name Southern Thule, which Cook "preserved." Beaglehole, though generally disposed to give the elder Forster short shrift, found this proposition credible because the German scholar was "more in touch with legend than was Cook."[16]

When the weather cleared, Cook stated that beyond Southern Thule "we could see no land." This was a relief because the strong swell behind him would have made exploration of a lee shore quite dangerous. Discovery-averse, Johann Forster hoped, for a different reason, that this "Southern Land" was not as large "as we are now afraid of." Standing north, Cook acquired "sight of a new coast," which he named Cape Montagu (Sandwich's family name), and another headland still twenty miles farther. Cook wrote: "We saw land from space to space between them, which made us conclude that the whole was connected." A major discovery seemed at hand, but he noted tellingly: "I was sorry I could not determine this with greater certainty, but prudence would not permit me to venture near a Coast, subject to thick fogs, on which there was no anchorage, and where every Port was blocked or filled up with ice and the whole Country, from the summits of the Mountains down to the very brink of the clifts which terminate the Coast, covered many fathoms thick with ever lasting snow."[17] On the third voyage, conventional opinion about Cook's fitness to serve notwithstanding, he would sail at the edge of the Arctic ice in fog without certain refuge to anchorages across fifteen degrees of longitude.

⚓

Now deep into his second voyage, Cook was becoming more careful with where he took *Resolution,* but his reflections on glaciation reached greater concentration and clarity. In the published account, he wrote that "several large ice-islands lay upon the coast" of Cape Montagu, but one in particular attracted his attention: "It had a flat surface, was of considerable extent both in height and circuit, and had perpendicular sides, on which the waves of the sea had made no impression; by which I judged that it had not been long from land, and that it might lately have come out of some bay on the coast, where it had been formed." Cook was inching closer to a correct understanding of how icebergs formed. North of Cape Montagu, he saw more terrain that he cautiously and nondescriptively named "Saunders" (solely) in his journal (after Charles Saunders, his commander in Quebec), though he surmised it was an island. Lastly, at 57° S, Cook sighted the Candlemas Isles (an appellation prompted by the liturgical calendar, an increasingly common pattern late in his second and third expeditions), one of which is now named Cook Rock. When the penguins disappeared, he concluded (correctly, it would be determined by later explorers) that "we were leaving the land behind us and that we had already seen its northern extremity."[18]

From the context of land, sea, and ice, Cook offered a hypothesis that Southern Thule, Cape Montagu, "Saunders," and the Candlemas Isles – in aggregate what he called Sandwich Land – were "either a group of islands, or else a point of the continent." He enlarged on this last thought: "I firmly believe that there is a tract of land near the Pole which is the source of most of the ice that is spread over this vast southern ocean," applying the now common name for that vast extent of water encircling Antarctica. He reiterated an earlier observation that it was "probable" this polar land "extends farthest to the north opposite the southern Atlantic and Indian oceans; because ice was always found by us farther to the north in these oceans than any where else, which I judge could not be, if there were not land to the south; I mean a land of considerable extent." The last clause precluded any confusion with South Georgia or Sandwich Land.[19]

Cook correctly deduced that a polar continent existed, but he did something more. By extrapolating from the variable distribution of ice in several southern oceans, he accurately forecast where polar terrain ranges farthest north. The Antarctic Peninsula and Wilkes Land both extend north of the polar circle into the Atlantic and Indian oceans, respectively. Conversely, the Ross Sea and McMurdo Sound, arms of the Pacific, nearly reach the eightieth parallel. Not coincidently, both Roald Amundsen and Robert Falcon Scott chose this side of the continent for their early twentieth-century landings in quest of the South Pole.

Cook now revisited sea-ice formation theory and speculated about what his hypothesis portended for Antarctic geography. He posed, for the sake of argument, that no considerable southern land was extant "and that ice

James Cook, *Chart of the Discoveries Made in the South Atlantic in His Majesty's Ship* Resolution *under the Command of Captain Cook in Jany 1775.* Though not as well known as his navigational accomplishments in the Pacific Ocean, Cook's exploration of the South Atlantic proved to be the ultimate laboratory for his contributions to the modern understanding of polar climatology. Given the Antarctic focus of Cook's second voyage, this chart's orientation is inverted from the normal projection so that south is to the top of the image.

may be formed without it." From this, it followed "that the cold ought to be every where nearly equal round the Pole and consequently we ought to see ice every where under the same parallel, or near it; and yet the contrary has been found." To demonstrate this, Cook recapitulated his experience: ice was rarely seen at 60° S in the Pacific, whereas in the Atlantic it was encountered as low as 51° S. Bouvet had reportedly seen it at 48° S, where the Atlantic and Indian oceans converge. These observations brought Cook to the first of several interpretive conclusions: "The greatest part of this southern continent (supposing there is one), must lie within the polar circle, where the sea is so pestered with ice, that the land is thereby inaccessible."[20]

This, of course, proved correct. Cook might have comfortably stopped there, but in an egotistical rush he added grandiloquent text, which made him appear foolish in retrospect: "The risk one runs in exploreing a coast in these unknown and Icy Seas, is so very great, that I can be bold to say, that no man will ever venture farther than I have done and that the lands which may lie to the South will never be explored."[21] This passage was a variant of the famous line announcing his highest latitude and that he had gone as far south as anyone ever would. By the process of elimination, Cook's criss-crossing of the Southern Ocean had indeed delimited the footprint that any polar continent might occupy. And though it was upwards of a half century before other explorers would travel in his tracks, *never* is a long time. His was the common mistake of not anticipating technological advancement.

George Forster shared his short-sightedness. His account, though occasionally critical of Cook, agreed with him in this instance. He observed that polar geography could "continue undetermined for ages to come, since an expedition to those inhospitable parts of the world, besides being extremely perilous, does not seem likely to be productive of great advantages to mankind."[22] The first Antarctic sightings occurred in the 1820s and coastal landings a decade later. Amundsen and Scott reached the South Pole in the austral summer of 1911–12.

In the Admiralty's narrative, Cook referred to Sandwich Land as "a country doomed by nature never once to feel the warmth of the sun's rays, but to lie buried in everlasting snow and ice." This last phrase, given current climate trajectory, is ironic overstatement. Cook may have prematurely precluded the possibility of someone proceeding farther south than Sandwich Land, but he well understood what risks such a voyage might entail. Should a high-latitude harbour "be so far open as to invite a ship into it, she would run a risque of being fixed there for ever, or of coming out in an ice island." A "heaving snow-storm attended with a sharp frost, would be equally fatal."[23] This text foreshadowed Cook's actual experiences on his third voyage off Alaska's Icy Cape, when the ice pack almost closed in on him and his experiences off the Siberian coast, when conditions dictated his departure from the Arctic Ocean.

Cook had an additional point to make, this one echoing his Great Barrier Reef discourse about postexpeditionary evaluations of explorers alternating between timidity and temerity. He asserted that after the foregoing explanation, "the reader must not expect to find me much farther to the south." This was not "for want of inclination," Cook protested, "but for other reasons." With a ship that was neither properly conditioned nor provisioned for further high-latitude exploration, he considered it rash "to have risqued all that had been done during the voyage, in discovering and exploring a coast, which, when discovered and explored, would have answered no end whatsoever, or have been of the least use, either to navigation or geography." South Georgia and the South Sandwich Islands were not the encounter with the austral lands that Dalrymple and others had idealized, but they were discoveries nonetheless, especially in the way they leavened Cook's analysis of polar oceanography.[24]

⚓

Veering northeast from the Candlemas Isles, Cook made one last attempt to tie down the existence of Cape Circumcision. *Resolution* left that precinct in a snowstorm so heavy Cook stated he was "obliged every now and then to throw the Ship up in the Wind to shake [the snow] out of the Sails."[25] Entering the western edge of the Indian Ocean, he initiated a reapplication of his South Pacific crossover strategy by laying a new track atop his run to the high latitudes south of the Cape of Good Hope two years earlier. In the event, from Sandwich Land, *Resolution* would sail nearly forty degrees of longitude to the east in another fruitless search for Bouvet's elusive cape. By combining the results of his opening leg of exploration with this follow-up near its end, Cook sailed two separate and discursive circles in the ocean south of the cape.

On February 13, 1775, Cook's ship was struck by "so sharp a frost that the Water in all our Water Vessels on Deck was in the Morning covered with a sheet of ice." The thermometer reached 29° F. The next day, *Resolution* crossed the prime meridian of Greenwich at 57° 50' S, an event otherwise unremarked on by Cook but that moment effectively completed his circumnavigation of Antarctica. Bearing northeast toward the supposed location of Bouvet's fugitive landfall, Cook was assured by the high seas from the south that there was no land to starboard. Locking onto the fifty-fourth parallel, where it was supposed that Bouvet had sighted land, he ran that line eastward.[26]

After crossing Cape Town's meridian and expending three more days in pursuit of Bouvet's cape, Cook began thinking that "no such land ever existed." To remove all doubt, he steered an additional thirteen degrees of longitude to the east, but by this point he had already passed the island's more westerly location. Of this last futile excursion, he wrote dismissively, if not derisively: "I was therefore well assured that what [Bouvet] had taken for land could be nothing but an Island of Ice, for if it was land, it is hardly possible we could have miss'd it, was it ever so small."[27] In fact, Bouvet had found land, but his coordinates were inaccurate, slightly for latitude and, less surprisingly, longitude. Bouvet Island, the most remote place in the world, the true Ultima Thule, is fourteen hundred miles southwest of Cape Town, not southeast of it where Cook was searching. The island was not sighted again until 1808, when a British whaler happened on it by accident.

By February 21, Cook was two degrees east of his track into the high latitudes of the Indian Ocean made in December 1772, which led him to conclude there was "no purpose to proceed any farther to the East under this parallel." It was also at this moment that Cook realized he had circumnavigated the globe for a second time. This awareness put him in a reflective mood. He had run out of discovery possibilities; the sails and rigging were well worn; and perhaps most importantly, the ship's provisions had decayed. Utilizing a leadership trope he popularized in travel literature, Cook stated in his published account that *Resolution*'s crew was healthy "and would have cheerfully gone wherever I thought proper to lead them; but I dreaded the scurvy laying hold of them at a time when we had nothing left to remove it."[28] He directed a course toward South Africa's Table Mountain and refreshment in Cape Town, the last major stop before heading home to England.

⚓

The discovery phase of his second voyage over, Cook turned to summarizing the expedition. First up was Terra Australis Incognita, a central facet of his career for seven years running: "I had now made the circuit of the Southern Ocean in a high Latitude and traversed it in such a manner as to leave not the least room for the Possibility of there being a continent, unless near the Pole and out of the reach of Navigation; by twice visiting the Pacific Tropical Sea, I had not only settled the situation of some old discoveries but made there many new ones and left, I conceive, very little more to be done even in that part." This was an elaboration on his earlier observation that he was done with the southern Pacific. Now he rendered a verdict: "Thus I flater my self that the intention of the Voyage has in every respect been fully Answered, the Southern Hemisphere sufficiently explored and a final end put to the searching after a Southern Continent, which has at times ingrossed the attention of some of the Maritime Powers for near two Centuries past and the Geographers of all ages."[29]

Rare is the explorer who admits to failing to find what is sought. Still, Cook was right about the underlying truth. There was no southern continent, as classically conceived. In his book, George Forster also chimed in on the question. The object of "our hazardous voyage," he stipulated, was "to explore the southern hemisphere to the sixtieth degree of latitude, and to ascertain the existence of a southern continent in the temperate southern zone." *Resolution's* various tracks "rendered it evident, that a continent does not exist in the temperate southern zone, but have likewise made it probable, by advancing into the frigid zone to seventy one degrees south, that the space within the antarctic circle is far from being every where filled up with land."[30] The obtuse construction of Forster's last clause was important because he and his father were now comfortably agnostic in their thinking about polar lands being required for the formation of sea ice.

At several points earlier in the voyage, Cook had surmised the existence of a continent in icy latitudes otherwise inaccessible to his vessel. He now readdressed the issue: "That there may be a Continent or large tract of land near the Pole, I will not deny, on the contrary I am of opinion there is, and it is probable that we have seen a part of it." This last was an exaggeration, probably borne of the inconclusive status of Sandwich Land. He then reiterated several facets of his polar terrestrial theory: "The excessive cold, the many islands and vast floats of ice all tend to prove that there must be land to the South and that this Southern land must lie or extend farthest to the North opposite the Southern Atlantick and Indian Oceans." This, he reasoned, was why a "greater degree of cold" was found in those seas "than in the Southern Pacific Ocean under the same parallels of Latitude."[31] To confirm this, Cook checked the ship's log. It showed that in the Pacific the thermometer had seldom reached freezing temperatures until *Resolution* transected the sixtieth parallel; in the upper ranges of the Indian and Atlantic oceans, however, it had reached 32 °F as low as 54° S, a pattern also consistent with the occurrence of floating ice.

Regarding the phenomena of ice islands, Cook offered a conclusive reflection heavily informed by the geomorphic encounter at South Georgia. Though generally negelected by historians because of the palm-tree paradigm, it is one of the most important of his career. The "coagulation" of icebergs, he stated, "has not, to my knowledge, been thoroughly investigated: Some have supposed them to have been formed by the freezing of the Water at the Mouths of large Rivers." But he now rejected scientific orthodoxy because his observations at sea would not allow him "to acquiesce in this opinion." Cook's reasoning was cogent. In *Resolution*'s many meetings with icebergs, no one ever detected any terrestrial matter – rock, soil, or vegetation – which would have been evident if the icebergs congealed on inland waters. The only polar terrain he had seen was mountainous, elevations that never enjoyed "heat sufficient to melt the snow in any quantity" necessary to create a flow of ice or water that could disembogue into the ocean.[32]

Having criticized current theory, Cook knew he was obliged to offer his antithesis. It took the form of an early and serviceable version of glacial calving. The valleys of South Georgia, he explained, "covered many fathoms deep with everlasting snow," terminating at the sea "in Ice clifts of vast heights." It is here that "the Ice islands are formed, not from streames of Water, but from Consolidated snow which is allmost continually falling or drifting down from the Mountains, especially in Winter when the frost must be intence." These accretions filled up "all the Bays be they ever so large ... which cannot be doubted as we have seen it so in summer."[33]

Cook then explained how icebergs entered oceanic currents. When the ice "clifts" created by continual snowfall and drifting "are no longer able to support their own weight ... large pieces break off which we call Ice islands." Tabular icebergs, those with "a flat even Surface," were "formed in the bays and before [that] the flat Vallies." Those with "a spired unequal surface must be formed on or under the side of a Coast, composed of spired Rocks and precepices, or some such uneven surface, for we cannot suppose that snow alone, as it falls, can form on a plain surface, such as the Sea, such a variety of high spired peaks and hills." He considered it "more reasonable to suppose that they are formed on a Coast whose surface is something similar to theirs."[34] Cook had correctly gauged that glaciers are rivers of ice and the origins of flat-top icebergs, but he missed the mark regarding "spired peaks," as he did not fully comprehend the erosive effects of weathering and gravity on the shaping of ice. He knew that icebergs deteriorated at sea, but his theory that they first floated into the sea as an embossed impression of what had previously held them fast was mistaken.

Clerke provided additional insights into how Cook's glacial theory had evolved and into the scientific dynamic among officers and naturalists on deck. He reported: "Untill this last Southern Campaigne we were various in our Opinions concerning the formation of an Ice Island, but the sight of those Lands at the bottom of the Atlantic render'd this matter very plain, and gave us a very clear idea of its Origin – increase, etc." Seeing icebergs break off South Georgia's "icy cliffs" had dispelled "all our doubts and clearly convinc'd us, that the Isles are form'd under the Cover of Lands either in Bays or wherever the water is so much shelter'd that the general purterbation of the Sea cannot much effect it." The "immense quantities of Snow which falls to the share of these happy Climes," Clerke added sarcastically, "freeze and embody itself with the Ice, 'till it becomes too enormous to bear its own weight and of course immerges into the Sea." He concluded, as did Cook, that the "myriads" of icebergs the expedition had encountered "cou'd not have been form'd but by a great deal of shelter and of course large quantities of Land." By this, he implied the existence of a larger tract of land closer to the pole than South Georgia and the South Sandwich Islands.[35]

Cook also returned to reflecting on the collateral issue of pack ice, but this portion of his pioneering analysis met with mixed results because he stayed bound to the mistaken notion that saltwater far from land did not freeze. Clearly, he acknowledged, saltwater surfaces froze over; he cited a number of locales north of the equator, including those he had seen personally, and vast oceanic extents around Antarctica. Cook then attributed (correctly, it would turn out) his earlier discovery of unfrozen seawater two degrees below the nominal freezing point for freshwater to "the Salts it contained." How, then, was pack ice created? The freeze over, Cook postulated, occurred "in Winter, when the frost is set in and there comes a fall of Snow." It was this snow, he hypothesized, that froze on the surface as it fell, "in a few days or perhaps in one night form[ing] such a sheet of ice as will not be easy broke up; thus a foundation will be laid for it to accumulate to any thickness by falls of snow, without it being attall necessary for the Sea Water to freeze. It may be by this means that these vast floats of low ice we find in the Spring of the Year are formed and after they break up are carried by the Currents to the North."[36]

Cook's analysis was partially right. Snowfall can add to the pack's thickness and will become ice if ambient temperatures are cold enough. But

Cook could never quite get past the old dogma about saltwater not freezing. Falling snow had been Johann Forster's early theory when they first encountered the immense ice fields in the Indian Ocean, and that idea seems to have been firmly implanted in Cook's mind. In their published accounts, however, the Forsters would correctly assert that low ice was formed in the sea. Cook was not the only member of the scientific team who continued to adhere to the old theory, as the postvoyage debate between George Forster and William Wales indicates. In response to Forster's narrative, Wales described overwintering on Hudson Bay in preparation for Venus's transit, where he saw solid ice near the coastline, but "farther out in the bay, it never formed at all." From this, Wales concluded there was no reason to believe "that ice can form upon the surface of the water in a large extended ocean." Forster, in turn, ridiculed Wales's naïveté.[37]

Wales was misled by a phenomenon known in modern glaciology as a polynya, an extensive opening in compact ice created by offshore winds. Some occur with such regularity within the Arctic ice pack that they have names, for example, Northwater in Baffin Bay. Inuit established communities near recurring polynyas because they facilitated the wintertime hunting of marine mammals such as whales, walrus, and seals, who come to the surface for air. In *On Sea Ice* (2010), W.F. Weeks claims that most modern glaciologists "do not know where the nearest polynya is located, the same cannot be said for polar bears," who hunt those same species plus humans.[38] Polynyas were undoubtedly the source of the myth that the North Pole was ice-free. In the end, Cook finessed his problem by concluding that freshwater snowfall formed a crust on saline surfaces when temperature and oceanic placidity were conducive. Thus, the sea could be frozen, but the saltwater could not; the thickness of ice after its formative stage was simply a function of accumulated snowfall.

Near the end of his dissertation on glaciology, Cook claimed that his treatment had been "written wholly from my own observation." But in his introduction to Johann Forster's journal, Michael Hoare asserts that Cook's conclusions bear "unmistakable signs of debate and exchange with" Forster. There is no way to be sure, but this seems fair. Hoare also praises "Cook's farsightedness" in making a case for a southern continent and his iceberg origination theory. He notes that Forster "did not introduce any of his reasonings and ideas on the subject" until he wrote his treatise. Furthermore, even when Forster did record the bulk of his thinking on glaciological issues in the icy latitudes of the Indian Ocean, he attributed them to the plural "we," which presumably included Cook.[39]

Closing his reflections on ice, Cook deployed one of his favourite figurative expressions, that of the adventurous, seemingly solitary explorer. With words only slightly more tempered than his famous "farther than any man" comment, he offered a mock invitation for others with sufficient "resolution and perseverance" to better his record or prove him wrong. Should another explorer "clear up" questions surrounding the origins of icy phenomenon "by proceding farther than I have done, I shall not envy him the honour of the discovery but I will be bold to say that the world will not be benefited by it."[40] Cook was wrong about this and the fate of Antarctica's ice, the great environmental question of the twenty-first century. It is also worth noting that notwithstanding Cook's negative characterization of the polar zone's frigidity and intimidating aspect, in just a year's time he would eagerly take up the challenge of exploring the opposite pole.

⚓

The freezing of seawater is, in effect, a desalinization process. The thermodynamics of this phenomenon are extraordinarily complex (far more so than for freshwater), but put simply, after initial crystallization and with sufficient and enduring cold, the frozen seawater gradually becomes salt free, or nearly so, because the saline minerals, which do not freeze, are pulled by gravity into the underlying ocean as a dense brine solution. This explains why Cook could occasionally haul low, flat pieces of field ice out of the southern oceans and, to his surprise, find that the melted liquid had no detectable salt flavour. The hydrated salts that precipitate from seawater brine are principally sodium chloride, sodium sulphate, and calcium carbonate. There are other minor compounds such as potassium chloride, magnesium chloride, and calcium chloride, but they precipitate out at temperatures below -36.8 °C and are thus rarely encountered in nature. Even lake and river water contain trace amounts of salts but taste fresh.[41]

Reflecting his scholarly community's common misunderstanding, W.F. Weeks, in the definitive textbook on the subject, asserts that "the scientific study of sea ice has essentially all been carried out during the last 100 years," starting with Fridtjof Nansen. After *Fram*'s icebound circuit of the Arctic, Nansen's *Northern Mists* (1911) came to be considered the foundational book for both polar exploration and ice science. Most knowledge about sea ice has accrued since 1945 as a result of a number of developments and concerns: Cold War strategic considerations, Arctic offshore oil and gas drilling, satellite-based remote sensing, and fears about greenhouse gases and climate change. In his chronicle of early polar explorers, Weeks makes a passing mention of James Cook but asserts that William Scoresby the Younger, son of an English whaler, qualifies as "the first sea ice scientist" on the basis of publications issued in 1818 and 1820.[42] That honour actually belongs to James Cook, and Johann Forster to a lesser extent. Both published extensive analyses of polar ice phenomena, but over two centuries of fascination with Cook's travels in Polynesia have obscured their groundbreaking work in polar hydrology.

Typically, seawater freezes at –1.8 °C. This compares to 0.0 °C (32 °F) for freshwater, the difference reflecting variations in the saline content of the two fluids. Because salt increases seawater's density (which is why floating objects are more buoyant in saltwater than in freshwater), lower temperatures are needed to reach the freezing point. When seawater's surface layer is consistently cooled, it begins to freeze, and crystals grow toward the warmer fluid underneath. As the cold endures or increases, the salts precipitate out at temperatures ranging from approximately –3 °C for calcium carbonate to –22 °C for sodium chloride (78 percent of the salt component in seawater). As cooling continues or deepens, more and more ice is formed, but as the ice grows, the salt brine separates from the increasingly fresh (though now crystalline) water until nearly all the salt dissipates into the seawater below the ice, sometimes taking the form of a stalactite at the base of the ice sheet. Thus, salinity values are lowest near the surface. The longer it takes a sheet of ice to achieve a given thickness, the less salt it will have near the surface. When it first crystalizes, there is little difference in the salt content of Arctic or Antarctic sea ice, even though seawater in the Northern Hemisphere is approximately two parts per thousand less saline because North American and Siberian rivers drain into the Arctic Ocean. Correspondingly, sea ice grows slightly faster in the Arctic than near the South Pole, though surface air temperature is still the principal factor in its accretion.[43]

Glaciologists have a nomenclature for the stages of ice formation in bodies of saltwater. In the most common pathway, sticky ice platelets (crystals) with interlocking needles form a thin skim on the surface. This is called frazil. Per Weeks, "the original ice nuclei do not form in or on the surface of the water but in the surrounding air," what he terms "the vapor cloud," which forms over open water during freezing conditions. When circumstances are calm but cold, continued freezing adds more frazil to the oily film, eventually forming a slushy viscous layer called grease ice. If cooling endures, grease ice thickens to become dark nilas, an elastic silky layer about two inches thick. At this density, nilas is nearly transparent and thus dark in colour like the water beneath it; as proto-ice, it is very saline because brine is still trapped in pockets between ice crystals. As the nilas thickens to about four inches in depth, it begins to lighten in colour. Since its reflective property has increased, it is known as light nilas or grey ice. All these phases, which occur to a thickness of one foot, are referred to collectively as young ice.[44]

Young ice that accretes beyond one foot in thickness becomes fully opaque and is called first-year ice. At this stage and with sufficient frigidity, the crystallization process expels the briny compounds through drainage tubes that permeate the ice's structure in vertical networks. Essentially, the brine is drawn down through the ice from the cold upper layer, which is exposed to the air, to the incrementally warmer (and more saline) portion of the sheet. It is assisted by gravity (saltwater is denser, or heavier, than freshwater). Over the course of a polar winter, first-year ice can become several feet thick.[45]

If a layer of first-year ice survives the subsequent warming season, it is known as second-year ice, which is now much harder and displays a blueish cast, a consequence of the brine draining out of the ice's upper layers. Ice that lasts beyond the second summer is considered multiyear ice or, more commonly, pack ice, and over time it may thicken to ten to twelve feet. At this point, the desalination of the upper ice is completed, and all but a trace amount of brine is gone. Multiyear ice can, once melted, become a source of fresh water, at least to human taste. Cook occasionally commented to this effect, though his more common source of potable water was chunks broken off from freshwater icebergs. The stoutest version of pack ice is called paleocrystic ice, once known to reach fifty feet thick. Historically, optimal growth conditions required snowfall that was sufficient to protect the ice from melting on the surface in the summer but not so thick as to form a blanket limiting growth in the winter. Created over the course of hundreds of years, this now rare type of ice may have been encountered by Cook in both the Antarctic and Arctic. But at the time, natural desalination was a mystery to Cook, and even now, Weeks avers, the process is "complex, and still not completely understood."[46]

⚓

Cook did not reach his conclusions about ice formation in a vacuum. They were informed by discussions with fellow officer Charles Clerke, Hudson Bay expert William Wales, and Johann Forster. Following the voyage, when Forster and Cook engaged in a dispute over who was responsible for writing the Admiralty's official account, Cook prevailed, but that did not prevent Forster from getting a book published. His stand-alone volume on scientific issues raised during the expedition, *Observations Made during a Voyage Round the World* (1778), was issued a year after Cook's report. Though not as celebrated in the literature on Cook as his son's book, *Observations* was a wide-ranging scientific treatise, remarkable in its erudition. Michael Hoare deems it "a path-breaking ... first-hand systematic study of a portion of the globe divided deliberately between lithosphere, hydrosphere, atmosphere and bio- and human geography" and "a worthy precursor" to the work of both Humboldt and Darwin. Unfolding in six chapters, Forster's book respected Cook's purview relative to the nautical account, rarely touching on events in *Resolution*'s chronology. Indeed, though historians have emphasized his frictions with Cook, in *Observations* Forster wrote about him in very complimentary terms. Forster was also careful not to tread on Wales's specialties in astronomy, tides, and magnetism because it was "improper, to attempt a business so ably discharged by others."[47]

Although Forster's dominant theme was ethnology, he focuses on "Ice and Its Formation" in the second chapter of his treatise. Forster opens his discourse by alluding to the aesthetic properties of icebergs. With language reminiscent of Cook's description of his first sublime encounter, he begins:

Nothing appeared more strange to the several navigators in high latitudes, than the first sight of the immense masses of ice which are found floating in the ocean; and I must confess, that though I had read a great many accounts on their nature, figure, formation and magnitude, I was however very much struck by their first appearance. The real grandeur of the sight far surpassed any thing I could expect; for we saw sometimes islands of ice of one or two miles extent, and at the same time a hundred feet or upwards above water.[48]

The amount of ice floating in the high latitudes astonished Forster. Besides icebergs, vast quantities of loose chunks drifted about backed by "large flat and solid masses, of an immense extent." The pack held still more icebergs captive, "as far as the eye could reach." Forster's differentiation of ice types echoed Cook's, as did his observation that they encountered ice at different latitudes in different oceans. Forster noted that sudden drops in temperature would prefigure the appearance of pack ice, and he was the source of information, published separately by his son, that the temperature on deck dropped when *Resolution* sailed by a large iceberg. This he considered proof "that the ice masses contribute considerably towards cooling the atmosphere."[49]

Forster marvelled at the ice pack's mobility. On the expedition's initial run from Cape Town into the Indian Ocean in December 1772, for example, a compact body of ice was seen between 50° and 51° S. But in February 1775, when searching for Bouvet's Cape Circumcision, he remarked that "we came to the same place, where about 26 months before, we had met with such an impenetrable body of ice, as had obliged us to run to the East, but where at this last time no vestige of it appeared." He made the more general point that "we never fell in with ice, which we would with certainty consider as stationary, but, on the contrary, found it commonly in motion."[50]

This understanding would fortify Cook's determination during the third voyage to implement the Admiralty's allowance for a return to the Arctic if he failed in his initial attempt. That is, after his disappointment in the summer of 1778, when the Arctic ice pack thwarted his intended run through the Northwest Passage, Cook had reason to expect that if he made a second try the ice might have moved in the meantime, potentially clearing a path toward the Atlantic. But this expectation proved to be naive. Unlike the Antarctic ice pack, which is surrounded by open oceans that allow ice to move about quite freely, the Arctic Basin is essentially land-locked, and the ice has, in effect, less room to manoeuvre.

In his treatise, Forster also addresses the composition and origination of ice. Like Cook, he notes that he found it "worthy of notice" that in all its forms the ice, once hauled out of the oceans and melted, yielded fresh water. Forster, however, makes a careful exception: "spungy and honey-combed" samples deformed by "agitation of the waves" held a "consider-able quantity of brine in the interstices." But, he notes, solid pieces of ice, protected from waves and therefore more integral in their constitution, were entirely potable. Forster then addresses the same key question that Cook puzzled over many times: Where and how was the ice formed? Judging it "undeniable" that the ice came from "regions lying beyond the tracks of ours ships," he carefully avoids the use of "lands," but he does not rule out that theoretical possibility either. He argues that ice had to originate "either near some land, or in the open ocean," but since they had encountered little of the former, he leans in the direction that it was created "far from any land."[51]

Forster then wades into a scholarly debate by referencing the works of George-Louis Leclerc, comte de Buffon; Mikhail Lomonosof; and David Crantz, all of whom had published on the subject of ice formation. These authorities, he stipulates, "were of opinion that the ice found in the ocean, is formed near lands, only from the fresh water and ice carried down into the sea by the many rivers in Sibiria, Hudson's Bay, etc. and therefore when we fell in with such quantities of ice in December, 1772, I expected we should soon meet with the land, from whence these ice masses had been detached." This expectation mirrored Cook's initial belief as well. Forster notes that the absence of land, revealed by several transects of the Antarctic Circle, "raised some other doubts concerning the existence of the pretended Southern Continent." Forster does not elaborate, but he was probably alluding to Cook's reading of ocean currents and swells. In sum, he concludes, *Resolution*'s voyage had "put it beyond doubt, that there is no land on this side of 60° in the Southern hemisphere, if we except the few inconsiderable fragments we found in the Southern atlantic ocean," referring to South Georgia and the South Sandwich Islands. Driving one more nail in the theoretical suppositions that had launched Cook's second voyage, Forster argues that if every part of the Southern Hemisphere not visited by *Resolution* was somehow "occupied by land, this would still be too inconsiderable to counterpoise the lands of the Northern hemisphere."[52]

Forster and Cook had both observed that pack ice, when melted, was fresh and that fragments of icebergs retrieved from the sea yielded potable water, but these observations raised the more contentious question: Could saltwater freeze? Forster cites Buffon's argument that "there is no example of having ever found a sea wholly frozen over, and at a considerable distance from the shores." But he corrects Buffon by noting "that the Baltic is sometimes entirely frozen," as is the North Sea between Denmark and Norway occasionally. Forster then contends that if these seas or others, frozen "from 360 to 420" miles out, were "not to be reckoned a great distance from the land, I do not know in what manner to argue." Notwithstanding legend and fugitive events to the contrary, he concludes that it is "more than probable, that the [Arctic] ocean is frozen in winter, in high Northern latitudes, even as far as the Pole." If "gentlemen" of science want to infer from broad sheets of ice adjoining the European mainland "*that the ocean does not freeze in high latitudes, especially where there is a considerably broad sea,*" Forster was confident he had put forward "instances to the contrary."[53]

Forster then addresses the old theory's corollary: that rivers conveyed freshwater ice into saltwater seas. He grants that Siberia's great rivers, which have sources in the temperate zone, flow from south to north into the Arctic. In the Southern Hemisphere, however, citing the same phenomenon Cook described, Forster recounts how South Georgia "in the midst of summer [was] ... entirely covered with immense loads of snow, the bottoms of its bays were choked up with solid masses of ice, of 60 or

80 feet above water, and we saw no vestiges either of rivers or of springs." If that was the case at 54° S, with the temperatures running from 30 °F to 34 °F, Forster pleads, "how can we then expect any springs or rivers in 60° or 71° South latitude, or rather still higher up to the South, where the sources of these imaginary rivers of the pretended Austral land, must be removed?" It would be impossible, he argues, for any such rivers to carry "those ice-masses into the ocean, which we met with in such stupendous quantities." Forster adds what he deems the clincher: never once did he or anyone aboard *Resolution* see any drift wood in "all the Southern seas." Places such as Novaya Zemlya in northern Siberia, Spitsbergen in the North Atlantic, and even the parts of Alaska visited by Vitus Bering were exposed to "a prodigious quantity of wood thrown" on their shores, "though none is growing there." On the other hand, he notes, the French had seen "not above one or two pieces of wood thrown up by the sea" in the Falkland Islands, "nor did we see any on the island of South-Georgia: all which sufficiently evinces the truth of the above assertion."[54]

Moving from geography to chemistry, Forster turns to the supposition that saltwater does not freeze. He starts with an anecdote from Dutch explorer Willem Barents, who in 1596 saw saltwater freeze to two inches thickness while he was at sea. Since this occurred in September, Forster asks: "What effect then must intense frost of a night in January not produce?" He then turns to the laboratory experiments of Edward Nairne and Bryan Higgins, conducted in 1776 after Cook brought *Resolution* home. Forster describes and footnotes their efforts at length. He concludes from the former's work that "salt-water does freeze," and the ice, once melted, yields water "perfectly fresh." The "briny particles" only adhere "to its outside." But for Forster the most compelling argument in favour of the view that seawater can freeze was "the immense bulk and size of the ice masses formed in the ocean." Considering that the Antarctic's "severe frosts are continued during six or eight months of the year, we may easily conceive that there is time enough to congeal large and extensive masses of ice." From this, he concludes that "the ocean does freeze, having produced so many instances of it."[55]

Forster then challenges Buffon's theory that temperatures are uniform in each hemisphere, another application of the ancient equipoise hypothesis. As for Cook, Forster's experience at South Georgia and "Sandwich Land" had been dispositive. These islands, he writes, were "covered with eternal snow to the shores of the sea, in the months of December and January, corresponding to our June and July." Forster then states that he believes "every unprejudiced reader will find it necessary

to allow the temperature of the Southern hemisphere, to be remarkably
colder than that of the Northern; and no one will, I believe, for the future,
venture to question this curious fact in the Natural History of our Globe."
(This argument was conducive to the prospect of reaching higher latitudes
in the Northern Hemisphere and Cook's ensuing quest for the Northwest
Passage.) As to why this was the case, Forster offers a tentative argument:
"The absence of land in the high latitudes of the Southern hemisphere,
creates this material difference in the temperature of the air, between the
corresponding degrees of latitude in the Arctic and the Antarctic hemispheres." Northerly parts of Siberia and North America, he states, were
inhabited and even bore agricultural produce in the summer. By contrast,
all *Resolution* had found near 60° S was "two little spots in the Southern
Atlantic ocean. The thermometer, in the height of summer, in these high
latitudes, was never five degrees above the freezing point, and we saw it
frequently pointing below it. We often had snow and sleet, and found our
water in the skuttled watercask on deck, frozen during some nights. If all
this happen in the midst of summer, what must the condition of these
climates be during winter?"[56]

The Northern Hemisphere is warmer than the south but not, as Forster
believed, as a function of holding a greater fraction of the globe's land
surface. Instead, northward-flowing cross-equatorial currents transport
oceanic heat in that direction. Oceans are actually warmer than land because they absorb more incoming solar energy. Scientists call this the albedo effect, the ability of a particular substance to reflect solar radiation
back into space. Ice, for example, has the highest surface albedo, meaning
that most of the sun's energy bounces off it. Water, on the other hand, is
less reflective than either ice or land, and therefore it is the most absorb-
ent. Forster understood this, to an extent. He notes that the sun's rays
were "swallowed up in the depth of the ocean." His mistake was in thinking of the great Southern Ocean as a sink when in fact the solar energy
it absorbs is eventually conveyed into the Northern Hemisphere. This
watery dynamic – not, as Forster thought, land radiating heat back into
the adjoining atmosphere – is what makes the northern half of the globe
warmer. In Forster's defence, all this was new to him. In the spirit of the
Enlightenment's Republic of Letters, he encouraged further observation,
investigation, and debate to "shew the causes of my error." It was only by
such discourse, he argued, that "science will be benefited and human
knowledge enlarged."[57]

Forster and Cook did not have a cordial relationship during the expedition, and afterwards he criticized Cook by observing that when

"sending out men versed in science," such as himself, "to remote parts of the world," their efforts should be "liberally supported and generously enabled." For this sin, Beaglehole characterized Forster as a "hack-writer" and "a more prominent than useful member of academic society." In Cook historiography, when Beaglehole leads, others usually follow – in this case into what Michael Hoare calls "the Beagleholian slough of Forster despond." As Nicholas Thomas has observed, most writers "generally have nothing to say about Forster beyond remarking about his notoriously fractious personality," which "has been blown out of all proportion." This pattern of interpretation also ignores the many encomiums Forster threw in Cook's direction in print: he calls him a "great," "experienced," and "able navigator" and refers to information that Cook had "obligingly" provided. A section of Forster's draft text that the Admiralty evaluated to see if he would pass muster as the author of the official account described the great navigator's southern accomplishments in glowing terms. "Cook's two Voyages at last triumphed over the Continent" of Terra Australis, he wrote, crossing the Pacific "in four different latitudes and examining it from the Torrid to the Frigid Zone." By proving "the Nonexistence of the famous *Cape Circumcision*" Cook, Forster argued, had abolished the last of "the little refuges that the favourers of this ill founded opinion might still have left."[58]

The erudition and thoroughness of Forster's *Observations* belies Beaglehole's disparagement and, as Thomas states, this vilification "has obscured ... the character and status of science on the voyage." Cook's and Forster's glaciological analyses were truly path-breaking but heretofore have remained unrecognized. Much like the privileging of the *Endeavour* voyage featuring Joseph Banks, Beaglehole's antipathy toward Forster reflects the antipodal bias of Cook scholarship. Hoare asserts that Beaglehole disliked contentious people, especially if they were not English. In the assessment of Karl Rensch: "Of all the early scientists who had accompanied exploring expeditions to the Pacific, Reinhold Forster had the best academic qualifications." In an implied criticism of Banks, Rensch finds that Forster's command of many languages, geography, botany, zoology, and literature made his observations "more reliable than the impressionistic and naïve descriptions" of an untrained amateur.[59]

⚓

After failing to find Bouvet's cape, Cook turned for Cape Town. The sails and rigging were worn and slowly giving away. Provisions had decayed to the point where "little more nourishment remained in them than just to keep life and Soul together." Applying a stewardship trope common in exploration literature, Cook added: "It would have been cruel in me to have continued the Fatigues and hardships [the crew] were continually exposed to longer than absolutely necessary, their behaviour throughout the whole voyage merited every indulgence which was in my power to give them." Cook was proud of his officers, who surmounted "every difficulty and danger which came in their way," including "being seperated from our companion the Adventure." All this, he concluded, "induced me to lay a side looking for French discoveries and to steer for the Cape of Good Hope."[60]

A ship flying Dutch colours was sighted on March 16, 1775. This was the first European vessel seen since *Adventure* had separated from *Resolution* seventeen months earlier. Approaching the cape, Cook "demanded from the Officers and Petty officers the Log Books and Journals they had kept, which were delivered to me accordingly and Sealed up for the Inspection of the Admiralty." All charts and drawings fell under a similar restriction. Cook enjoined everyone "not to devulge where we had been till they had their Lordships permission to do so."[61] This embargo was a routine naval directive that, given the ever-increasing popularity of travel accounts in the eighteenth century, was intended to limit unauthorized publications that might inadvertently compromise discovery claims or violate diplomatic niceties.

Cook and the crew first learned of *Adventure's* melancholy fate in New Zealand from a Dutch Indiaman they hailed off the African coast. Now back in the channels of intercontinental trade, the *Resolution* was also favoured by a British merchantman, who provided tea and other fresh provisions. Its captain, who did not intend to stop in Cape Town, was prevailed on by Cook to carry a letter to the secretary of the Admiralty "just to acquaint him where I was." According to the ship's log, *Resolution* reached Table Bay on March 22, but by the local calendar, it arrived a day earlier, having gained a day by circumnavigating the globe from west to east. The ship anchored amidst a host of other French, Dutch, and British vessels. The captain of a British ship agreed to take the junior officers' journals plus some charts and drawings to London to provide an advance read on the voyage's accomplishments. In port, Cook found a letter from Furneaux describing the affray at Grass Cove and his reason for a quick flight home from the South Pacific (spoiled provisions, ostensibly). Furneaux had searched for Bouvet's Cape Circumcision as well, but he too failed to find it at the location prescribed by French charts. Cook took this as "another proof that it must have been Ice and not Land which Bouvet saw."[62]

Cook set in motion the process of procuring bread, meat, greens, and wine for the crew and, for himself, "some real repose after the fatigues of so long a voyage." He sent only three men ashore for "the recovery of their health." Cook also directed the ship's repair. Based on his experience at Batavia during *Endeavour's* voyage, he believed the Dutch, old rivals that they were, would take "a sham[e]full advantage of the distress of Foreigners."[63] Cook was briefed on the latest exploration news about French and Spanish voyages in the Indian and Pacific oceans. Flush with his own success, he was nonplussed by Spanish additions to the chart of the Society Islands; Polynesia now mattered little to him. The Forsters were likewise busy, dispatching letters to fellow scientists and patrons such as Linnaeus and Daines Barrington and readying their drawings and journals for what they expected to be a quick path to publication once they reached home.

After five weeks stay in Cape Town, *Resolution* put to sea on April 27. Fully confident in the ability of "Mr Kendals Watch" to calculate longitude, Cook set a course for the meridian of St. Helena. He reported that the chronometer "did not deceive us and we made it accordingly on the 15th of May." At the British colony, Cook had the "mortifying" experience of having to explain away some of the local elements in Hawkesworth's account of *Endeavour's* voyage, which had been published after Cook set sail on *Resolution*. In his journal, Cook started rescuing his reputation by avowing: "I never had the perusal of the [Hawkesworth] Manuscript nor did I ever hear the whole of it read in the mode it was written." He took particular exception to the charges that *Endeavour's* crew had mistreated St. Helena's slave population and that he had mocked tools used on the island. Both charges were drawn from Banks's journal but attributed to him by Hawkesworth.[64]

Resolution left for Ascension Island, where the ship took on more provisions and remained until May 3. At this point, Cook was no longer keeping a daily log, a sure sign that the voyage had entered its anticlimactic phase. Nevertheless, Cook was captivated with the idea of establishing the exact longitude of Brazil's coast, which in his estimation most charts placed too far west. This was the kind of behaviour the Forsters found obsessive, but Cook was not "willing to give up every object which might tend to the improvement of Navigation and Geography for the sake of geting home a Week or a fortnight sooner."[65]

On June 11, *Resolution* crossed the equator. The sails of other ships now became a common sight. After a brief stop in the Azores, "the better to enable us to fix with some degree of certainty the Longitude of these Islands," Cook was relentless – *Resolution* headed for home. She made land near Plymouth on July 29, 1775, "and the next morning anchored at Spithead." Charles Clerke quickly dashed off a letter to Joseph Banks reporting on his return from the "Continent hunting expedition." Being aboard *Resolution*, he wrote, was akin to spending "3 Years underground."[66]

Cook ended his published account by extolling the expedition's dearth of casualties and sickness, and, given the high-latitude orientation of the voyage, cited "the singular felicity we enjoyed, of extracting inexhaustible supplies of fresh water from an ocean strewed with ice." An expedition of three years and eighteen days' duration, during which only four men had been lost and just one of those to illness, was indeed a remarkable accomplishment. Beaglehole suggested that a seaman was better off "with Cook in the Antarctic than ... in London." Cook apologized that his account "has been more employed in tracing our course by sea, than in recording our operations on shore." This emphasis, he explained, was mission-driven. The purpose of his voyage "into the Southern Hemisphere" was to look for Terra Australis Incognita. Absent that discovery, Cook was left to credit "our persevering researches," which yielded "less room for future speculation about unknown worlds remaining to be explored." He took satisfaction in circumnavigating the globe in icy latitudes over several years "in such varieties of climate, and amidst such continued hardships and fatigues." In the account's last line, Cook stipulates that "the disputes about a Southern Continent shall have ceased to engage the attention, and to divide the judgment of philosophers."[67]

Cook's letter to the Admiralty, drafted in Cape Town and carried to London by another ship, offers a noteworthy recapitulation of *Resolution*'s track. It yields the unmistakable impression that the most significant segment of the voyage was in the high latitudes of the Atlantic Ocean, not the vaunted Pacific embedded in Cook lore. In the letter, Cook references the discovery of South Georgia. This was a "coast which from the imence quantity of Snow upon it and the vast height of its Mountains, we judged to belong to a great Continent, but we found it to be an isle of no more tha[n] 70 or 80 leagues in circuit." This passage faithfully reflects the real-time perceptions recorded in the journals of both Cook and Clerke. That is, when it was least expected (during the final year of the voyage in the last quarter searched), perhaps the fabled southern land had been stumbled on. Cook recounts proceeding toward Southern Thule and his sketchy

"Sandwich Land." This terrain, he writes, was "so buried in everlasting snow that it was necessary to be pretty near the Shore to be satisfied that the foundation was not of the same composition." He candidly admits that the ship had been carried away from this coast, "which we could not afterwards regain, so that I was obliged to leave it without being able to determine whether it belonged to a Continent extending to the South or was only a group of isles." (This is the foundation for Cook's mistaken belief that he had seen a facet of the polar continent.) The appearance of "Sandwich Land," he avers, replenished his lagging faith in the existence of Cape Circumcision, so "I quited this Horrid Southern Coast with less regret." As we have seen, Bouvet Island also eluded Cook.[68]

Daniel Solander, Banks's assistant on the *Endeavour,* was privy to Cook's Cape Town letter after it arrived in England. In a quick jot to Banks, he wrote: "Abstract: 260 new Plants, 200 new animals – 71° 10' farthest Sth – no continent – Many Islands, some 80 Leagues long," this last a reference to South Georgia. After Solander saw Cook in person, he dispatched another note to Banks. It briefly discusses *Resolution's* stops on the shores of Tahiti, Tonga, the New Hebrides, and New Caledonia but makes nary a mention of Cook in the high latitudes, ice, or the possibility of a polar continent. Nevertheless, Solander deems it a "Glorious Voyage." It has been the common view of historians ever since that the second voyage was Cook's paramount achievement.[69]

Although Cook's third voyage was in fact a greater accomplishment (recognition for which has eluded him because he died in the execution of it), his second expedition nonetheless added 30 degrees of comprehension to the high southern latitudes across 360 degrees of longitude. With the exception of South Georgia and Sandwich Land, this space contained no terrain. Cook was the rare explorer whose reputation could prosper in the absence of actual discovery. As Nicholas Thomas puts it, "Cook's history-making ... lay in discovering nothing." The great navigator's "sense of accomplishment did not hinge on the presence or absence of the continent, but on the distance he ventured and the precise record he kept of his route and his findings." Put another way, Cook's discoveries may have been negative, but they were complete.[70]

Tony Horwitz states that whereas "most explorers might dream of the glory attending discovery of a new continent, Cook revelled in the opposite. A born skeptic, he prized fact over fantasy."[71] This is true to a point, but the naming of Cape Disappointment on South Georgia seems to indicate a psychological turn in Cook's mind that took place late in his second voyage and informed his enthusiasm for a third expedition with

Nathaniel Dance, *James Cook,* oil on canvas, 1775–76. This is generally considered Cook's official portrait, painted in 1775–76, of his public acclaim in Great Britain. Given the dominance of the palm-tree paradigm that defines the great navigator's career, this image is noteworthy because Cook's index finger is aimed at the vicinity of New Georgia (later South Georgia) and Southern Thule, the element of the South Sandwich Islands closest to the pole. This chart was the only cartographic specimen that Cook's memorialist referred to in the prefatory section of the Admiralty's account of the third voyage.

positive discovery – that is, the Northwest Passage – as an objective. Cook had no false modesty about the Antarctic voyage's accomplishments, but near its end he began signalling a lack of fulfillment. But for his choice of approaches to the far south, both of which aimed at Antarctica's most southerly recesses, Cook might have encountered the continent more directly. As it was, he was the first to glean an intimation of a land mass at the South Pole.

⚓

Chagrined by Hawkesworth's treatment of the *Endeavour* voyage, Cook was more determined than ever to have control over the official narrative of the Antarctic expedition. As he eventually phrased it in the official account of the second voyage, "what I have here to relate is better to be given in my own words, than in the words of another person; especially as it is a work designed for information, and not merely for amusement." Cook believed his "candour and fidelity" would "counter-balance the want of ornament," a credible statement but one that was also a common trope in discovery literature. In a letter to his old mentor from Yorkshire, Captain Walker, Cook projected that his book would lack "those flourishes which Dr. Hawkesworth gave the other, but it will be illustrated and ornamented with about sixty copper plates, which, I am of opinion, will exceed every thing that has been done in a work of this kind ... As to the Journal, it must speak for itself. I can only say that it is my own narrative, and it was written during the voyage."[72]

As a journalist and writer, Cook did not resort to the common conceit of pretending, through a manipulative recasting of journal entries, to always know where he was or how the story was going to turn out. As Tony Horwitz notes, Cook had the very unusual habit for an explorer of "telling the reader what he did *not* know." Rarely reaching for the grandiloquence one associates, for example, with the proto-Romantic Meriwether Lewis, Cook was, instead, in Frank McLynn's phrasing, a "master of understatement," content to follow his instructions, always professional, "the ultimate duty-bound navigator."[73]

George Forster offered an influential if self-interested perspective on Cook's evolution from explorer to writer. According to Forster, it was not Hawkesworth's exaggerations and mistakes but "the news of the prodigious profits of the compiler [that] inspired him with the desire of becoming an author." Forster asserted it was only at Cape Town that Cook determined "to obtain a share in the emoluments of the history of the voyage." Cook was indeed conscious of the royalties from book sales that might be coming his way. In his letter to Wilson, Cook professed that he would "have the sole advantage of the sale." In the contest between the Forsters and Cook over authorial prerogative, both fame and monetary reward hung in the balance. But George Forster was wrong in his assertion. The vast amount of redrafting Cook engaged in while writing his journal (three concurrent versions were in progress at all times) shows an intention to compose a postexpedition narrative from the outset of the voyage. Cook's fabled ambition, notes Nicholas Thomas, "had turned from geographic to literary accomplishments."[74]

A desire for royalties was not the only aspersion George Forster cast in Cook's direction. He also seeded what would become conventional historiographical criticism by noting Cook's failure to take a senior naturalist on his third voyage. In a public letter to the Admiralty's first lord, the earl of Sandwich, dated June 1, 1778, when Cook was already in Alaskan waters, Forster asserts that the scientific community will "not fail to reflect, that if you had been actuated ... by the love of knowledge, you would have sent at least *one* man of letters with Capt. Cook, when he sailed on his third great voyage in July, 1776."[75]

The Cook-Forster authorship dispute would spill over into the planning for Cook's northern voyage. Daines Barrington, who had sponsored Johann Forster as Banks's replacement, figured prominently in this denouement. On his own volition, Barrington seems to have privately assured Forster that he would author the official narrative, presumably as an added enticement, given Banks's late withdrawal. Forster returned to England armed with this assumption, but Cook fiercely resisted, fearing a triple loss: control of the narrative, literary renown, and financial reward. Having cajoled Forster into the undertaking, Barrington initially took Forster's side, insisting that the naturalist write a more prestigious geographical memoir and not just an addendum on natural history, as Cook envisaged.

Various proposals for coauthorship were discussed, but in the end, Forster was undone by his own text. His writing samples were not deemed appropriate by the Admiralty, under whose auspices the book was to appear. More saliently, Forster's new glaciological theory had precipitated a falling out with Barrington, the principal public advocate for the existence of a Northwest Passage. The proposition that saltwater did not freeze was essential to the practicability of sending someone, ultimately Cook, in search of the Arctic gateway. When Forster lost his patron, the contest was over. Cook wrote the *Resolution* account, as he had always hoped and expected would be the case. In response, Forster assisted George (who was not bound by the Admiralty's literary restrictions) with writing *A Voyage Round the World,* a narrative that competed with Cook's, then he published his own *Observations* in the summer of 1778. Coincidentally or not, Cook's less insistent treatment on how sea ice was formed, as opposed to Johann's more fervent view on the subject, happened to serve the navigator's career aims.

Avidly awaited by readers all over Europe and America, Cook's *A Voyage towards the South Pole* was published in May 1777, ten months after he had left on his next and last expedition. (The rigour of preparing a complex text for publication heavily informed Cook's approach to keeping the third voyage's journal, in which narrative storytelling, not nautical data such as location, weather, and distances, dominates the text.) Copy-edited by Anglican Canon John Douglas (after Cook segmented the journal into chapters to meet literary conventions), *A Voyage towards the South Pole* approached the popularity of Hawkesworth's *Endeavour* digest. In literary terms, it may have been more influential than its predecessor, as it inspired other writers such as Mary Shelley and Samuel Taylor Coleridge. Bernard Smith's brilliant deconstruction of the latter's *Rime of the Ancient Mariner* leaves no doubt that the epic poem derived much of its literary power from the combined influence of Cook and William Wales, *Resolution*'s astronomer who later became Coleridge's instructor.

One line from Coleridge is especially pertinent: "And through the drifts the snowy clifts, Did send a dismal sheen." This imagery was drawn from the second voyage's encounter with the ice blink phenomenon, on which subject, Wales "was something of an authority," having been introduced to it on Hudson Bay in 1769. Wales's experience had served the expedition well during *Resolution*'s first encounter with pack ice in December 1772, southeast of Cape Town, when a "whitish haze in the horizon" augured an imminent encounter with a field of floating ice. Wales shared his surmise with those on deck only, as he phrased it, to be "laughed at for my information."[76] He was, of course, completely vindicated. One person who definitely absorbed this lesson was James Cook. The great navigator's discernment of ice blink on the quest for the Northwest Passage would prove to be the pivotal moment of exploration during his third and final voyage.

PART THREE *A Third Voyage*

EIGHT

An Ancient Quest, a New Mission

After *Resolution* returned from its rigorous three-year voyage, Daniel Solander reported to Joseph Banks that James Cook looked "as well as ever," indeed, "better than when he left England." Solander's remark contradicts the pervasive view that Cook was used up or otherwise ill-prepared for leading another expedition. For example, in *Captain Cook's World*, John Robson writes: "Cook had been a tired man when he returned to Britain at the end of his Second Voyage in 1775. He had ... fallen seriously ill halfway through the voyage. He was ready for some rest and the opportunity to recover." More ominously, in *Captain Cook*, Vanessa Collingridge states: "Something happened to Cook on his second voyage"; he "grew tired" not only physically but mentally, resulting in "cracks that would ultimately lead to his death." In reality, only one member of the expedition confessed to being physically defeated. Near its end, Johann Forster confessed: "I am no more young."[1] Cook was thankful to have some respite from the stresses of global circumnavigation, but his physical reserves were not tapped out. As Nicholas Thomas states: "We do not, in fact, know that he was desperately tired or strained. No one who met him at this time suggests that he was."[2] The Solander anecdote suggests the opposite.

After returning from the Antarctic, Cook took an administrative position at the Royal Hospital in Greenwich, a rest home for disabled and elderly seamen. When he accepted this light-duty sinecure in August 1775, he reserved the right "to quit it when either the call of my Country for more active Service, or that my endeavours in any shape can be essential to the publick." This was no secret. Solander told Banks that Cook had secured a promise from the Admiralty that he could return to active duty "whenever he should ask for it."[3] Cook knew himself and anticipated the need for what in modern parlance is called an exit strategy.

Word was circulating in London that *Resolution*, under the command of Charles Clerke, would soon return to the Pacific to return Mai, a passenger Tobias Furneaux had taken aboard at Huahine, to his homeland.

With the foreshortened *Adventure* voyage, Mai had reached Great Britain in July 1774. He became a favourite in London's parlours, much like Ahutora, who was feted in Parisian society at the conclusion of Bougainville's expedition. But as time wore on, Mai's sponsors grew weary of managing his appearances, which laid the groundwork for his return. One week after taking the Greenwich position, Cook referenced this plan in a letter to John Walker. *Resolution,* he stated, "will soon be sent out again, but I shall not command her." He already had misgivings about his posting, telling Walker he had gone from one extreme to another. Cook stated that a few months earlier, "the whole Southern hemisphere was hardly big enough for me and now I am going to be confined within the limits of Greenwich Hospital, which are far too small for an active mind like mine." It was a "fine retreat" with a good income, but he was uncertain "whether I can bring my self to like ease and retirement, time will shew."[4]

Cook was not interested in *Resolution*'s next voyage when it was limited to returning Mai, but there were other active minds in play. Even before Cook had liquidated the Terra Australis fantasy, there had been ferment in British intellectual and commercial circles regarding the other great geographic puzzle of the age – the supposed existence of a Northwest Passage through, or around, North America. Every European power desired a continuous waterway from the Atlantic to the Pacific that could serve as a shortcut to Asian markets. No one in Great Britain had explicitly stated that Cook would take up this challenge, but he loomed over the proposition nonetheless. On January 19, 1773, two days after Cook first crossed the Antarctic Circle, Daines Barrington informed the Royal Society that he had met with Lord Sandwich to discuss the practicability of a voyage to the North Pole and, on the presumption that it was ice-free, find a passage to the East Indies. The society endorsed Barrington's idea, which later that year resulted in an unproductive voyage into the North Atlantic. In 1774, Barrington shifted focus by proposing an attempt from the Pacific side of North America. Sandwich told the society that an expedition that year was out of the question but that one "will be undertaken after the return of Capt Cook in 1775."[5]

Not coincidentally, Parliament expanded eligibility for the £20,000 prize for finding this shortcut between Europe and Asia to include officers in the Royal Navy. It had formerly been limited to captains from the Hudson's Bay Company (HBC) or other privateers. The act, which passed in December 1775 after Cook reappeared, amended the 1745 statute in one other respect. It repealed the provision that specified Hudson Bay as the qualifying outlet for the prize, stipulating only that the expected passage would reach the Atlantic north of the fifty-second parallel. Knowledgeable observers favoured prospects in the upper sixties, but 52° N crosses the southern extremity of Hudson Bay, so out of either inertia or wishful thinking its prize-winning potential was maintained.

⚓

The search for the passage had a long history, and as J.C. Beaglehole observes in his introduction to Cook's journals, it was "an illusion just as sedulously nurtured as that of the great southern continent," invented and propagated by people "who may be called geographical romantics."[6] The proposition was not so much whether the passage existed. It only waited to be discovered, and it would be navigable. Explorers from several nations figured in the centuries-long search for the route, both before Cook and afterwards. Nevertheless, the passage was a particular obsession for Great Britain, and to a preponderant degree, it is a British story. The openings of the passage at either end were established by English navigators: Baffin in the east in 1616 and Cook in the west in 1778. Sir John Franklin's fatal expedition in the north central Arctic in the nineteenth century is the most famous episode in the corridor's history.

Russian voyages, French theoretical cartography, and a heavy dose of Spanish legend add considerable spice to the saga, but the efforts of those countries, or mere rumours of them, only spurred the British on. Struggling to maintain control of its American colonies and with its West Indies sugar economy under duress, Britain had added incentive in the 1770s. In the age of sail, a Northwest Passage was thought capable of cutting a twelve-month voyage, through either the tumultuous seas surrounding Cape Horn or the longer route around the Cape of Good Hope, in half. A more direct path would not only cut travel times but also eliminate vulnerabilities to nature and other nations. That a British discovery of the passage would torment its chief rival, France; thwart Russian expansion into North America; and outflank Spain's presumed dominance in the Pacific only provided greater motivation.[7]

Columbus's discovery of a "new world" had led to the parcelling of the Western Hemisphere by various European powers, but the original impulse behind his voyage never disappeared. Once the continental dimensions of North America were hazily discerned, the multinational search for routes through and ultimately around it took hold and endured until the middle of the nineteenth century. A saltwater route through the ice-choked straits in the northern Canadian archipelago would finally be sketched by the 1850s, but it was not until railroad men stitched together a latter-day version of the Northwest Passage across the continent near the end of the nineteenth century that the idea reached practicability. In the first decade of the twentieth century, the classic version from Europe to the Pacific basin was laboriously transected. In the twenty-first century, the dream of a quick shortcut has been resurrected, inspired by global warming and the concomitant recession of the Arctic ice pack. Seasonally, ships now make the passage in a few weeks' time.

⚓

A century after Columbus's discovery and two centuries before Cook set out on his voyage, North America was still perceived as much a geographic obstacle as a colonial destination. The several voyages of Martin Frobisher in 1576–78 (like Columbus's, thought to be destined for Cathay, or China), touched at the southeastern corner of Baffin Island. Frobisher conceived the idea that he had found a northern strait through the continent analogous to Magellan's, but the supposed corridor proved, after subsequent exploration, to be the deep bay that now bears his name. John Davis led a voyage in 1586 into the waters between Greenland and Baffin Island. He was directed to probe as far as 80° N in pursuit of a passage. This latitude was thought to be practicable because of the durability of the medieval idea that the Arctic Ocean was navigable and the North Pole ice-free. In the event, Davis turned back at 66° N, and in a follow-up voyage the following year, he reached 72° N before ice foreclosed his options in the strait now named for him.

In 1616, William Baffin reached Smith Sound at 78° N, a record for the Canadian Arctic that would stand until the nineteenth century. His most practical find was Lancaster Sound, at 74° 20' N, which in time proved to be the main gate to the passage from the Atlantic side. Hudson Bay, because of its more southerly aspect, was always the idealized portal, and in Cook's era there was still a glimmer of hope it would yield a strait to the Arctic Ocean and thence westward to a Pacific outlet. (Such a connection exists, via Fury and Hecla Strait, emanating from the Foxe Basin north of Hudson Bay proper, but this ice-choked gap would never serve as a shortcut because it joins the western extent of Lancaster Sound, a corridor that can be reached more expeditiously in the manner first pointed out by Baffin.)

The bay's exploration by its namesake, Henry Hudson, is a story in itself. In 1607, on behalf of the Muscovy Company, which had ambitions for reaching the Chinese market by discovering a Northeast Passage atop Russia, Hudson slightly exceeded 80° N east of Greenland before being thwarted by ice. He was the first to attain that northern benchmark. Over the next two years, Hudson conducted additional runs to the northeast, the second time as an employee of the Dutch East India Company. Stymied by the ice, Hudson bounced off the pack and in a historic redirection headed west across the Atlantic, bringing to an end his search for what is now commonly referred to as the Northern Sea Route. This corridor presented its own challenges, and it would briefly tempt Cook, but the immanent dominance of the czars and Empress Catherine crimped the attractiveness of finding a sailing route above Eurasia.

On his rebound, Hudson sailed into the river that now bears his name. He had again failed to find a shortcut to Asia, but the gateway he did discover was, until the railroad era in the mid-nineteenth century, one of the most productive routes into the interior of North America, especially after the establishment of the Erie Canal in 1825. In 1610, Hudson entered his northern strait and bay, but emblematic of future difficulties searching for the Northwest Passage's portals, he was set adrift by a mutinous crew. Navigable some years for as little as two months, Hudson Bay would prove to be a dead end for two centuries' worth of dreamers. Thus, at the end of the first generation of Arctic exploration, the consensus was that if there was a shortcut heading west from the Atlantic, it probably lay at 66° N or higher, and access to it might only be seasonal at best.

The pioneering navigators (their names are synonymous with the region – Frobisher, Davis, Baffin, Hudson) recorded their findings so that others might follow them, but their disconnected discernments caused purely speculative geographers to jump into the breach. One of the earliest and most fantastical tales, published in 1626, involved the fictional Lorenzo Ferrer Maldonado, who was rumoured to have reached 75° N via Davis Strait in 1588 before proceeding west to a strait that separated Asia and America, a place where he met Lutheran traders.

Even in the less credulous times of Cook, armchair geographers still imagined that an outlet from Hudson Bay would realize the ancient vision, but in the end, it only served the fur trade. The HBC and its competitors gradually pieced together a commercial network of rivers and lakes to the west, so that by the time Cook knocked on North America's Pacific door, traders had closed much of the terrestrial gap between the bay and the west coast of the continent. The key figures were James Knight, William Stuart, Moses Norton, and, especially, Samuel Hearne.

In 1715, Knight, the chief factor of Fort York, dispatched Stuart on an intertribal peace initiative to further the company's commercial prospects, including searching for the allied desiderata of straits, rivers, or precious metals. Accompanied by Thanadelthur, a Denesuline (Chipewyan) woman from the interior, and Cree from the fort's neighbourhood, Stuart travelled seven hundred miles across the forbidding terrain known since as the Barrens to the country southeast of, but not quite to, Great Slave Lake. Stuart and Thanadelthur returned in 1716 having facilitated a rapprochement between the Cree and the Denesuline. They also brought back stories of a river whose banks were lined with chunks of pure copper metal so heavy that several men could not lift them. This was not the gold Knight most desired, or an ice-free saltwater strait trending toward the Pacific, but the Coppermine River would loom large in Cook's saga.

In the wishful thinking that lay at the heart of the reigning cartographic paradigm, the large body of water Stuart and Thanadelthur heard about (eventually denominated Great Slave Lake) was transmogrified into "the sea." Here, one Arthur Dobbs enters the story. A member of Parliament, Dobbs's great cause was ensuring the appropriateness of the HBC's monopoly charter and its corporate effectiveness. He had a passion for geography, too, perhaps not coincidentally since his neighbour was Jonathan Swift. Dobbs was aggravated by the company's reluctance to invest in exploration for exploration's sake. He succeeded in pushing for a conclusive survey of the bay's western shore, a search that was expected to yield the Northwest Passage. The Royal Navy commissioned a voyage in 1741, commanded by Christopher Middleton, formerly a captain in the company's trading operations. Middleton was able to make significant cartographic refinements but, of course, he was unable to find a saltwater link north or west that could carry sail to the passage's Pacific portal: the legendary Strait of Anian.

This elusive and entirely mythical gateway floated up and down the west coast of North America on maps issued by cartographers of many countries over several generations. The fluidity of the strait's location was a function of particularized national objectives, the incorporation of information from mythical voyages, and pure speculation, all fuelled by the fact that the North Pacific was, for Europeans, the most obscure part of the world, excepting the poles. Only Francis Drake's cursory effort in 1577–80 had shed any light on this long coastline, and even his account was replete with quasi-fictional elements. The Strait of Anian first appeared cartographically in the middle of the sixteenth century as a narrow channel separating Asia and America, analogous to the actual Bering Strait. By the early seventeenth century, the concept had shifted southward to a passage that subdivided North America, or at least provided access to its interior.

In one projection, the strait was shown running inland in the vicinity of northern California. Well into the eighteenth century, French cartographers (always the most imaginative) continued to depict this opening in the mid-latitudes.

Dobbs refused to accept the objective outcome of Middleton's voyage. Picking up the thread of speculation, he introduced a major complication in the grand narrative of the Northwest Passage in *An Account of the Countries Adjoining to Hudson's Bay* (1744). In the map accompanying this publication, Dobbs outmanoeuvred the French projection by eliminating the Strait of Anian altogether. He drew the continental coast from Cape Blanco (in today's southern Oregon) in a northeasterly direction, cutting conveniently across twenty degrees of latitude toward Hudson Bay in a gentle curving line.

Dobbs supplemented his theoretical cartography by dredging up the mythological voyage a Spanish admiral, Bartholomew de Fonte, had supposedly conducted in 1640. According to legend, Fonte had sailed from Lima to 53° N, where he entered the Strait of Anian. This opening led to a river, across an extensive lake, and past falls to another lake, which steered toward another strait and a mid-continent encounter with a ship from Boston that had travelled west via Hudson Bay. This was not a passage in the classic sense but more of a corridor; it anticipated what would become, post-Cook, the second iteration of an idealized Northwest Passage – a membranous "communication" between the Pacific and the Atlantic. The Fonte cartographic image was patently nonsensical and, as Barry Gough writes, it "should perhaps have remained hidden for all time," and it would have had it not served Dobbs's agenda.[8] His geographic propaganda held sufficient currency that both Cook and George Vancouver would reference, nominally research, and ultimately ridicule it.

Criticisms of the Fonte myth abounded in the run-up to Cook's final expedition, but the most graphic evisceration appeared in Johann Forster's history of northern explorations, the English-language edition of which was published in Dublin in 1786. Though Forster had a complicated relationship with Cook, in this book, which appeared at a time when the thoroughness of Cook's third voyage was coming under fire, he came to the great navigator's defence. Forster determined that the Fonte story had originally appeared in 1708 as "the production of some idle visionary." This "improbable" report, which told of tides reaching beyond "water-falls or cataracts" and "salt water even beyond a cataract," was so absurd to Forster that it was more in keeping with Daniel Defoe's "New Voyage round the World" than any "authentic records." As Beaglehole quipped: "Rarely can disbelief have been so willingly suspended."[9] In any event, Dobbs was able to create sufficient churn around the notion of a practicable Northwest Passage that Parliament authorized a £20,000 prize in 1745 for any merchant captain who could find it. This act, once amended to qualify officers in the Royal Navy, would help lure Cook to the Arctic.

⚓

Over time, Hudson Bay would prove to be a cul-de-sac. However, its extension deep into North America allowed it to become the jumping off point for the single most important development guiding the course of Cook's third voyage: the overland expedition of Samuel Hearne, an HBC trader posted at Churchill on the inland sea's western shore. In the constant readaptation to geographical reality – which was the subtext of the quest for the Northwest Passage from Martin Frobisher off Baffin Island in the 1570s to Meriwether Lewis on Lemhi Pass in 1805 – Hearne's trek to the Arctic Basin in the early 1770s was pivotal.

Hudson Bay's prospect as a gateway to the Pacific endured because it was thought to be the only practicable option. All the inlets pointing west from Baffin Bay were in latitude 70° N and higher and often inaccessible because of ice, even in summer. Within this context, Hearne's venture was in effect a flanking manoeuvre to get on the back (western) side of any Atlantic opening to the passage, thereby simplifying the search. Since the width of North America had not yet been scientifically determined (in time, it would be another Cook legacy), it was thought that an overland mission northwest of Hudson Bay, following in the footsteps of Stuart and Thanadelthur decades earlier, might lead to an inlet off the Pacific analogous to Hudson Bay, if not the main ocean itself.

Why would Hearne's patron, Moses Norton, think that such a body of water existed? The concept of a second North American inland sea – what George Vancouver later mocked as the new Mediterranean – was a favourite notion of French cartographers. Accessible from the Pacific, the Mer de l'Ouest dominated North America's northwest quadrant. The principal propagators for this image were the map-making brothers Guillaume and Joseph-Nicolas Delisle and Phillippe Buache. Its promise was that if navigators could make their way to this sea from the Pacific's temperate latitudes, they would gain access deep into the continent's interior, if not almost to Hudson Bay itself. This legend had such deep-seated currency in French and British intellectual circles that it would dog Cook's reputation posthumously.

The voyages of Vitus Bering, a Danish explorer in Russia's employ, precipitated a separate torrent of geographic speculation. Bering was presumed to have sailed around the eastern end of the Asian land mass in 1728, when he reached 67° N through what seemed to be a strait separating the continents, but without actually sighting the American shore. This circumstance left speculators with a lot of room to manoeuvre. When news of his exploits reached western Europe, it was thought that the Russians might have discovered the opening to both the Northeast Passage over the top of their own country and the access point to seas surmounting North America. A map published in 1735 by J.B. du Halde hinted at the terminus of Asia but finessed the issue of what lay to the east by pushing the image of Siberia to the extremity of the right margin. Bering's followup voyage of 1741 took a more southerly course in an overt attempt to reach North America. He coasted the Aleutians, Kodiak Island, and the Kenai Peninsula, and he sighted Mount Saint Elias, but the expedition shed no further light on the putative strait and its juxtaposition with the North American mainland.

In 1753, Buache published a noteworthy map that served up an admixture of fantasy and reality. Siberia and North America were indeed separate and not a contiguous land mass, but the exact location and dimension of "Bering's strait" was hazy and would remain so until confirmed by Cook. Nevertheless, the mere intimation of this strait suggested a strategic pairing with Lancaster Sound, sighted by Baffin in 1616. For the balance of the eighteenth century (specifically until Vancouver's journal was published in 1798), the French school of speculative geography peppered its maps with imaginary inland seas reached via fanciful straits heading east off the Pacific. These openings included Fonte's supposed strait, resurrected from obscurity by Dobbs in 1744, and what Glyn Williams terms "the only slightly less improbable story of Juan de Fuca's voyage of 1592."[10] Samuel Purchas, in his history of exploratory voyages, published in 1625, relates a second-hand report about a Greek pilot, Apostolos Valerianos, who sailed the Pacific Coast for Spain under the name of Juan de Fuca. He alleged having reached a huge inlet at 48° N and reputedly sailed deep into this opening over the course of twenty days. Like Fonte's story, Dobbs repopularized the Fuca legend, and it was these tales that helped inform the idea of the Mer de l'Ouest.

Historian Barry Gough has attempted to establish the plausibility of the Fuca story. He does not assert that Fuca made the voyage of legend, but he does argue that such an explorer existed and found an opening to the strait that others later plumbed. Gough maintains that because Vancouver

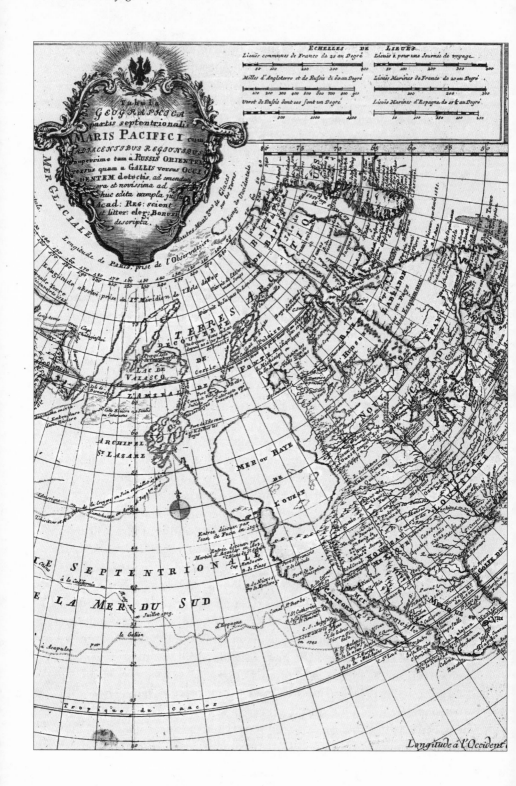

◄ Leonard Euler, *Tabula geographica partis Septentrionalis Maris Pacifici*, 1760. Based on Philippe Buache, *Carte general des nouvelles decouvertes au Nord de la Mer du Sud*, 1752. This image was included in Euler's atlas, designed for classroom use, and issued in Berlin under the auspices of the Royal Prussian Academy of Sciences. Euler's engravers replicated the work of other cartographers, in this case Buache. In the wake of Vitus Bering's recently completed voyages, which drew the interest of geographers to the North Pacific, both Buache and his occasional collaborator and nephew, J.N. Delisle, issued duelling projections for the location of the Northwest Passage. Both versions included the long-standing French notion of the Mer de l'Ouest, but Buache gave the concept its fullest expression. This image outlasted Cook, who thought his northern voyage had demolished the idea. It was not fully vanquished until George Vancouver's three-year survey of the Northwest Coast in 1792–94.

dismissed Fuca even more vehemently than Cook did historians have piled on.[11] Suffice to say, mid-eighteenth-century French cartographers, energized by Bering's discoveries, normalized the geographic lore dredged up by Dobbs. Their serial projections depicted temperate-zone openings in North America's Pacific Coast that reached inland seas proximate to Hudson and Baffin Bays.

The French imagining of North America's west coast was certainly more enticing to a would-be explorer such as Hearne than the image emerging from St. Petersburg. In 1758, the St. Petersburg Academy of Sciences belatedly published Gerhard Müller's account of Bering's second voyage. An English edition in 1761 included a map locating Bering Strait, close to its actual location, but its most salient feature was a conjectural and exaggerated extension of the Alaskan subcontinent farther north and west than where the French typically placed America's Pacific shore and thus profoundly more distant from Hudson and Baffin Bays. South of Alaska, Müller's map contained more inaccuracies (including a "River of the West" emerging at 45° N from headwaters at Lake Winnipeg), but as Beaglehole opined, the image was not a "rash collation" and could have logically served as the premise for a British exploratory venture. Anticipating Cook's cartographic temperament, Müller was dismissive of French speculative theory; instead, he considered it "better to omit whatever is uncertain, and leave a void space, till future discoveries shall ascertain the affair in dispute." But armchair geographers, such as Dobbs, relished filling vacuums with optimistic projections. None of these, says historian Glyn Williams, was more enchanting than the "beguiling but fanciful maps of the Delisle/Buache school" and their Mer de l'Ouest, whose convenient openings provided a temperate-zone alternative to the Russian strait close to the Arctic Circle.[12]

▶ Gerhard Friedrich Müller, *Nouvelle carte de decouvertes faites par des vaisseaux russes,* 1754. Published subsequent to the purely speculative works of Buache and Delisle, which gave credence to the supposed discoveries of Admiral de Fonte, Müller's chart represented the St. Petersburg Academy of Sciences' official view of Bering's discoveries. Müller was the first cartographer to depict the Alaskan subcontinent (albeit inaccurately in size and relationship to Asia), an image that Cook refined.

⚓

Hearne's patron, Moses Norton, was chief factor for the HBC fort at Churchill from 1762 to 1773. The post was the company's northernmost (58° 46') on the bay's west coast, and it is there that the narrative embers of the Stuart-Thanadelthur story describing rich mineral deposits inland were revived. In 1764, when Norton wrote company directors in London, he asserted that Hudson Bay likely had no outlet to the Pacific, but he argued that the country to the northwest was still well worth exploring because of the possibility of finding the Northwest Passage, in addition to commodities such as furs and minerals. In 1768, Norton travelled to England to pitch his plan for a major expansion of HBC operations, one that would include full incorporation of the vast Denesuline district first discerned by Stuart and Thanadelthur a half century earlier. Norton apprised his vanguard agent Hearne, only twenty-five when he departed Churchill in 1770, that the main objectives were improving tribal relations and determining the suitability of the new country to sustain trade, followed by procurement of copper from the "Far Off Metal River." Lastly, Hearne was told to determine "either by your own travels, or by information from the Indians, whether there is a passage through this continent."[13]

Hearne's journal was not published until two decades later, but word of his expedition filtered back to London after he returned to Churchill in June 1772. (His account eventually became a classic in travel literature because of its grisly description of the massacre of an Inuit clan perpetrated by Denesuline and Cree over Hearne's objections, a gothic tale since debunked by Ian MacLaren.)[14] Guided by Matonabbee, a Denesuline Chief, Hearne's party headed west in the middle of the winter of 1770–71. In May, it struck the headwaters of the Coppermine River and then reached the "long wished-for spot" of its confluence with the Arctic Ocean on July 17. At its termination, the river was "scarcely navigable for an Indian canoe," Hearne wrote, "being no more than one hundred and eighty yards wide, every where full of shoals."[15]

Though the Coppermine looms large in the history of northern exploration, particularly Cook's third voyage, at 525 miles in length it is a relatively minor stream, certainly in comparison to the Mackenzie farther west, and Canada's twenty-sixth largest. Hearne found only but a single lump of copper but, more importantly, that the river emptied into saltwater that was ice-free three-quarters of a mile from shore. The geographic import of his expedition came into the Admiralty's cognizance just as

Cook returned from his second voyage. Hearne's lasting contribution was the datum from which a latitude for the river's mouth was subsequently calculated to be 71° 54' N. Hearne was not practised in the use of a quadrant, and he took only one reading earlier in his expedition. Motivated by thoughts on the planning needs of Cook's northern voyage, the Admiralty extrapolated from Hearne's notes to establish this faulty reckoning. Polar expeditions in the nineteenth century proved that the mouth of the Coppermine is at 67° 48', two hundred miles farther south.

In addition to locating a segment of the Arctic coastline, Hearne also shed light on several important facets of continental geography. Having crossed vast tracts of land northwest of Churchill, he would have struck a saltwater outlet from Hudson Bay if there was one. He did not. Perforce, the coast Hearne saw at the mouth of the Coppermine had to extend a very long distance to the west. By itself, that should have ended Hudson Bay as a prospective element in any latent Northwest Passage corridor, but it did not. Furthermore, Matonabbee's circuitous route home included a long southwesterly sweep toward Great Slave Lake, which showed that there was no intrusion of saltwater into the middle of the continent west of Churchill. At the narrowest point of this lake, which Hearne was the first European to see, Indigenous informants from still farther west told him they "knew no end to the land in that direction." Nor, Hearne continued, "have I met with any Indians ... that ever had seen the sea to the Westward." Instead of affirming the existence of the Mer de l'Ouest, he learned of an "immense chain of mountains" running "North to South" and blocking access to the Pacific.[16]

This was a treasure trove of geographic information. Hearne had essentially proved to the satisfaction of those planning Cook's third voyage that there was no Northwest Passage *through* America. After all, Hearne had crossed the heart of the continent all the way to the Arctic without finding a large river let alone an inland sea. Instead of finding the mythical passage, he gleaned two essential truths about North America: the Rocky Mountain cordillera ran nearly to the Arctic Circle and, as Hearne inscribed in his narrative: "The Continent of America is much wider than many people imagine."[17] These discoveries impeached Dobbs's fanciful speculation and the French geographic school, which had hypothesized a mid-latitude passage through the continent. On the face of it, Hearne's findings seemed to be the end of an era dating to Frobisher, Davis, Baffin, and Hudson in the North plus La Salle and Cabeza de Vaca in more southerly latitudes. Combining French, Spanish, and now British exploration of

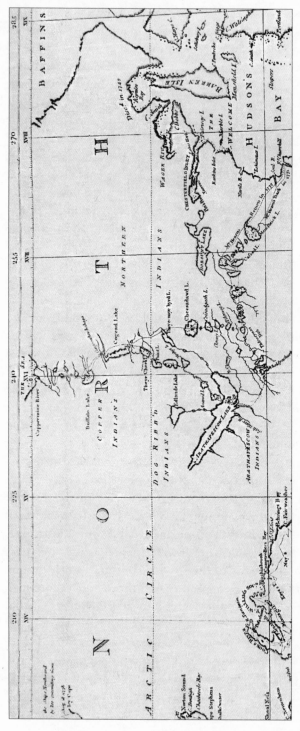

Hearne's Coppermine River route to the Arctic, *A General Chart Exhibiting the Discoveries Made by Captn. James Cook, in This and His Two Preceding Voyages; with the Tracks of the Ships under His Command, by Lieut. [Henry] Roberts of His Majesty's Royal Navy* (detail), 1783–84. Roberts's chart, intended principally to display the tracks of Cook's ships during his three voyages, contains an underappreciated section showing interior detail of Canada west of Hudson Bay. Drawing on information provided to the Admiralty by the Hudson's Bay Company prior to the start of Cook's third voyage, Roberts depicted the route of Samuel Hearne to the mouth of the Coppermine River at "the Sea," otherwise undifferentiated. Hearne's geographic insights guided the Admiralty's instructions to Cook to begin his search for the Pacific portal to the Northwest Passage at 65° N.

the continent's mid-section, a line drawn from Hearne's northing in the Arctic to the Gulf of Mexico showed that there was no saltwater passage between the Atlantic and Pacific.

After the HBC slipped a copy of Hearne's field notes to the Admiralty, the voyages of Fuca and Fonte should have been irreversibly relegated to the status of myth, and within the best-informed circle of geographic thinkers, including Cook, they were. Simultaneously, Hearne's expedition supplanted speculative theory and informed Cook's third voyage instructions. The guiding intelligence was clear: any passage, presuming it existed, was in the Subarctic latitudes in seas along the northern edge of America that were partially frozen in mid-summer. Remarkably, even after Cook's voyage, no element of which contradicted the thrust of Hearne's findings, the legends of Fuca and Fonte would be resurrected by a new generation of speculators.

⚓

While Cook searched for Terra Australis, interest in the boreal zone was also reaching another of its periodic crescendos. These enthusiasms tended to peak in thirty-year increments, and in the early 1770s Daines Barrington, after influencing Cook's second voyage through his placement of Johann Forster, stepped onto the northern stage, reprising the role of Arthur Dobbs. In addition to his friendship with Lord Sandwich at the Admiralty, Barrington, as a member of the Royal Society and "citizen" in the Republic of Letters, was a regular correspondent with Swiss geographer Samuel Engel. Barrington, a lifelong bibliophile with an interest in natural phenomena, had popularized Engel's 1765 publication on ice formation, which reiterated the long-held view that saltwater did not freeze at oceanic depths. On the eve of Cook's final voyage, Barrington asserted, citing Engel's authority, "it is very probable, there is little or no ice" near the North Pole. The premise was that ice only formed as landfast in shallow bays and in rivers "during winter, and does not break up and get into the sea till the latter end of March or the beginning of April, when it begins to thaw upon the shores." According to Barrington, "accounts agree that in very high latitudes there is less ice" and perhaps "nothing but sea" because "if there be no land near the Pole, then there can be no bays in which ice can be formed to interrupt the navigation."[18]

When Barrington wrote these lines early in 1775, Cook and Johann Forster had already come to the opposite conclusion. But the die was cast for new voyages to the Far North before Cook returned. Barrington had already forged the North Atlantic expedition of Captain Constantine John

Phipps in 1773. In response to the Royal Society's pleadings, prompted by Barrington's enthusiasm, the Admiralty dispatched Phipps with two ships for an imagined voyage to China that reprised Hudson's failed transpolar voyage more than 150 years earlier. Entering the Barents Sea on his eastern turn toward the Siberian coast, Phipps's path was blocked by the ice pack off Spitzbergen. Phipps, like Hudson, reached 80° N, which, except for the most fantastical reports and speculations, was emerging as the northern limit of seasonal navigation. Phipps might have failed to reach the pole, but this did not thwart Barrington's zeal. Indeed, it seems to have energized him.

After the Phipps setback, Barrington collected stories from HBC sailors, Greenland whalers, and "a very eminent merchant of Hull." He conformed his findings into papers that were read at meetings of the Royal Society in May and December 1774. These, in turn, formed the core of a pamphlet, *The Probability of Reaching the North Pole Discussed*, published in the late spring or early summer of 1775, just as Cook was returning from Southern Thule. Barrington regaled his readers with apocryphal accounts of more than two dozen navigators who had reached parallels ranging from the more customary 80° N to a Dutch vessel that was supposed to have sailed to an astonishing 89° 30' N, where it was as hot "as in the summer at Amsterdam." In a further instance, a "Capt Mac-Callam" made it to 83° 30' N. When he turned for home, "the sea was not only open to the Northward," there had not been a "speck of ice for the last three degrees, and the weather at the same time was temperate." In yet another, a Dutch ship "between fifty and sixty years ago" reached 88° N in water "rolling like the bay of Biscay."[19]

Referring to himself as the "unworthy proposer of the Polar voyage," Barrington wrote that the "unfortunate barrier of ice" that Phipps had encountered "was only temporary, and not perpetual." In short, Phipps's experience was written off as simply bad timing since many others had reportedly reached farther north. The ice that stymied Phipps, Barrington hypothesized, had travelled in a southeasterly direction from the "*Frozen Sea* into which the great rivers of Siberia and Tartary empty themselves." Since "it is so difficult to freeze any large quantity of salt water," he reasoned, the ice was simply "dispersed by winds, tides, and currents" from their point of emergence in Siberia to the "high Northern latitudes."[20]

This was a succinct recapitulation of classical polar hydrological theory, but Barrington developed a corollary to explain why Phipps had only proceeded as far as 80° N while others had, allegedly, reached the high eighties or nearly the pole itself. The "sudden assemblage of ... an accumulation of ice" created only the "appearance ... of forming a perpetual barrier, when perhaps, within the next 24 hours, the wall of ice might entirely vanish." Among other examples, he offered a report on the harbour at Louisburgh on Cape Breton Island. One May 1, it was "entirely open; but ... the [next] morning, it was completely filled with ice, so that a waggon might have passed over it in any direction."[21]

Barrington threw other stray notions into his tract, such as the fact that whales, which thickly populated the northern seas, needed to gain access to the surface for breath. Two of the voyages he researched would have made it to the pole, but "common sailors," hearing of their captains' intentions, "remonstrated, that if they should be able to proceed so far, the ship would fall into pieces, as the Pole would draw all the iron work out of her." Barrington's sources did provide some valid information of tactical value, such as the following: "The most likely time for such discoveries to be made, is in the months of July and August, when the ice is most commonly furthest from the land." Two ships were a necessary precaution for polar voyages, Barrington was told, and should be "entirely under the command of some expert, able, and experienced seaman, who has frequented those seas for some time past." This perfectly described Cook and was followed by the advice that a northern voyage had to be staged for discovery opportunities "about the 10th of May," when the ice began to break up.[22] Perhaps not coincidentally, Cook would begin to display anxiety about his discovery timetable in mid-May 1778 as he probed Alaska's Prince William Sound.

In the concluding section of the tract, "Thoughts on the Probability, Expediency, and Utility of Discovering a Passage by the North Pole," Barrington stated that the time for "philosophical speculation" was over and that "an object of the greatest importance to the public welfare ... should be no longer delayed." No country but Britain was "so well situated for such an enterprize ... New channels of commerce would be thereby opened, our navigation extended, the number of our seamen augmented, without exhausting our strength in settling colonies, [or] exposing the lives of our sailors in tedious and dangerous voyages through unwholesome climates." This idealization of the Northwest Passage was followed by Barrington's conclusive call to action: "A project which has dwelt in the mouths and memories of some, and in the judgement and approbation of a few, from the time of Henry the Eighth, should be revived, and at length, for the benefit of his subjects, carried into effect, under the auspices of GEORGE the Third."[23]

Barrington's advocacy rallied Parliament and the Admiralty. There were skeptics, but as Beaglehole phrased it, "the only way to settle the argument was to go and look."[24] But Barrington's novel twist, given the outcome of the Phipps expedition, was to switch the focus for a new northern voyage to the Pacific, where the Russians had recently been active. Their garbled accounts became the last piece of the puzzle that Cook would try to figure out.

⚓

Reports of Russia's plans for further exploration in the North Pacific had begun to circulate in London and other European capitals late in 1773. The Spanish were so alarmed by reports of Russia's eastward expansion that Juan Pérez sailed from Mexico in 1774 on a Pacific Coast reconnaissance. He reached 54° 40' N (a parallel that resonates in continental history because the United States later inherited the Spanish claim) but did not encounter any Russians. The next year, Bruno de Hezeta and Juan Francisco de la Bodega y Quadra again reconnoitred the region, the latter attaining 58° N. Adding to the international intrigue, sources in Paris suggested that Bougainville was about to sail in search of the Northwest Passage, a report that Barrington trumpeted in his promotional pamphlet. The rumour proved unfounded, but the mere spectre of the possibility, combined with Russian incursions into North American waters, was the principal reason Parliament liberalized the Northwest Passage prize rules by allowing navy commanders to qualify. Royal assent was necessary to implement this change, and it was not secured until Cook returned from his second voyage.

The Russians had resumed formal discovery expeditions with the voyage of Lieutenant Ivan Sindt, conducted during 1765–67. This venture did not take on currency in Great Britain until 1774, when Matthew Maty, secretary of the Royal Society, published an English-language version of Jacob von Staehlin's account of Sindt's and other recent voyages. Maty's connection to this publication would loom large in the course of future events. Staehlin's book had originally appeared in Germany in 1773 under the auspices of the St. Petersburg Academy of Sciences, which had been founded in 1725 by Peter the Great to emulate the learned societies in London and Paris as part of his larger campaign to westernize Russia. None of the academy's original sixteen members were actually Russian, reflecting the czar's problem. One of them, French geographer Joseph-Nicholas Delisle, authored the instructions that guided Bering's second voyage. Czar Peter also relied on Baltic German scholars (from Estonia, Lithuania, and Latvia) for his intellectual initiatives. Cartographer Gerhard Müller had emerged from this nexus, as did Staehlin. Magnus von Behm, who became the Cook expedition's benefactor in Kamchatka, also represented this demographic.

The chart that accompanied the Staehlin publication, *An Account of the New Northern Archipelago, Lately Discovered by the Russians,* proved to be far more important than the narrative of Sindt's expedition. Whereas Müller's map depicted the oversized westward projection of the Alaskan subcontinent, Staehlin's new image pulverized that land mass, leaving in its stead the much smaller island of "Alaschka" in the middle of a strait between Siberia and a truncated North American mainland. Alascha (the first time the progenitor of "Alaska" appeared in print) was the largest element of an archipelago that extended south and west. It was a grotesque mischaracterization of the Aleutian chain, which had been reasonably projected beforehand by Müller. Staehlin's image implied that Bering had sailed to the eastern (American) side of Alascha at 65° N, the same parallel for the strait Müller depicted. However, Staehlin's tantalizing strait was both fifteen degrees farther east than Müller's projection – meaning hundreds of miles closer to Baffin and Hudson Bays; and unlike Müller's map, the route above North America was no longer encumbered by the gigantic northward projection of the Alaskan subcontinent. In short, Staehlin's map suggested that both Hearne's Coppermine River outfall and the Atlantic were far more accessible than previously thought. As historian John Norris observes, Staehlin conveniently reduced the Northwest Passage to "a minor sub-polar concern."[25] Though newer and presumptively more accurate than Müller's depiction, in truth Staehlin's map took Russian cartography backward, which a frustrated Cook would soon learn.

By 1775, all the cartographic parameters guiding the strategic direction of Cook's third voyage were in place. Going east to west, Phipps had recently reached 80° N east of Greenland, and Baffin had long ago reached 78° N, still the record for the Canadian Arctic at the time Cook sailed from England. Hearne was thought to have found a sliver of open Arctic sea just shy of 72° N. Bering had traced the Asian coastline to 67° N, and now Maty and Staehlin had pointed to a strait between "Alascha" and North America at 65° N, from which the continental trend line bent easily toward the east. Capping it all, Barrington had propagated a glaciological theory that was conducive to sailing between some of these points.

Every previous attempt to find a Northwest Passage (other than legendary ones) had approached from the Atlantic, only to be thwarted by ice-choked bays or straits. The sea Hearne saw appeared now to be easily reachable by way of Bering's (or what might be better termed Staehlin's) strait via the North Pacific. It was at this stage in mission development, as the calendar turned to 1776, that British planners brought Cook into the conversation. After all, who had more practical experience and knowledge about resupply stations for a long voyage into the Pacific, its prevailing winds and currents, or (presuming it was encountered) navigating around sea ice? The logic was summed up retrospectively by John Douglas in his introduction to Cook's third-voyage account:

> It was wisely foreseen, that whatever openings or inlets there might be on the East side of America, which lie in a direction which could give any hope of a passage, the ultimate success of it would still depend upon there being an open sea between the West side of that continent, and the extremities of Asia. Captain Cook, therefore, was ordered to proceed into the Pacific Ocean ... and having crossed the equator into its Northern Parts, then to hold such a course as might probably fix many interesting points in geography, and produce intermediate discoveries, in his progress Northward to the principal scene of his operations.[26]

Cook's discovery zone was a corridor at least six degrees of latitude in width, possibly as much as thirteen to fifteen degrees in a "good" ice year.

There was still one obvious problem facing the Admiralty's planners. Presuming Cook proceeded through Bering Strait and past the Coppermine River, how would he find his way into the Atlantic if every attempt from that side had failed to find an aperture? The thinking seems to have been that Cook would simply find the gateway to Baffin Bay if it was there; if not, pursuant to the Barrington-Engel school of climatology, he had only to sail sufficiently far away from land to find open water toward the pole and thence proceed home. Nevertheless, a backup plan of sorts was devised. Lieutenant Richard Pickersgill, who had sailed twice with Cook, was dispatched in the brig *Lyon* to the waters west of Greenland two months before Cook was slated to leave England. Pickersgill's first duty was to protect the English whaling fleet in Davis Strait from depredations by "American Ships in those parts belonging to the Inhabitants of His Majesty's Colonies now in Rebellion." Afterwards, he was to proceed to Baffin Bay "for the purpose of making Discoveries."[27] This was code for keeping a look out for Cook.

Pickersgill's mission was entirely unrealistic. *Lyon* was not strengthened for service in the ice, the crew was not equipped with proper cold-weather gear, and Pickersgill did not have the proper temperament. He left the Thames in May 1776 (Cook departed in July), far too late to provide any protective benefit for the whalers. Pickersgill was also easily defeated by the ice; he only reached 68° N, ten degrees shy of Baffin's record. By October 1776, he was back in England and was relieved of command in advance of the 1777 season, when, according to the Admiralty's schedule, Cook was supposed to be approaching the passage's Pacific portal.

Lieutenant Walter Young succeeded Pickersgill in command of *Lyon*. Young sailed in March with a précis of Cook's instructions. He was directed to look for openings to the Arctic and even winter over if need be to await Cook's appearance. Young was no more durable than Pickersgill. Beaglehole called him "one of the most inefficient explorers that ever existed." Reaching 72° 45' N, he met ice and promptly turned for home.[28] Thus, the Atlantic complement to Cook's venture ended in June 1777 when Cook was still staging for his northern voyage in Polynesia. The Admiralty's failed Baffin Bay stratagem emphasizes the rarity of Cook's fidelity to mission and the spirit of endurance he brought to these same latitudes and from half a world away besides.

⚓

As Cook prepared to leave England, he faced the prospect of effectuating a positive discovery of historic magnitude – solving the three-hundred-year-old Northwest Passage puzzle – not merely the negation of someone else's theory. His previous voyage in the trackless latitudes of the Southern Ocean had been spent pursuing fugitive and ephemeral sightings of lands and legends propagated by other explorers, demolishing them all. That was a significant accomplishment, but in a way it was serially anticlimactic. As Glyn Williams expresses it, "instead of the looming haystack of a southern continent, Cook was searching for the slim needle of a Northwest Passage."[29] Cook was cautiously optimistic about his new mission, and he knew that if he succeeded he would secure a fortune in the form of the £20,000 prize and fame in Columbian proportions.

Cook's instructions, like those of the second voyage, were heavily informed by his own outlook toward the matter at hand. They contained a shorthand version of the customary directions for Enlightenment-era voyaging. As for the sailing agenda, Cook was to take the now familiar route to provisioning stops in Cape Town, New Zealand, and Tahiti (where Mai was to be dropped off). He would then proceed to North America's west coast at the forty-fifth parallel, where he would commence the discovery phase of the voyage. This latitude approximated Drake's sighting of "New Albion," and it strategically avoided any risk of diplomatic entanglements in Spanish settlements. Spain was openly hostile to any British incursion in the Pacific and, had their ships encountered him, its colonial officials would have had Cook "detained, imprisoned and tried in accordance with the Laws of the Indies."[30]

After taking on "wood and water" in the first convenient North American harbour, Cook was to "proceed Northward along the Coast as far as the Latitude of 65°, or farther, if you are not obstructed by Lands or Ice." This parallel was another conscious selection guided by presumptively authoritative geographic knowledge. It dismissed any theoretical value in the voyages of Fuca and Fonte and the French cartographic school but plainly validated Bering's findings and the more recent information about Russian voyages in the Arctic contained in Staehlin's account. In 1784, as he prepared the narrative of Cook's third voyage for publication, John Douglas anticipated a criticism of Cook that would reach such a fervour later that decade that Vancouver had to be dispatched to the Northwest Coast to address it. Douglas said that Cook wasting his time in the range of Fonte's and Fuca's supposed discoveries in the high forties and fifties was akin to asking the great navigator "to trace the situation of Lilliput and Brobdignac." The voyages of Fuca and Fonte were, like Gulliver's, "mere objects of imagination" and "destitute of any sufficient external evidence"; they bore "so many striking marks of internal absurdity, as warrant our pronouncing them to be fabric of imposture." This was true, but the maritime fur traders who followed in Cook's wake found a number of inlets, channels, and other apertures on the Northwest Coast forced to a reopening of the matter.[31]

The line of text in the voyage instructions most frequently overlooked by historians advised Cook "not to lose any time in exploring Rivers or Inlets, or upon any other account, until you get into the before-mentioned Latitude of 65°." Only when he reached that line was Cook to "search for, and to explore, such Rivers or Inlets as may appear to be of a considerable extent and pointing towards Hudsons or Baffins Bays."[32] The designated latitude approximated Bering's northernmost position off the Siberian coast as reflected in the maps of Müller and Staehlin, but to relate the strategic underpinnings of this guidance, Douglas expounded on Hearne's overland trek to the mouth of the Coppermine River. His text was the first public disclosure of Hearne's expedition.

As reported by Douglas, when Hearne had first spied the Coppermine's delta in mid-July 1771, rain and fog drifted in over him. Because of the river's restricted navigability, he deemed his discovery to be of no practical value and, consequently, did not wait for clearer weather to determine the location's latitude. Accordingly, Admiralty officials extrapolated distances from Hearne's last known fixed point and concluded "that the mouth of the copper-mine river lies in the latitude 72°."[33] This overshot reality by several degrees and no doubt contributed to Cook's surprise when he encountered the Arctic ice pack south of this parallel.

Summing up the consequences of the Hearne effect, Douglas said the Admiralty came to a self-evident conclusion: "We now see that the continent of North America stretches from Hudson's Bay so far to the North West, that Mr. Hearne had travelled near thirteen hundred miles before he arrived at the sea. His most Western distance from the coast of Hudson's Bay was near six hundred miles; and that his Indian guides were well apprized of a vast tract of continent stretching farther on in that direction." Given the "indubitable proofs that no passage existed so far to the South as any part of Hudson's Bay ... if a passage could be effected at all, part of it, at least, must be traversed by ships as far to the Northward as the latitude 72°, where Mr. Hearne arrived at the sea." Drawing on Bering's trace of the Asian coastline to 67° N, Douglas sensibly concluded that any separation from North America had to approximate that latitude. Indeed, fastening this down was posited as a foundational element for Cook's voyage, because determining "the relative situation and vicinity of the two continents ... was absolutely necessary to be known, before the practicability of sailing between the Pacific and Atlantic Oceans, in any Northern direction, could be ascertained."[34]

The only nonpolar exploration authorized in Cook's instructions was in the Indian Ocean, where "some Islands" had "been lately seen by the French." These were to be investigated "for a good Harbour, and upon discovering one make the necessary Observations to facilitate finding it again." This was the land that had been discerned by Kerguelen, a development Cook had first learned about at Cape Town outbound on the second voyage and inquired about further when he stopped there again near its end. But Cook's instructions also stated: "You are not however to spend too much time in looking out for those Islands, or in the examination of them if found, but proceed to Otaheite" to drop off Mai. The instructions made no allowance for the insular follow-up in the South Pacific that historians from Beaglehole forward have decried as a shortcoming of the third voyage, nor did they allow for excursions in search of the Great River of the West or Juan de Fuca's supposed strait. The opposite was true. Cook was specifically advised that to get to the northern discovery zone at 65° N on a timely basis, he was "not to lose any time in search of new Lands, or to stop at any you may fall in with, unless you find it necessary to recruit your wood and water." Douglas, drawing on Cook's own statements from his account of the second voyage, averred that there was "little more to be done" in the tropics of the South Pacific and that the Southern Hemisphere had been "sufficiently explored."[35]

The instructions contained a timeline from which Cook would deviate in a significant fashion. The expectation was that Cook would leave Cape Town in October or November 1776, deliver Mai by February 1777, and reach the discovery zone of the Pacific's high northern latitudes in June. In the event, Cook ran an entire year late. But he did follow the seasonal schedule by arriving on the Northwest Coast in spring 1778. So, by mid-summer, he was positioned for discovery in the icy latitudes.

The instructions anticipated that Cook's route might be "obstructed by ... Ice."[36] It is difficult to measure the weight that should be lent to this text given the prominence of the countervailing idea that the Arctic might be ice-free at the pole. There is no evidence that Cook himself subscribed to this theory (which would have enthusiasts for another century); indeed, everything he said about voyaging toward the South Pole suggested otherwise. Cook's ships were not strengthened for ice, but this proves little; his ships aimed toward Antarctica were similarly unarmed. Unlike whalers or nineteenth- and early-twentieth-century explorers, Cook never countenanced the prospect of embedding himself amidst the ice and riding it out to some undefined exit, as Amundsen did. Cook considered himself too capable a navigator to get stuck in the ice, as he later proved off the Alaskan coast.

Essentially, Cook's prescription was this: enter Bering's or (per Staehlin) some adjacent strait in the North Pacific at the seasonally optimum moment, reach the Arctic coastline that Hearne had espied, and then look for an exit into the Atlantic. If, on the basis of his own observations or intelligence gleaned from an encounter with Inuit ("the Esquimaux"), he discerned "a certainty, or even a probability, of a Water Passage" into Baffin or Hudson Bay, he was to use his "utmost endeavours to pass through." If Cook found "no Passage ... sufficient for the purposes of Navigation," he was to "repair to the Port of St Peter and St Paul in Kamtschatka." (Bering's ships, *St. Peter* and *St. Paul*, had sailed from this harbour in Avacha Bay on the peninsula's Pacific Coast, inspiring its name. It remains a major port for the Russian navy.) Cook was extended the latitude of proceeding elsewhere to pass the winter if he judged it "more proper." If the first season of discovery was not successful, he was to try again the following spring by proceeding "as far as, in your prudence, you may think proper, in further search of a North East, or North West Passage, from the Pacific Ocean into the Atlantic Ocean, or the North Sea." That is, if a route over North America was impracticable, Cook was authorized to attempt to reach Great Britain atop Siberia and Russia. (He would try both.) Should Cook fail in the attempt in either direction, he was to head home "by such Route as you may think best for the improvement of Geography and Navigation." From the course set by his successors, we can infer that, in this event, Cook made plans to survey the coasts of Korea and Japan.[37]

⚓

Resolution weighed anchor on July 12, 1776, and stood out of Plymouth Sound. On his departure, Cook faced a daunting list of unknowns, many more than his previous voyages had presented at a comparable stage, starting with the premise that access to Hearne's Arctic coast was possible from the Pacific. Though Bering's first voyage had strongly hinted that Asia and North America were separated, the details were sketchy. As an anonymous eulogist (subsequently identified as the admiral of the fleet, John Forbes) phrased it in the postvoyage account, the object of Cook's "last mission was to discover and ascertain the boundaries of Asia and America, and to penetrate into the Northern Ocean by the Northeast Cape of Asia."[38]

For Cook's European contemporaries, the coastlines adjoining the North Pacific were, as Glyn Williams observes, "the least-known areas of the inhabited globe."[39] This was especially true on the American side, where Cape Blanco marked the northern limit of geographic comprehension near the forty-second parallel. (This line constitutes the modern California-Oregon border.) To the north lay the longest unexplored stretch of coast in the world. As Douglas phrased it in the introduction to Cook's third-voyage account, "the Southern hemisphere had, indeed, been repeatedly visited, and its utmost accessible extremities been surveyed. But much uncertainty, and, of course, great variety of opinion, subsisted, as to the navigable extremities of our own hemisphere; particularly, as to the existence, or, at least, as to the practicability of a Northern passage between the Atlantic and Pacific Oceans." Northern exploration promised more in return than southern exploration because, "if such a passage could be effected, voyages to Japan and China, and, indeed, to the East Indies in general, would be much shortened; and consequently become more profitable, than by making the tedious circuit of the Cape of Good Hope." Pointedly observing that "so much was done toward exploring the remotest corners of the Southern hemisphere," Douglas found it inconceivable "that the Northern passage should not be attempted."[40] These sentiments echoed Cook's active disdain for further exploration in the South Pacific.

And Cook was the only man with the stature to take on this ancient quest. "The operations proposed to be pursued," wrote Douglas in retrospect,

> were so new, so extensive, and so various, that the skill and experience of Captain Cook, it was thought, would be requisite to conduct them. Without being liable to any charge of want of zeal for the public service, he might have passed the rest of his days in the command to which he had been appointed in Greenwich Hospital, there to enjoy the fame he had dearly earned in two circumnavigations of the world. But he cheerfully relinquished his honourable station at home; and, happy that the Earl of Sandwich had not cast his eye upon any other Commander, engaged in the conduct of the expedition.[41]

NINE

Southern Staging Grounds

James Cook took on the challenge of finding the Northwest Passage – the Holy Grail of Enlightenment-era discoveries – in the aftermath of a London dinner party with senior Admiralty officials on January 10, 1776. This occasion, which involved First Lord Sandwich, Comptroller Hugh Palliser, and First Secretary Philip Stephens, was ostensibly designed to secure Cook's counsel on how to conduct the forthcoming voyage. It ended with him indicating an interest in the command. The Admiralty resorted to tempting Cook with the lead because they did not dare ask outright. Sandwich and his colleagues pumped the scheme in the hopes Cook would offer, which he did. His motivation for coming out of retirement has puzzled historians. Glyn Williams argues that Cook had "surely had his fill" of "routine surveying work" by that point. That is certainly correct regarding the South Pacific, but Cook was not tired of discovery, or meaningful work more generally. His stated concern to John Walker about being "confined within the limits of Greenwich Hospital" is dispositive on this point.[1] At the Admiralty dinner, it probably occurred to Cook that the prospect of another explorer finding the fabled Northwest Passage was too difficult to countenance.

Royal Society dinners occupied a lot of Cook's time between the second and third voyages. He had been formally proposed for membership at a meeting on November 23, 1775. His nomination was sponsored by Joseph Banks and seconded by Daniel Solander, Constantine John Phipps, Henry Cavendish, Philip Stephens, and Nevil Maskelyne. Their citation references his skill in astronomy and "two important voyages for the discovery of unknown countries, by which geography and natural history have been greatly advantaged and improved." Cook attended society functions several times while his nomination hung (gathered signatures) prior to his election on February 29, 1776. He attended meetings or dinner parties at least six more times before departing on his final voyage, including one encounter with the famous diarist Thomas Boswell.[2] Given subsequent problems with *Resolution*'s seaworthiness, it has been argued that Cook should have spent less time socializing and more time supervising the refitting of his ship for the ensuing voyage.[3] But truly, the entire British fleet was in a deteriorated state at the beginning of the war with the American colonies.

Hobnobbing with the leading lights of British intellectual life was clearly not sufficient to keep Cook in retirement. A man of modest means, he mentioned to Walker potential monetary rewards, but the simplest explanation for his undertaking another voyage is that he had become accustomed to living at the edge of the world. He thrived there, notwithstanding the risks and stresses inherent in such enterprises. Richard Henry Dana Jr., in his classic *Two Years before the Mast* (1840), wrote of sailors who at homeport or in the absence of the bustle of life at sea began to feel unhappy. It was only when they were "busily at work" that their "mind and body seemed to wake together." Within a month of taking the northern mission, Cook told Walker: "It is certain I have quited an easy retirement, for an Active, and perhaps Dangerous Voyage. My present disposition is more favourable to the latter than the former, and I imbark on as fair a prospect as I can wish."[4] He was both ready and eager for a new venture and at age forty-seven had a vigorous outlook on life.

The only thing Cook was tired of was negative discovery, of disproving the fugitive findings and geographic notions propagated by others. With this new voyage, he had the chance to make an epochal find. After *Endeavour*, Cook had told Walker that he had "made no very great Discoveries."[5] This statement is usually written off as an example of Cook's modesty, but there is no reason to think it was dishonest. Tahiti, New Zealand, and Australia had all, in some measure, been previously explored. Historians marvel at the scope of Cook's second voyage, but he did not find the object of that mission, and his return to England was anticlimactic. The dour Johann Forster could not replicate the excitement in elite circles that Banks had generated. Even lacklustre Tobias Furneaux upstaged Cook by bringing Mai to Great Britain, thereby feeding the country's appetite for Polynesian exoticism. And then three months before Cook reached England in July 1775, the initial skirmishes with the American colonists had occurred at Lexington and Concord, a turn of events that dominated conversation in London's salons and meeting rooms.

Historian Richard Van Orman asserts that although an explorer's work "was usually toilsome and sometimes dangerous, few were happy doing anything else. In fact, the more difficult the undertaking, the more they enjoyed it." Felipe Fernández-Armesto describes this phenomenon as "Faustian yearnings for knowledge and fame."[6] The comparatively muted response to the Antarctic voyage created a context where Cook sought fulfillment via an opportunity for positive discovery and for one more appeal to public acclaim. Thus, after completing his final draft of *A Voyage towards the South Pole and Round the World* – which was published in May 1777, nearly a year after he left on the third voyage – he set off in search of the Northwest Passage.

⚓

Cook's principal adjutant on the third expedition was Charles Clerke, commander of *Discovery*, the smallest of all the Cook-voyage ships. Genial and loyal, Clerke had been on the first two journeys and was far more competent than Furneaux. He might have led the expedition but for Cook's interest. Clerke was late reporting because he had guaranteed the finances of his brother John, also a captain in the Royal Navy, who had absconded to the East Indies. Charles, with insufficient resources of his own, was temporarily consigned to debtor's prison. He wrangled his way out of jail but not before contracting tuberculosis, a bacterial infection that affects the lungs and slowly saps their ability to function. Before antibiotics, the only known treatment was a resort to fresh air in a warm sunny climate. Clerke's condition would create a poignant moment between the commanders when their ships left Polynesia for the Arctic. Clerke died from this condition before the voyage was over.

Resolution's first lieutenant was John Gore, an American by birth and the expedition's commander after Cook and Clerke died. Gore was a durable figure, having sailed with Wallis before accompanying Cook on *Endeavour*, but he was not a bright light in the discovery business. His gullibility regarding Terra Australis would prefigure his naive enthusiasm about various prospects for the Northwest Passage's western gateway and stands in stark contrast to Cook's dispassionate ability to read both sea and land.

Also on *Resolution* was the young, sensitive, and erudite second lieutenant James King. After the voyage, he would complete the expeditionary narrative begun by Cook. Midshipman James Trevenen later recalled King's "cheerfulness and sweetness of disposition." Marine corporal John Ledyard, who published an unauthorized account of the voyage in which he is oft-critical of Cook, admired King for the "humanity and goodness" he showed Indigenous Hawaiians during the stressful days after Cook's death.[7] *Resolution*'s sailing master was the technically capable but abrasive William Bligh, soon to be famous as *Bounty*'s captain. The best known of several midshipmen, George Vancouver, was on his second tour with Cook and assigned to *Discovery*. Midshipmen George Gilbert (whose father had been master on *Resolution* during the previous voyage) and Trevenen kept journals that contained certain stray comments that were later seized on to frame much of the interpretive orthodoxy surrounding Cook's supposed lack of diligence in the South Pacific.

Other notables on Cook's flagship were landscape artist John Webber and Ledyard, the American-born marine and later would-be terrestrial explorer of northern climes. Clerke's officers on *Discovery* were first lieutenant James Burney (who had sailed on the Antarctic voyage), second lieutenant John Rickman (who also published an unauthorized account of the third expedition), and several future fur-trade explorers who figure in the evolution of the Northwest Passage as a cartographic image after Cook's death: master's mate Nathaniel Portlock (an American), able seaman Joseph Billings, and armourer George Dixon.

In Cook historiography, the most pointed criticism about the third voyage's complement is the absence of a naturalist with the stature of a Banks or Forster. Johann Forster was the first to make an issue of this publicly, attributing the expedition's difficulties after the fact to this deficiency and presumed lack of judgment on Cook's part. Many authors since have subscribed to Forster's thesis, according to which it is thought that Cook needed a friend to confide in or, alternatively, that the expedition required someone outside the naval chain of command to check his worst tendencies. Bernard Smith deduced that even the on-board library "was less well equipped than either the first or the second voyages." Nearly every modern account mentions Cook's supposed aside to King, made shortly after his assignment to *Resolution:* "Curse all the scientists and all science into the bargain!"[8]

Cook's alleged exclamation, first and rather suspiciously reported in Forster's preface to Rickman's unauthorized book, is usually interpreted as Cook's overreaction to previous shipboard difficulties with the German polymath. It is easy to imagine Cook voicing a negative comment about Forster, but it seems implausible that King, having returned to London after the voyage to help with the publication, would have betrayed the now dead captain's confidence on such a sensitive point. Lost in this indictment is that for his third voyage, Cook envisaged the exotic Southern Hemisphere as a mere staging area and not a destination for scientific discovery. In the more familiar Northern Hemisphere, botany and zoology were not expected to be central elements of Enlightenment inquiry, only geography would be. Nevertheless, Banks secured a spot for Kew Gardens's David Nelson as a botanical collector. Officially, Nelson was an aide to William

Bayly, who had been astronomer on *Adventure* but was now attached to *Discovery*. (King served as the principal astronomer on Cook's flagship.) *Resolution*'s surgeon, William Anderson (who had been on the second voyage), was also an able-minded, self-taught naturalist who Cook held in high regard as an ethnologist. Anderson was aided in his official role by the roguish extrovert and well-educated David Samwell and by William Ellis, who also published an unauthorized narrative of the voyage.

Thus, contrary to the self-flattering impression Forster imprinted, there was considerable scientific talent aboard Cook's ships. His "curse," presuming it occurred, was probably a theatrical outburst borne of a desire for shipboard serenity and not indicative of a shift in Cook's basic sensibility away from Enlightenment principles. The contributions to science and art made by King, Bayly, Bligh, Anderson, Samwell, Ellis, and Webber bear this out. What Cook wanted to achieve on his third voyage, and succeeded in executing, was an expedition without conflicting agendas. One need only look to the tensions between late-eighteenth-century French navigators La Pérouse and Antoine Raymond Joseph Bruni d'Entrecasteaux and their naturalists to see how common Cook's sentiments were.

Of the aforementioned figures, James King was the most important. Fourth in the line of succession and far less known to students of exploration history than the more junior Bligh or Vancouver, he played a greater role in shaping the perception of Cook's final voyage than any person other than Douglas and Beaglehole. A scion of a well-connected, upper-class family and having studied in Paris and Oxford, the cerebral King, though a naval officer, approximated the scholarly ideal of the Banks-Forster model in his knowledge across a range of subjects. He also had a navigator's eye for land and sea, which endeared him to Cook. King was polite, gracious, charming, popular (with everyone except Bligh), and his skill with a pen made him the obvious choice to complete the postvoyage account (which listed him as a contributing author). He was a confidant of the commander, as was made evident by several passages in the Admiralty's narrative and by observant Hawaiians who thought he was Cook's son. Several of his entries helped create the impression that Cook was somehow "different" on the third voyage, when in fact it was King's genteel sensitivities that were the differentiating variable between the early Cook and the late one.

⚓

After leaving England with a crew of 112 men, Cook set a course for Cape Town. When *Resolution* crossed the equator on September 1, 1776,

Anderson mocked "the old ridiculous ceremony" of ducking those who had not crossed the line before, and he criticized Cook for allowing it. On October 17, *Resolution* anchored in Table Bay, where she awaited Clerke in command of *Discovery.* Since Cape Town was the last port where they could engage in regular communication with Great Britain, correspondence with officials, friends, and family was dispatched before jumping into the unknown. Samwell provided insight into opinion on the prospects for discovering the Northwest Passage: "If we find it we shall be in England next Winter. We have various Opinions about it[;] some think we shall and others that we shall not find it." King, writing to no less a figure than Edmund Burke, reported on the state of his health and a trip into the South African countryside; most importantly, he stated that "C. Cook and I are on a perfect good footing, and that every thing goes on as I could wish."[9]

Leaving the cape on November 30, Cook directed the ships toward the sites of recent French discoveries in the Indian Ocean's roaring forties. His journal, now shorn of daily technical seafaring data, reflects a more relaxed approach toward voyaging in the Southern Hemisphere. It reads like a first draft of what he clearly envisaged as the published account. He polished it continually during the course of the expedition.

Beaglehole observed that Cook could have had no idea during the *Endeavour* voyage that he was "writing for a large public, or was doing anything more than compose the usual document that a commander in the naval service was directed to prepare for the official inspection of his superiors." Cook began to see himself as an incipient author during the second voyage and now on the third he presumed to have the latitude to simply carry the story along, comfortable in the knowledge that his subordinates would keep track of specialized information such as the expedition's latitude and longitude. As Bernard Smith states in *Imagining the Pacific,* "Cook, more than ever before, was aware that he was making contemporary history and would have had the publication of the new voyage in mind from the time he departed."[10] Accordingly, early in this expedition, he only wrote when something essential was pending. For example, his journal notes the departure from Cape Town, but he does not resume writing until Marion Island is encountered nine days later.

On his eastward course to Kerguelen's discovery, Cook ran the ships in the high forties, at no time exceeding 50° S. Nevertheless, the animal menagerie on board (intended as presents to Polynesian leaders) started dying off from the cold. The crew also felt the chill, notwithstanding that

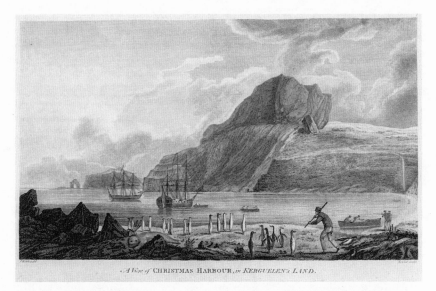

A View of Christmas Harbour, in Kerguelen's Land. Engraving by Taylor after a painting by John Webber, 1783–84. Kerguelen Island was the sole geographic feature within the Southern Hemisphere that Cook was directed to explore during the Northwest Passage voyage. Cook had learned of the exploits of French navigator Yves-Joseph Kerguelen in Cape Town on the outbound leg of his second voyage. Geographic enthusiasts heralded Kerguelen's discovery as a prospective headland for Terra Australis Incognita. Cook failed to find the island on his second voyage but successfully charted it on his third.

it was these same latitudes that Cook had sailed to during the Antarctic voyage for warmth after his freezing forays along the ice. The ships reached Kerguelen's islands on Christmas Eve at a latitude reckoned to 48° 29' S near the meridian that runs through India's southern tip. There, he found a "great plenty of fresh Water, Seals, Penguins and other birds on the shore but not a stick of wood." The last phrase, a graphic expression of vegetal scarcity that Cook had used before, dotted exploratory accounts in the era and demonstrates the prominence of botanical science during the Enlightenment and the practical need of wood for heat and cooking. Samwell documented the island's wildlife, validating the strength of the expedition's unofficial team of naturalists. Christmas Harbour "was lined with Penguins which standing in regular Rows all upright and having white Breasts which gave them a perfect uniformity, they did not appear at a distance much unlike a Regiment of Soldiers drawn up." On December 28, Cook conducted a sovereignty rite on shore where tokens of similar

French activity had been discovered. Anderson, not much of an enthusiast for nationalist tradition, considered the ceremony "ridiculous" but more fitting "to excite laughter than indignation."[11]

After completing his survey of Kerguelen's small archipelago, Cook reflected on his previous voyage and its mission. Clearing the eastern cape of the main island, he noticed a telltale swell from the southwest, proof there was not an extensive tract of land in that direction. Drawing on the sole insight gleaned from *Adventure's* track after her first separation from *Resolution*, Cook observed that if land extended "any farther to the South it would been scarce possible" for Furneaux to miss it. He chided the French explorer, who had imagined these islands "to be a Cape of a Southern Continent," which, he humbly asserted, "the English" had since proved nonexistent.[12]

The balance of the voyage across the Indian Ocean was uneventful, at least for the laconic captain who likened sailing in its "thick foggy weather" to running "in the dark." This experience unnerved Anderson. His concern about sailing in "considerable foggs as these" was that "should any small island or even rock be plac'd in our track no fortitude or dexterity could save a ship running perhaps six or seven knots from instant destruction." He added: "We are perhaps every moment in the situation just mention'd or we may be at a vast distance from danger, but equally ignorant of both." These were risks that Cook was long used to in high-latitude, fog-ridden voyaging, but the scenario Anderson painted would actually play out later in the Aleutian Islands.[13]

⚓

Cook then conducted a cursory exploration of Tasmania's coastline at Australia's southeast corner. The brevity of his effort, which failed to discern its insular nature, has come in for criticism. Historians, following Beaglehole, have magnified all presumed exploratory oversights by Cook on his last voyage, and the truncated Tasmania survey is typically seen as a harbinger of future failures. As John Robson phrases it: "This was, perhaps, the first indication that, on this voyage, Cook had lost some of his drive and curiosity." Vanessa Collingridge writes that the "old Cook" would have fully explored Tasmania, but he was now "a tired man" having "lost the lust for discovery." She then jumps to the more extraordinary conclusion that Cook had "tipped over into irrationality" and that his "lack of curiosity was as ominous as it was bizarre."[14]

These interpretations do not account for Cook's awareness that he was already tardy in reaching the Pacific and behind schedule for reaching

North America. His instructions envisaged him leaving Tahiti in early February, but he only reached Tasmania in late January and had a stop for refreshment in New Zealand still ahead. The Admiralty's guidelines allowed for following up on French discoveries in the Indian Ocean but made no mention of surveying any part of New Holland. If anything, Cook showed more initiative with this quick scan of Tasmania than was expected.

Resolution and *Discovery* reached Queen Charlotte Sound during the second week of February 1777. The crews welcomed the moderate weather of the austral summer in New Zealand, which was, as Anderson phrased it, "never disagreeably warm." Cook had a brewery erected at Ship Cove for making spruce beer. To Samwell, this beverage was "of infinite use to the Ships Companies by keeping them free from the Scurvy, and reflects great Honour on the Commander who paid due Attention to this important Article."[15] Cook had learned about the curative power of pine needles and juniper berries from his time in Canada. Knowledge about *arbor vitae* – trees of life, later discovered to contain four times the concentration of vitamin C as citrus juices – had been passed to Europeans by the First Nations of the St. Lawrence Valley.

Cook also conducted an informal inquiry into the affray at Grass Cove, where some of Furneaux's men had been killed four years earlier. He eschewed taking revenge on Kahura, the presumed ringleader of the massacre. There was rampant disaffection over this decision among Cook's men. Even Mai, who had been on *Adventure*, could not understand it. To the crew, Martin Dugard avers, Cook's leniency resembled what "the out-of-touch London poofs" back home preferred, not what the requirements of empire required. Without offering specifics, Cook's biographer Frank McLynn argues that Cook should have addressed "the men's concern about revenge" more adroitly.[16] These interpretations document the no-win situation Cook found himself in at the time and in retrospect too. If he was predisposed toward violence against Polynesians (as the post-colonial interpretation holds), his generous treatment of Kahura was a pronounced anomaly. On the other hand, Cook's reputation would have suffered if he had turned his back on the Royal Society's "hints" and failed to restrain a crew intent on vengeance.

This episode prefigured a raft of such ruptures. Cook's last voyage seemed to be populated by a more fractious element than his previous expeditions, a fact little commented on. The first scholar to note this was Sinclair Hitchings, who called *Resolution*'s detachment of marines an "unruly lot." Most of them regularly got into trouble for neglect of duty, refusing to stand sentry, quarrelling, drunkenness and theft, and other acts of misfeasance. McLynn allows that Cook was dealing with "sailors far more strident, assertive and disaffected than any he had known previously."[17]

The last quarter of the eighteenth century was a time of great social upheaval in Western civilization, as the situation in the American colonies proved. The hallmarks of the next decade were the far-reaching French Revolution and, on a microcosmic level, the mutiny on the *Bounty*. In 1797, almost the entire Royal Navy mutinied. As historian Richard Van Orman notes: "The age of the explorer was also the age of the revolutionary."[18] Some figures, such as Thomas Jefferson, were devoted to both causes. The antiauthoritarian ethos of the emerging Romantic era, with its passions and excesses, explains the increase in disciplinary issues on the third voyage. The conceit of Cook scholarship is that the only variable between the final expedition and the two that preceded it is Cook alone. But if the cultural zeitgeist influencing the crew's temperament was the greatest difference, not the captain's demeanour, then the whole argument about Cook's deportment during the third voyage is turned on its head. In the present case, Cook's restraint went unrequited. Even the Māori thought he should have exacted revenge on Kahura and taunted him for failing to do so.

⚓

Before leaving New Zealand, Cook allowed Mai to bring aboard two young Māori boys as a retinue. The gesture showed the captain's sentimental side, but it was one that all parties would later regret. He aimed the ships northeast toward Tahiti, imagined as the last stop before the discovery phase of the voyage would begin. The plan was to drop off Mai, make presents of plants and livestock thought to be useful, secure provisions, and then head north of the equator. Not anticipated were the stiff headwinds the ships faced on this tangent. When he sailed to Tahiti from New Zealand during the southern winters of his second voyage, Cook had gentle winds at his back, but he had never sailed in this direction in the austral summer or early fall. Cook has been criticized for not foreseeing this but, as Anderson stated at the time, it was impossible to know the wind patterns after "sailing through a sea only once or twice."[19]

On March 29, 1777, the expedition encountered Mangaia, the southernmost in a group of islands southwest of Tahiti, later named after Cook by Russian explorers. Up to this point, Cook had mapped the coasts of Kerguelen Island, as directed in his instructions, and briefly scanned

Tasmania, a legacy from his earlier voyages. His inspection of this new chain was cursory at best. As recorded in the published account, he had the island of Takutea assessed by Anderson, who walked about and "guessed that it could not be much more than three miles in circuit."[20] Consistent with his stated views from the end of the second expedition, Cook saw no strategic purpose in cataloguing the near infinitude of islands in the South Pacific. He did not fail to explore these islands, he refused to explore them because they were incidental to his current mission and, for that matter, his previous voyage as well.

The Cook Islands figure in the story of the third voyage principally because it was here that Cook altered his course and, consequently, the expedition's timetable. The supply of water on these islands was disappointing. On April 6, he wrote of "the summer in the northern Hemisphere" being "already too far advanced for me to think of doing anything there this year." He deemed it "absolutely necessary to persue such methods as was most likely to preser[v]e the Cattle we had on board in the first place, and save the Ships stores and Provisions in the second the better to enable us to pro[s]ecute the Discoverys in the high northern latitudes the ensuing summer." Instead of beating against the prevailing easterlies, Cook veered west toward Tonga and the Friendly Islands. He had spent only eleven days there on his second voyage, but from that brief exposure he was "sure of being supplied with every thing I wanted."[21]

What might be called Cook's Tongan Diversion is a foundational element within the standard negative casting of the third voyage. Frank McLynn, for instance, argues that when Cook left New Zealand, he failed to make proper allowance for the hydration of his crew and "a menagerie with a colossal thirst for fresh water." In a worse cut, he writes that Cook "decided to play safe and head for Tonga instead of Tahiti."[22] Cook had not played it safe getting to New Zealand by crossing hundreds of miles of the Indian Ocean sailing blind (according to Anderson) in an attempt to make up time. Nor should we discredit Cook's instinct that being in a hurry precipitated its own set of risks.

Cook reached Nomuka, one of the largest of the Friendly Islands, in May 1777. Though he had just made the most significant change to the course and timetable of any of his voyages, Cook was initially in good humour about it. That changed when the ships were increasingly subject to theft. No doubt the Tongans viewed their actions as a form of reciprocal giving or, if you will, entitled taking. Their thinking probably went along the following lines: these visitors have so much stuff, surely they can afford to share. As historian John Gascoigne points out, in most Pacific societies, cultural contacts were cemented by extravagant gift exchanges, but the Englishmen on Cook's ships were acculturated to a strict market economy, and they presented themselves to Indigenous peoples as "ungenerous when it came to giving such presents." Occasionally, Cook viewed theft with philosophical detachment. He could chastise island men of high rank in sufficient measure to get them to hold off from stealing, but since the underlying desire to acquire European materials was unabated, he surmised that the leaders simply directed their servants to do "this dirty work." Flogging the actual perpetrators had no effect, because it "made no more impression than it would have done upon the Main-mast."[23] Nevertheless, a pattern of fractiousness was established, and it lasted for the duration of his time in the Friendly Islands; a month later, Cook found himself shooting at someone who had stolen a trivial item.

Cook intended to head south from Nomuka to the main island of Tongatapu. But, Cook wrote, Finau, a notable Tongan leader, vehemently "importuned" him to deviate from this course "as if he had some particular interest to promote by diverting me from it." Indeed, unbeknownst to Cook (and as would be discovered several decades later by a British beachcomber adopted by the Tongans), Finau was intent on capturing the British ships. His plot fell apart internally in a dispute over timing, adding one more to the many instances where Cook could have died earlier than he did. In any event, instead of heading south, Cook visited the Haapai Group to the northeast, assured by Finau that there "we could be supplied plentifully with every refreshment, in the easiest manner." Though he did not entirely understand the internal logic of Finau's purpose, Cook knew he was being manipulated, probably by opposing factions. After leaving Nomuka, a report came to his attention that ships like his had arrived in his former port. This dubious news, Cook averred, was "artfully contrived, to get us removed from the one island to the other."[24]

Cook eventually reached Tongatapu, the chain's largest island, on June 10, 1777. Several weeks were marked off the calendar to observe a solar eclipse on July 5. In the meantime, there were ceremonies to witness. Cook joined one procession of Tongan Chiefs shirtless, since no one was allowed to participate in the ritual covered above the waist. Some of Cook's men were scandalized by what they perceived as a lack of propriety, but his participation can be explained as the act of a proto-anthropologist not wishing to give offence. Athletic competitions also filled the time, but they predictably led to roughhousing and conflict. Samwell reported "that in all disputes between our People and the [Tongans], Captn Cook ever acted with the utmost Impartiality, being as ready to hear the

Complaint of a [Tongan] and to see justice done to him when injured, as he was to any of his own Men, which equitable way of proceeding rendered him highly respected and esteemed."[25]

But all was not well. Cook's earlier detachment about pilfering was soon replaced with exasperation. Stealing became a constant, and the thieves' tactical ingenuity in creating diversions to enable the stealing of low-value items probably created the impression in Cook's mind that the thefts had become a game intended to embarrass him. In his journal, he never details the increasing severity of the punishments he doled out, but others were shocked. Midshipman George Gilbert thought Cook had callously punished a Tongan thief "in a manner rather unbecoming of a European," that is, by cutting off his ears. In another instance, he ordered a man's arms cut with crosses to mark him as a thief. William Anderson held contradictory views on this state of affairs. He thought binding a man in a painful posture after being flogged was not humane, but he later referred to the "peaceable terms that we liv'd on with the natives while here," excepting "a few trifling quarrels."[26]

Cook himself wrote of leaving "the *Friendly Islands* and their Inhabitants after a stay of between two and three Months, during which time we lived together in the most cordial friendship." There were "accidental differences," which he attributed to "their great propensity to thieving," but he claimed it was a pattern encouraged "by the negligence of our own people." Cook highlighted that "these differences were never attended with any fatal consequences, to prevent which all my measures were directed." This passage clearly indicates that Cook believed any punishment for theft short of lethality was appropriate. The voyage might have been extended by a full year by diverting to Tonga, but he was pleased to have "expended very little of our Sea Provisions" by living off "the produce of the islands." Besides which, he had dropped off some cattle, and those intended for Tahiti "received fresh strength." In sum, Cook concluded, "the advantages we received by touching here was very great, when it is considered that the Season for proceeding to the north was lost before I steered for these islands." Douglas emphasized this point in his editorial treatment of Cook's journal by adding a line in print: "The time, employed amongst them was not thrown away."[27] His defensiveness about Cook's itinerary presciently anticipated future criticism.

⚓

The Tongan Diversion was that portion of the third voyage when a presumably fatigued Cook passed on opportunities to explore nearby Fiji and

Samoa. This common interpretation was propagated by Beaglehole when he suggested that Cook, "on his second voyage, if he had heard of the existence of large islands so close to any of his anchorages ... would have ... fastened them down securely on his general chart." Even since Nicholas Thomas published his iconoclastic view criticizing this orthodox understanding, the best Cook scholars have continued to echo Beaglehole's thesis that a tired Cook from the third voyage passed on discovery opportunities that the presumably more diligent Cook from the preceding voyages would have pursued.[28]

A close reading of the Cook record provides considerable evidence contradicting such certitude. Indeed, Cook is subject to an ironic double standard in this regard. He is criticized for diverting to an unintended destination – Tonga and the Friendly Islands – but then taken to task for not conducting an even greater digression by exploring islands adjunctive to those he had not intended to visit in the first place. Contradicting his thesis that the captain was becoming an angry man with a taste for violence, Frank McLynn implies that Cook refused to engage with Fiji's warrior culture because he was afraid of a fight. Incongruous interpretations about Cook's conduct abound. Vanessa Collingridge calls Cook's "lack of ... usual geographical curiosity" about Fiji and Samoa as "nothing less than shocking," yet she writes off his dutiful survey of Kerguelen's islands, a destination specifically called out in his instructions, as "no great find" and a "pointless" enterprise. Thomas posits that Cook demurred on Fiji because sailing there would delay depositing cattle and horses in Tahiti, which he regarded "as a central objective and a dignifying one of his mission." For Thomas, the Fiji proposition "is a non-problem; neither a temper, nor a tiredness so mysterious that Cook was unaware of it himself, had any bearing on his" exploratory interests.[29]

The idea that a diminished Cook was tiring of discovery was originally seeded in Gilbert's journal. He found it "somewhat surprising that Capt Cook did not go in search" of these neighbouring islands "according to His usual practise." Gilbert could not account for this, "as we certainly had time while were lying at Tongataboo." The midshipman's awareness of what had transpired during the first two Pacific voyages was necessarily based on second-hand knowledge. He had not sailed with Cook before and was thus unable to witness the numerous instances on those expeditions when – for reasons of schedule, unfavourable circumstance, or simply caprice – Cook passed on visiting or charting islands in the South Pacific. Since this was Gilbert's first voyage to the South Pacific, one can understand his disappointment in not visiting as many tropical destinations as possible. Nevertheless, his criticism was seized on by Beaglehole as the principal element in an evidentiary pattern that led to his questioning of Cook's vigour. Heretofore, Lynne Withey is the rare scholar to consider the possibility that Cook thought he had explored the Southern Hemisphere sufficiently, asserting that maybe "one more small island was not worth the trouble." But in the end, she reverts to the norm by stating it was "more likely Cook was tired, and ill," perhaps not to the point of impairment, but he was still "lethargic about pursuing new discoveries."[30]

Anyone who reads Cook's second voyage journal and its stated position on the point of further exploration of the South Pacific, combined with his third-voyage instructions, should not be surprised that he passed on Samoa and Fiji. But we need not rely only on Cook's journal to support the view that the South Pacific had been "done." Several years later, a contemporary of his, La Pérouse, while meeting with British officers in New South Wales and describing his discovery goals to them, would complain that Cook had "left nothing to those who might follow in his track to describe, or fill up." Another French explorer from the post-Napoleonic era, Dumont d'Urville, confessed to being haunted by Cook because the main features of the Pacific basin had been so well delineated that only mop-up work – a stretch of insular coastline here, a few islands there – remained to be explored. Even Johann Forster once wondered: "What good can arise from seeing 2 or 3 Isles more in the South Sea?" During his second voyage, it only appeared that Cook was cataloguing islands; his charts of them were largely a corrective effort that better located and defined the extent of islands discovered by others.[31]

Cook chimed in on this matter indirectly. After returning from Antarctica, he conducted correspondence with a young would-be South Pacific explorer from France named Louis Latouche-Tréville. Cook's first letter, written in September 1775, is well known for its bold directive that "a man would never accomplish much in discovery who only stuck to his orders." Less scrutiny has been given to Cook's second letter, written in February 1776, after he had determined to conduct his third voyage. In it, he counsels that the only objectives worthy of further clarification were the south coast of Australia, the Solomon Islands adjoining New Guinea that Bougainville had discovered, and the equatorial zone between 10° N and 10° S, an area so little explored that he thought some islands might be discovered there.[32] Presciently, this comment anticipated Cook's inadvertent discovery of the Hawaiian Islands, which lie slightly north of this corridor.

⚓

Cook finally arrived in Tahiti on August 12, 1777, just when the Admiralty's plan would have had him making a track toward Baffin Bay. That mission was very much on Cook's mind, because two days later, according to David Samwell, the captain "had the Ships Company called together on the Quarter Deck, where he told them that he supposed they all knew the Destination of the Voyage, which was to try to find a Northwest passage on the Western Coast of North America the next Summer." William Anderson also chronicled the moment, noting that if the ships could not make it "into the Atlantic sea by navigating to the northward of America," then Cook intended to direct a course "round the northern parts of Asia and Europe." Amidst the most fabled tropical paradise, Cook immunized himself from potential shipboard concerns that the voyage was taking too long to conduct. He reminded the crew, according to Samwell, that the ships had left Great Britain more than a year earlier and "what a long time it would probably be before we returned there or to any European Port." The purpose of that statement was to explain "how necessary it was to be frugal in the use of the Provisions we had brought with us from England." Samwell pointedly noted "that this was more particularly necessary with regard to the Article of Spirits which if served out at full allowance would not last the Voyage." Cook was of a mind to cut the daily ration in half, though it "would be much more agreeable to him if during our Stay at these Islands," Samwell recorded, the men "would be satisfied to go without their Grog entirely except on Saturday Nights." Anderson reported that the seamen consented to this "immediately without an objection, and indeed it could not be consider'd as the smallest hardship where were we sure of being plentifully supplied with fresh provisions and abundance of cocoa nuts, which ... are certainly superior to any artificial beverage." Their readiness to comply with this plan, Samwell concluded, "afforded Capt. Cook much pleasure."[33]

The 1777 visit to the legendary isle of Tahiti was Cook's fourth. He gifted three cows, a bull, a horse, and some sheep, and wrote that his ship had been "lightened of a very heavy burden." This expression was perhaps a double entendre, because Cook took satisfaction "in having been so fortunate as to fulfill His Majesty's design in sending such usefull Animals to two worthy Nations," referring also to the previous deposit at Tonga. This, he felt, was suitable recompense "for the many anxious hours I had on their account."[34] With this passage, Cook transcends his typecast role as navigator to take on the role of imperial emissary and propagator. This prosaic duty stood in lieu of finding more islands in the Southern Hemisphere, and from his journal it is clear Cook revelled in this work.

After the rigging was overhauled and the ships caulked, Cook left Tahiti for its neighbouring islands. Though the Cook legend trumpets the thoroughness of his first two voyages in contrast to his supposedly lackadaisical approach to the third, he had never visited nearby Moorea (nor Bora Bora, arguably the archipelago's major island). Cook stated that Moorea's principal harbour, "for security, and goodness of its bottom," had few equals with "any of the islands in this ocean." It had the particular "advantage over most of them, that a ship can sail in and out, with the reigning trade wind; so that the access and recess are equally easy." Cook was sheepish about this discovery, observing that "it is a little extraordinary, that I should have been three times at Otaheite before ... and yet not know, till now, that there was a harbour in" Moorea. He ordered a sketch made "for the use of those who may follow us."[35]

Dallying near Tahiti not only filled time, it also filled Cook's narrative. As Bernard Smith cogently observes in *Imagining the Pacific*, the *Endeavour* voyage with Banks was oriented toward astronomy and botany. The second – with the stellar team of Cook, Wales, Bayly, and the Forsters – was oriented toward climatology. "By the third voyage," Smith notes, "Cook had come to realise that both scientific and popular interest had shifted to the native peoples of the Pacific; to the nascent science of ethnography."[36] Accordingly, in the absence of a Banks or Forster, Cook spent more time recording Polynesian practices to meet readers' demands.

But the sojourn in Moorea was not without severe consequences for Cook's reputation, both during his lifetime and after. Two prized goats, the source of valuable milk for flavouring tea and potential gifts, were stolen in separate incidents. The first, an act of Moorean retribution for a sailor's misdeed, was quickly resolved. The second goat, however, was secreted away to another island, delaying its return. In the literature on Cook, the reports of several crew members (Gilbert, King, Williamson, Samwell, and Ellis) frame the standard account of Cook's punitive expedition. Drawing on them, Vanessa Collingridge writes that the second theft and its lack of immediate resolution threw Cook "into a monstrous fury."[37] Indeed, Cook directed what can only be called a rampage across the island's shore, burning houses and canoes.

Gilbert could not "account for Capt Cooks proceedings on this occasion as they were so very different from his conduct in like cases in his former voyages." But Gilbert was working with second-hand information since he had not sailed with Cook before. With similarly constructed wording, suggesting a shipboard conversation, King said he was also unable "to account for Captn Cooks precipitate proceeding in this business." He framed the issue in geopolitical terms, observing that Cook's retribution "will be a very strong motive" for the islanders "to give a decided preference to the Spaniards, and that in future they may fear, but never love us."[38]

Cook was defended by Charles Clerke, who chastised the perpetrators for the "asburd obstinacy" that had precipitated his wrath. It was as if "the Devil put it in their Heads, to fall in Love with the Goats" and, Clerke added, the consequence of not returning the animals was thoroughly explained by Cook "before any destructive Step was taken." Cook, not surprisingly, has the least to say about this event. He referred to this "rather unfortunate affair ... which could not be more regreted on the part of the Natives than it was on mine." Nicholas Thomas, in explaining Cook's angry deportment, argues that his latter-day role in animal husbandry made the goat's theft anything but a trifling loss.[39] Cook's repentant phrasing in the aftermath of the incident shows his conscience, not its absence.

Most historians see no reason for such ambiguity. Beaglehole's original theory about Cook's fatigue might have started as a rhetorical question, but Frank McLynn had recently stated, "quite confidently and without fear of contradiction ... that Cook by the time of his third voyage had undergone a personality change." His "irritability, loss of initiative, and depression" were caused by – take your pick – Vitamin B deficiency, roundworm infection, psychosexual repression, or even opiate dependency. The "marked change" in Cook was "noted by everyone who had sailed with him before," McLynn adds. But the severest critics were in fact a handful of first-timers. Cook veterans such as Clerke said nothing of the sort. Without proof, McLynn argues that the captain's supposed disillusionment with exploration and erratic behaviour were functions of a subconscious realization "that the Northwest Passage would turn out to be a chimera and that he was engaged in a fool's errand." The scholarly consensus, as phrased by Michael Hoare, is that "on the third voyage Cook became tired, lost his self-control, got more cruel and wanton, and perished."[40]

This trope was nourished by an incident at Cook's next stop, Huahine. A man from Bora Bora, whom even the local Chiefs considered a ne'er-do-well, stole a sextant, an important piece of equipment. Not wanting to replay the Moorea scenario, wherein a community would pay the price for an individual's transgression, Cook tried to get his message across by shaving the culprit's head and having him flogged. When this man broke away from his shackles, Cook was both incredulous and apoplectic, emotions prompted less by the escape than the breakdown in his system of command. He suspected active collusion by some of his crew. Two seamen were shackled and flogged, and a midshipman, Robert Mackay, was disrated to the rank of able seaman.

As inhumane as flogging appears to modern sensibilities, Cook's orders were not outside the bounds of normal practice in the Royal Navy at that time, and the practice would not be outlawed by naval regulation for another sixty years. Writing in 1840, Richard Henry professed an abhorrence to flogging but avowed he would not "wish to take the command of a ship" himself without recourse to "chastisement." The "power of the captain," Dana averred, should not be "diminished an iota. It is absolutely necessary that there should be one head and one voice, to control everything" because a captain was "answerable for everything; and is subject to emergencies which perhaps no other man exercising authority ... is subject to." Apart from tradition, noted anthropologist Anne Salmond points out that although there were more lashes applied on Cook's third voyage than the first two combined, "the worst punishments happened on board the *Discovery*, which suggests that Clerke was at least as much to blame."[41] If the junior officers got lax in their exercise of discipline, the captain, any captain in that era, would be forced to draw the reins tighter to maintain order, no doubt chafing the men all the more.

All the same, it is hard to reconcile Cook's pattern of behaviour on Moorea and Huahine with those who prefer to see him, in the words of Ganannath Obeysekere, "as the humane embodiment of the Enlightenment," a perspective propagated by British apologists intending to offer up a contrast to "earlier violent Spanish expansionism." John Gascoigne, comfortable with the original Beaglehole hypothesis that Cook was merely "a tired man" and not mentally unhinged, says Cook's demeanour on the third voyage "was different more in degree than kind" from what he had exemplified previously. Reflecting the historical consensus, Gascoigne states: "His men plainly thought something had changed," and the difference was that "Cook became increasingly sensitive to any challenge to his authority, whether it came from his own men or from the native peoples he encountered."[42]

Still, it is important to remember that the premise of a "changed Cook" is based on the comments of those who were sailing with Cook for the first time. Salmond asserts that "the *Resolution*'s contingent included many first and second-voyage men." The term "many" is subject to interpretation but is generally conceived to mean a large number in relative or absolute terms. According to the ships' companies for all three Cook voyages, *Resolution II* had a crew of 112 men, only 13 of whom had been on a Cook voyage before, and of those 13, only 4 had been on both *Endeavour* and *Resolution I*. *Discovery*, with a complement of 70 men, had 10 Cook veterans, of whom only 2 had been on both expeditions, 1 of them being Clerke. Thus, there were only 6 people on both ships who had sailed on the two previous voyages, just 3 percent of the combined crew, and most of them were able seamen or craftsman. Only two kept journals, Clerke and William Harvey. Another 6 had been on the first *Resolution* voyage and were now on the third expedition, but, again, a majority of these sailors were from the ranks. Only the surgeon, William Anderson, and the master's mate, Henry Roberts, were in a position to comment knowledgeably on the pattern of Cook's exploratory conduct. Indeed, there was a higher percentage of carry-over from *Endeavour* to *Resolution I* than from the latter to *Resolution II*. None of the propagators of the "changed Cook" narrative came from either the single voyage or even the smaller two-voyage subset, which in itself ought to raise questions about the reliability of the Cook-was-now-a-changed-man hypothesis.[43]

It is clear that Cook had tired of brokering the three-way cross-cultural dynamic between himself, Indigenous peoples, and his men, much as he had tired of counting islands. To say this, however, does not mean that he was tired of discovery, as subsequent events would prove. Cook's true failure, pertinent to each of his voyages and not just the last one, was never realizing that the endemic theft he railed against was not a moral flaw inherent in South Pacific Indigenous peoples so much as a statement about the exotic appeal of the material culture he brought to their shore. "The wonder," nineteenth-century African explorer Mungo Park wrote in a parallel circumstance, "would be, not that the stranger was robbed of any part of his riches, but that any part was left for a second depredator."[44]

If the Royal Society's hints about dealing with Indigenous peoples gradually fell to the wayside, it happened because, renowned cultural historian Bernard Smith cogently argues, Cook had absorbed enough of the Enlightenment ethos "to become aware by his third voyage that his mission to the Pacific involved him in a profound and unresolvable contradiction." According to Smith, Cook saw himself as King George's "Enlightenment Man," desirous of spreading "the blessings and advantages of civilised Europe." But it was not the beneficence of Europe he was spreading but rather its "curses," venereal and pecuniary. The import of the former is readily understood but the latter is a more complicated topic. Enlightenment principles of cultural engagement and his own crew's survival depended on, Smith writes, the establishment of "markets among people who possessed little if any notions of a market economy." When commerce failed, Cook's only alternative, if the expedition was to survive, was the use of force in the same manner the Portuguese and Spanish had always applied. This, Smith concludes, hurt the Englishmen's pride, not only the more sensitive members of Cook's crew, who commented on the violation of humanistic ideals in their journals, but Cook himself. He witnessed the death of an important aspect of the Enlightenment's discovery agenda, but more dispiritingly, he brought it about by his own hand.[45]

On Huahine, there was also great pathos associated with the final deposit of Mai and the two Māori boys who had been taken aboard in New Zealand. Cook was fond of Mai, going so far as to secure an ensign for the stern of a canoe he gifted him. Cook said Mai made "a very affectionate farewell of all the Officers; he sustained himself with a manly resolution till he came to me, then his utmost efforts to conceal his tears failed." But Mai was not the only one feeling strong emotions. According to Lieutenant John Williamson, Cook was "much affected at this parting."[46]

The separation anxiety exhibited by the Māori youths was even more heart-rending. James King said Koa and Te Weherua showed "the most violent and poignant grief at parting with us; they had flattered themselves that we shoud have carried them to England, and we were all truly sorry at their disappointment." Cook professed: "If there had been the most distant probability of any Ship being sent again to New Zealand I would have brought [them] ... home with me." According to Gilbert, Te Weherua, the eldest, "display'd a nobleness of spirit above the common rank of people; and never associated with the sailors, but always kept with the Gentlemen." Koa "was always full of Mirth and good humour." The crew was fond of "his mimicry and other little sportive tricks." Cook said the older boy had "resigned himself" to spending the rest of his life far from home, but Koa "was so strongly attatched to us that he was taken out of the Ship and carried a shore by force." The tragedy of this scene, which Cook characteristically underdramatized, was fully captured by William Bayly, who said the boys "cry'd greatly at parting with us, and one of them," impliedly Koa, "jumped over board twice out of the Canoe" in an attempt to swim back to the ship. (When William Bligh returned to Tahiti as captain of *Bounty* in 1789, it was learned that Mai had died of natural causes in or near 1781. The fate of the young Māori boys was even more dreadful. James Morrison, *Bounty*'s boatswain, reported that Koa and Te Weherua had "greived much for ... their Native Country, and after OMai died, they gave over all hopes and having now lost their chief friend, they pined themselves to death.")[47]

Several of Cook's men would have gladly traded places with the Māori youths. Midshipman Alexander Mouat and gunner's mate Thomas Shaw attempted to desert. Charles Clerke and William Anderson, both of whom would die in the North from the effects of tuberculosis, asked to stay at Huahine in the knowledge that the polar latitudes would kill them. Cook's refusal has been found to be consistent with the depictions of a new, callous captain, fearful of censure if something should go awry on *Discovery*.[48] But he had legitimate reasons for keeping his leadership team together. The discovery phase of the expedition was finally about to begin. Having dragooned low-ranking seamen who wanted to desert, Cook could hardly have allowed an officer and a gentleman to stay behind. Clerke himself, after Cook was dead, could have directed the expedition anywhere he chose, but he returned to the North, and because of the self-imposed demands of leadership, he died there.

⚓

With departure for northern latitudes imminent, Cook drew up an elaborate set of instructions for Clerke. Cook observed that "the Passage from the Society Isles to the Northern Coast of America is of considerable length both in distance and time, and ... part of it must be performed in the very depth of Winter when gales of Wind and bad Weather must be expected and may possibly occasion a Seperation." Should that happen, Clerke was to find a convenient port at 45° N or still farther north to "recrute your wood and Water and to procure refreshments" while keeping "a good look out for me." If there was no meet by April 1, 1778, Clerke was to proceed to 56° N, "at a convenient distance from the coast," and stay there until May 10. If Cook did not appear then, Clerke was told "to proceed Northward and endeveour to find a passage into the Atlantick Ocean."[49] As it turned out, in a remarkable record of seamanship under very difficult circumstances, *Resolution* and *Discovery* never lost contact in the Pacific's high latitudes.

After these instructions were disseminated, *Resolution* and *Discovery* made a quick stop at Bora Bora, an island Cook had heard reports of during his earlier voyages but never visited. The rationale for doing so now was the prospect of recovering an anchor that Bougainville had lost off Tahiti a decade earlier but that had since been removed to Bora Bora. Author and broadcaster Vanessa Collingridge sees here evidence that "the old and curious Cook was back again," but a simpler North Pacific mission-driven explanation avails in his journal: "Not that we were in want of Anchors, but after expending all the Hatchets and other iron tools we had to procure refreshments, we were obliged to make others out of the iron we had on board to continue the trade." That is, Cook wanted to restock his supply of metal in anticipation of future needs in the North. Otherwise, he would never have visited Bora Bora.[50]

Before leaving the tropical paradise, Cook wrote: "So much, or rather too much, has been published of Otaheite and the neighbouring islands already that there is very little room for new remarks." This rhymes with Cook's comment near the end of the second voyage that he was done with the South Pacific, and it fully supports the view that his third voyage to Polynesia was, for him, a mere staging area, not a venue for extensive exploration. After securing the anchor and lifting the shore boats aboard the ships on December 9, 1777, Cook "steered to the Northward."[51]

TEN

Terra Borealis

James Cook was unsentimental when he left Tahiti, but George Gilbert probably captured the emotions of most when he wrote: "We left these Islands with the greatest regret immaginable; as supposing all the pleasures of the voyage to be now at an end: having nothing to expect in [the] future but excess of cold, hunger, and every kind of hardship, and distress." Many, like David Samwell, whose journal revealed a pronounced taste for sexual adventurism, surely signed on mostly for the part of the voyage now ending. He regretted bidding "adieu to the Delights and Luxuries of these Islands."[1] For Cook, on the other hand, the whole point of the voyage was about to begin. His goal was to get to 65° N by June, and he had six months to cover eighty-two degrees of latitude.

Cook planned to sail north of the equator to catch the westerly winds that would take him to the American coast, but on Christmas Eve day, 1777, land was sighted. It would become known as Christmas Island, one of the Pacific's largest atolls. Cook had asked the Tahitians if they knew of any islands to the north, and although they said they did not, he was not surprised to find it because, he said, lowlands such as Christmas Island were "common in this area." Cook intended to linger, Samwell reported, to commemorate "the old laudable Custom of keeping Christmas" and to observe an eclipse of the sun. The eventual departure was delayed when two seamen got lost. Cook maintained that "the Ships masts were to be seen" from any part of the island, but the men later told him they never thought to look for them, nor could they recall which direction the ships lay. Other journal writers on the voyage delighted in this episode, which has *Robinson Crusoe*-like overtones. Lieutenant John Williamson, who found them, described holding up a "cocoa nut shell" to the man who was still ambulatory and filling his mouth with "rum and water, not suffering him to drink much." The other sailor was found "almost dead," having survived by drinking the blood of some birds and his own urine.[2]

⚓

On January 2, 1778, Cook's ships weighed anchor and resumed their course northward. Crossing the equator, cold-weather gear was taken out of the hold and distributed to the crew at the astonishingly low latitude of 7° N, indicating that Cook was expecting to reach the mid-latitudes quickly. However, seabirds were soon encountered, and between 10° and 11° N turtles became a common sight. "All these are looked upon as signs of the vecinity of land," Cook stated, and on January 18 the first of several Hawaiian Islands was detected. Spanish ships had been traversing the North Pacific from Mexico to Asia and back for the preceding two centuries, but they had missed these islands. Prevailing winds and the North Pacific Current had carried Spanish vessels north of this chain eastbound and the North Equatorial Current south of it westbound. Since the islands ran parallel to these opposing tracks, they were never encountered. Cook was the first European to transect the central Pacific's equatorial latitudes on a north-south tangent. His perpendicular line intersected an archipelago that is more remote from a continental mass than any other sizable chain. As Samwell phrased it: "This appearing to us to be a new Discovery excited our Curiosity much, expecting to meet with a new Race of People distinct from the Islanders to the Southward of the [equatorial] Line; we were sometime in Suspense whether it was inhabited or not."[3]

Cook spent three days plying in the wind, surf, and occasional squall before establishing an anchorage off the south shore of Kauai, a sequence that prefigured his supposedly dilatory manoeuvres off the Big Island of Hawaii a year later. He discerned that it was populated by the "same Nation as the people of Otahiete and the other islands we had lately visited," a demographic insight he later amplified on. Cook seized this windfall to enhance his supply of pig meat, remarking "we again found our selves in the land of plenty." He had never seen islanders "so much astonished at the entering a ship before, their eyes were continually flying from object to object, the wildness of thier looks and actions fully express'd their surprise and astonishment at the several new o[b]jects before them." With some of his crew reporting "venereal complaints," Cook attempted a prohibition of "all manner of connection" in furtherance of Royal Society and Enlightenment ideals. For incentive, he restored the full allowance of grog, which had been suspended in Tahiti, but doubted his restraining order would be effective: "Inducements to an intercourse between the sex[es] are ... too many to be guarded against." What Samwell called the "lascivious gestures" of the Hawaiian women did not help matters. Cook led three boats ashore to look for a supply of fresh water and to "try the desposition of the inhabitants, several hundreds of whom were assembled on a sandy beach before the village."[4]

Cook's next sentence looms portentously in accounts of his career: "The very instant I leaped ashore, they all fell flat on their faces, and remained in that humble posture till I made signs to them to rise." Scholarly debates have raged over the meaning of this and similar gestures, Cook's interpretation of the same, and the culminating phase of this cultural dynamic a year later when the Hawaiians abandoned their deferential attitude in the run-up to his death. Whether this behaviour betokened an attribution of divinity to Cook or whether Cook's ego subsumed this outlook to the detriment of his judgment (a question that is a central element of the palm-tree paradigm that straitjackets Cook studies) has been addressed by other authors. Suffice to observe here that this phenomenon was not rare in the history of European exploration. Felipe Fernández-Armesto, the foremost modern historian of exploration, asserts: "The claim that sea-borne explorers' hosts supposed their visitors to be heaven-borne is, indeed, such a widespread topos as to be barely believable."[5]

Unbeknownst to Cook at the time, the first watering party, commanded by Williamson, killed a man in a tragic incident layered with ironic ramifications. Williamson thought the Hawaiian victim was attempting to steal his firearm when he was probably just reaching for something to grab onto to help pull the boat ashore. This might have been seen as another unfortunate event except Williamson, who earlier had been one of Cook's severest critics for his treatment of islanders, hid the incident from the captain until after they left the island.

Cook called this archipelago the Sandwich Islands in honour of his patron and briefly reconnoitred its westernmost elements: Oahu, Kauai, and Ni'hau. For many an explorer, and certainly those glancing at fear-nought jackets stacked on deck, the temptation to stay longer would have been strong. Not Cook. After two weeks, he determined they had spent "more time about these islands than was necessary to have answered all our purposes," meaning water and food, which his instructions allowed, as opposed to discovery work, which they did not. Curiously, the Admiralty's account, published after Cook's death, incorporated text at this narrative juncture at odds with Cook's terse determination to proceed on. John Douglas (as Cook) added that it was "to be regretted, that we should have been obliged, so soon, to leave a place, which, as far as our opportunities of knowing reached, seemed to be highly worthy of a more accurate examination."[6] This is further testament to the disjunction between Cook's immediate valuation of Polynesia's relevance to his third voyage versus retrospective judgments rendered by third parties from Cook's time to the present.

As *Resolution* and *Discovery* "stood away to the Northward," Cook pondered how to account for the Polynesian expansion "so far over this Vast ocean," a dispersion that encompassed a triangle running from New Zealand to Easter Island to Hawaii and spanned a space that he measured to "60° of latitude ... and 83° of longitude." King, serving ably in the Banks-Forster role, studied the expanse "thro which this Nation has spread" and translated its limits to "exceed all Europe, and is nearly equal to Africa," calculated to 4,200 miles on an east-west axis and 3,315 miles on a north-south axis. The triangle covered 10 million square miles of open water. (A North American–centric equivalent to the Polynesian Triangle would encompass a space running from its apex at Alert, Nunavut, to Mexico City, and then across the Atlantic to the Azores.) King determined that the language spoken in "all the Isles in the intermediate space are by their affinity or sameness in Speech to be reckon'd as forming one people," although "provincial dialects" occurred.[7] Scholars have legitimately subjected Cook's observations of Polynesian culture to criticism, but rare is the anthropologist who has made a field discovery as potent as Cook's ethnographic image of the Polynesian Triangle, a term of art firmly fixed in ethnological studies. He documented the last great divergence of humans from their original African roots. The dispersion from Indonesia and the Philippines, starting in late antiquity and reaching the steady state observed by Cook circa 1200 CE, was the greatest exploratory triumph in human history.

The Hawaiian Islands are Cook's most significant exploratory legacy, a point worth emphasizing because his discoveries during the first two voyages hold a privileged status among historians. Vanessa Collingridge's statement that his second voyage "marked the apogee of his contribution to his profession and to the world at large" is emblematic of the conventional interpretation. Martin Dugard employs a popular trope among scholars that Cook "stumbled upon" the archipelago as if this unexpected discovery on the third voyage is of a lesser rank than such serendipities on the first two. Cook did not conduct a sovereignty ceremony in Hawaii, evidence that he did not see Polynesia as strategically important to Great Britain. Conversely, he thought "Spain may probably reap some benifit by the discovery of these islands, as they are extremely w[e]ll situated for the Ships sailing from New Spain to the Philippine Islands to touch and refresh at."[8] Cook would conduct a tactical winter retreat to the chain after his northern summer in the icy latitudes, but at this point his fidelity to the voyage's mission directive dictated his departure.[9]

⚓

Cook stood north from the Hawaiian Islands on February 2. Nine days later, Charles Clerke mused that the crew had been inhabiting the Torrid Zone so long "we are all shaking with Cold here with the Thermometer at 60." He expected "a few good N: Westers to give us a hearty rattling and bring us to our natural feelings a little, or the Lord knows how we shall make ourselves acquainted with the frozen secrets of the Arctic." Restoration of the full grog allowance helped the seamen cope. The ships uneventfully covered vast stretches of open water, and Cook's journal remained empty until February 12, at 30° N, when he made note of steering to the northeast.[10]

Kelp and wood debris were detected on February 25, but no birds were sighted, which sent mixed signals about North America's proximity. (Part of the confusion, George Gilbert pointed out, was that the Northwest Coast of America was found "8 degrees to the East of what was layed down in the best Charts.") At dawn on March 7, 1778, at a recorded latitude of 44° 33' N, Cook took in a view of "the long looked for Coast of new Albion." His instructions had directed him to reach North America by February, so he was on schedule in that sense, though a year late. Cook's first glimpse of what one might call *terra borealis* yielded anticlimactic text: "The land appeared to be of a moderate height, deversif[i]ed with hill and Vally and almost every where covered with wood. There was nothing remarkable about it, except one hill, whose elevated summit was flat."[11] Cook's chart denominates this feature as "Table Mountain," no doubt inspired by the distinguishing feature of Africa's Cape of Good Hope, one of his favourite ports. This was the first terrestrial aspect described by Cook on the west side of North America, and based on J.C. Beaglehole's surmise, it is commonly understood to be Mary's Peak, which is twenty-three miles inland and the scenic backdrop to Corvallis, Oregon. However, another Table Mountain, located on virtually the same parallel as Mary's Peak, but only eight miles inland, may have been what Cook actually espied. In any event and contrary to local legend, Cape Foulweather was not the first place name Cook lent to Oregon's geography.

Since the expedition was still thousands of miles from the mission's target, Cook was not yet in descriptive mode. The more substantive commentary fell to King. He started by casting doubt on both legendary explorers such as Juan de Fuca and authentic ones such as his country-man Francis Drake: "This part of the Continent of America has not so far as we know, ever been seen." He also made a vital climatic determination, asserting that the conditions were seasonably "milder than it was on the Eastern Coast of the Continent." (Halifax is at a comparable latitude on the Atlantic.) The higher elevations of the Oregon coast range, he observed, were "free from Snow."[12] Finding a moderate climate in late winter halfway to the North Pole implied that perhaps Daines Barrington had been right: it was going to be easier to reach farther north on the western side of America.

Needing refreshment, Cook started looking for a harbour. Finding none in the foul weather (thus the place name), he stood away. Squalls drove the ships southward almost to the modern California-Oregon state line. Cook glimpsed the numerous headlands that are a distinguishing feature of the southern Oregon Coast. After giving two places names that were physiographic (Table Mountain) and atmospheric (Cape Foulweather), Cook returned to his curious habit, for an Enlightenment man, of naming places by drawing on the liturgical calendar, a practice he occasionally employed on the first two voyages but that became increasingly common on the third. He named two capes after ancient Christian saints: Perpetua and Gregory. The former has endured, but the latter is now known as Cape Arago. Cook had been voyaging so long he had begun to run out of English admirals, captains, and noblemen to honour.

The unsettled weather continued unabated. On March 14, a frustrated Clerke jotted: "We can neither forward our Matters by tracing the Coast, nor have the Satisfaction of getting into a Harbour to take a look at the Country." On March 21, a fortnight after sighting Table Mountain, Cook finally fell in "with the land beyond where we had already been," and the next day, in clearing conditions, he calibrated their location at 47° N. To the northeast, he observed an opening in the coast "which flatered us with hopes of finding a harbour." This inlet appeared to be too small to enter, so Cook now drew on an emotive sentiment for a toponym and named the point "Cape Flattery" because of its misleading promise. He charted the cape at 48° 15' N, which was within seven minutes of its actual location. This difference is well within the margin of error for shipboard computations in less than perfect conditions, but, mysteriously, Beaglehole erroneously asserted that Cape Flattery is located at 46° 22' N and characterized this as an unusually large miscalculation for Cook.[13] This supposed mistake fell comfortably within the now conventional view, which Beaglehole did much to propagate, that during the third voyage Cook was off his game.

The error was Beaglehole's, who was thinking of Cape Disappointment, the headland at the mouth of the Columbia River at 46° N.

Historians of northwestern exploration have reproached Cook for the lack of inquisitiveness that led to his "missing" the Great River of the West, later known as the Columbia, and the Strait of Juan de Fuca at 48° N. These "oversights" are typically written off as a function of Cook being, in the words of one historian, "tired and in diminished health." Stephen Bown calls Cook's denomination of Cape Flattery, which guards the southern gate to the Strait of Juan de Fuca, "a timeless monument to the great mariner's error." In a particularly egregious example, Peter Stark states that Cook's voyage "probing for the Passage or the mouth of the long-rumored Great River of the West" simply "didn't work out as planned." It is true that the Columbia and Fuca's strait are vital to any understanding of regional geography, but in the first instance Cook was well offshore and had no sense of passing by the outfall of a large river. Even if he had seen the Columbia, his instructions stipulated that exploring a river at that latitude was a needless diversion. In regard to legendary continental inlets, off Cape Flattery, Cook made passing mention of being in the vicinity of "where geographers have placed the pretended *Strait of Juan de Fuca,* but we saw nothing like it, nor is there the least probability that iver any such thing exhisted."[14]

Cook got the big picture right but was wrong on the local particulars. Informed by Samuel Hearne's revelations, he could comfortably dismiss the notion of a continent-piercing strait at such a low latitude. Fuca's passage of legend was not the strait we know today – an indent in the west coast of America – but the supposed Pacific gateway to the Northwest Passage, which carried to or near the Atlantic. Unlike the Columbia, had Cook seen the strait, he might have entered it because that would have yielded opportunities for the harbour he sought. George Vancouver would later find them in plenty, establishing the insularity of the adjoining terrain that bears his name. Beaglehole asserts that Cook saw relatively little land between 44° N and 55° N and was perhaps "too contemptuous" of nautical lore. Glyn Williams shares this view, calling Cook's dismissal of Fuca "dogmatic" because the atmospheric conditions and poor visibility precluded him from making an objective assessment.[15] Nevertheless, Cook's conduct was consistent with his instructions advising him not to begin searching for a Northwest Passage until he reached 65° N, a specification sensibly derived from the explorations of Bering and Hearne. The latter proved that any opening on the Pacific slope in such low latitudes as Fuca's strait would inevitably turn out to be a dead end, as Vancouver later documented.

⚓

On March 29, Cook was back in sight of land at 49° N. Discerning an inlet, he entered "to get some Water, of which [we] were in great want." Dozens of canoes enveloped the ships. This locale has come to be known as Nootka Sound, a place that would resonate deeply in regional history for the next decade and a half as the centre of the coastal fur trade and an imperial flashpoint between Great Britain and Spain. Once inside the harbour, Cook observed "many Canoes filled with the Natives" plying "about the Ships." A "trade commenced betwixt us and them, which was carried on with the Strictest honisty on boath sides." In exchange for hatchets, stray pieces of iron, and some beads, the sailors acquired bentwood visors, decorative masks, baskets, and hats, but their principal interest was in peltries. Bear and wolf skins were made into coats, while the fur from smaller land animals such as foxes and martens became lining for caps and gloves. Particularly valuable, Cook stated, was "the Sea Beaver, the same as is found on the coast of Kamtchatka."[16]

Cook was referring to the sea otter (*Enhydra lutris*), a small marine mammal found in shallow water close to shore, a species that would be hunted to near extinction over the next half century. His allusion to Kamchatka betokens a knowledge of the animal's importance to the North Pacific fur trade the Russians were conducting, but it was naive as to the future British trade nexus. Cook is generally credited as the originator of the ensuing British and American commerce in furs along the North-west Coast, but his role has been exaggerated. It would not become apparent until two years later, long after Cook was dead, that the crew's pelts proved to be worth a fortune when the ships stopped in Asian ports on their way home. In the moment, George Gilbert's reason for "procuring those skins, was for clothing to secure us against the cold." This view was shared by Samwell, who said: "To us who were bound for the North Pole," they were "valuable articles and every one endeavoured to supply himself with some."[17]

Trade in furs and other items with the Mowachaht – in whose territory the ships had anchored – proceeded apace and without major complication. Here, on the American shore, trade was keen, and the usual sort of pilferage common in such situations continued. A serene Cook stated it was "better to put up with these tricks than to quarrel with them." He joked that "hardly a bit of brass was left in the Ship, except what was in the necessary instruments." Whole suits of clothes were stripped of their buttons, and copper kettles, tin cups, and candlesticks were carried overboard, yet he humorously reported that the people of Nootka Sound "got a greater middly and variety of things from us than any other people we had visited." Was Cook conscience-stricken by recent events in Polynesia, or had he simply reconciled himself to a certain amount of roguish behaviour? Historian John Gascoigne argues that the longer Europeans stayed around a new culture, the severity of punishment for theft increased. As newcomers to the Northwest Coast, Cook and his crew still had plenty of cultural cushioning in reserve.[18]

Resolution was "very leaky in her upper works," so Cook set the caulking crew in motion. A squad went ashore to clear space for watering and for the forge for repairing the iron works in the masts. The rigging was decayed, so new sets were fitted. An observatory, essential to the voyage's navigational requirements, was set up, and at the quotidian end of the operational spectrum, so too was a brewery. What Samwell called "small beer," flavoured with fir boughs, was "served to the Ship's Company instead of Brandy," a batch being sufficient "for two or three months."[19] Cook often traded this beverage for grog during the course of the voyage, a routine development until the so-called sugar-cane beer mutiny occurred when Cook returned to Hawaii.

By April 20, most of the repairs had been effected, so Cook took "a view of the Sound," captain-speak fleshed out by George Trevenen, one of the midshipmen who served as a rower. In the picture Trevenen paints, we see a captain finally, at the verge of the mission's operative moment, letting down his guard. "We were fond of such excursions," Trevenen wrote, and "altho' the labour of them was very great ... this kind of duty, was more agreeable than the humdrum routine on board the Ships ... Capt. Cooke also on these occasions, would sometimes relax from his almost constant severity of disposition, and condescend now and then, to converse familiarly with us. But it was only for a time, as soon as we entered the ships, he became again the despot."[20] The concluding phrasing, which simply reflected the reality of the shipboard command structure, has loomed large in the historiography of Cook's third voyage. For the most part, Trevenen wrote about Cook in reverential terms, and even this story was, largely put, intended as a compliment.

Near the end of the Nootka sojourn, Cook was ferried to nearby Friendly Cove to secure provender for the few remaining sheep and goats on board. Cook stated: "The Inhabitants of this village received us in the same friendly manner they had d[o]ne before, and the Moment we landed I sent some [men] to cut grass not thinking that the Natives could or would have the least objection, but it proved otherways for the Moment

John Webber, Resolution *and* Discovery *in Ship Cove [Nootka Sound], 1778*. After reaching the Northwest Coast in the spring of 1778, Cook finally found refuge in a fine anchorage on the west coast of what was later perceived to be Vancouver Island. Here, Cook replenished the expedition's supply of wood and water and otherwise readied the ships for the discovery phase of the voyage. A productive trading regime was established with the Mowachaht people of Nootka Sound, including the expedition's first exposure to sea-otter pelts.

our people began to cut they stoped them and told them they must *Makook* for it, that is first buy it." This was novel. Author James Zug notes: "In all three voyages, never once had a native demanded payment for something that grew wild." After paying a dozen men, "who all laid cla[i]m to some part of the grass," Cook believed he was at "liberty to cut wherever I pleased, but here again I was misstaken, for the liberal manner I had paid the first pretended pr[o]prietors brought more upon me and there was not a blade of grass that had not a seperated owner." Cook realized that he had "emptied my pockets with purchasing, and when they found I had nothing more to give they let us cut where ever we pleased."[21] In Cook's estimation, the exchange concluded amiably.

John Ledyard, writing after the voyage (although, more to the point, he was writing after the American War of Independence and as a proto-entrepreneur hoping to raise capital to compete with British fur traders), regaled readers with a different version of this incident. His story pricked Cook's imperiousness while simultaneously burnishing his new nation's identity. According to Ledyard, Cook was shocked when he learned of the demand for payment "and went in person to be assured of it, and persisting in a more peremptory tone in his demands, one of the Indians took him by the arm and thrust him from him pointing the way from him to go about his business." Cook was reportedly struck with astonishment at this turn and exclaimed, "with a smile mixed with admiration ... 'This is an American indeed!' and instantly offered this brave man what he thought proper to take."[22]

This purple passage does not overtly contradict Cook's account, but it certainly conveys a different tone. The image of Cook being manhandled (an incident no other journal-keeper mentions) stands out. It may have been an embellishment otherwise consistent with Ledyard's postwar strategy, which sought to enhance both a personal and an American fur-trade "brand," to employ modern terminology. For Cook, this North American "grass cove" incident prompted a salient ethnographic observation: "I have no w[h]ere met with Indians who had such high notions of every thing the Country produced being their exclusive property as these; the very wood and water we took on board they at first wanted us to pay for." What he had not allowed for, Gascoigne points out, was that the Mowachaht's "long experience of trade with other groups along the Pacific north-west coast meant that they generally drove much harder bargains than the Polynesians."[23] It also shows that Cook assumed he could buy his way out of any predicament.

By late April 1778, Cook was eager to sail into the true zone of discovery. When the ships were towed away from their moorings, he stated: "Our Friends the Indians attended upon us till we were almost out of the Sound." The Mowachaht encouraged him to return and promised to "lay in a good stock of skins" in anticipation of his reappearance. Cook reported that he did not have "the least doubt but they will."[24] They never saw him again. But following the epiphany in China and James King's treatise in the Admiralty's account on how these "skins" might form the basis of a profitable commerce, Euro-American traders rushed to Nootka Sound. None of them behaved anywhere near as well as Cook and his men did.

Leaving Nootka Sound, Cook observed: "When ever it rained with us Snow fell on the Neighbouring hills, the Clemate is however infinately milder than on the East coast of America under the same parallel of latitude," calculated to 49° 33' N, which was equivalent to Newfoundland. The temperature never got below 42 °F, even overnight, and often approached 60 °F in the daytime. There was no frost in the morning, and grasses never stopped growing. King complemented his earlier observation about the region's comparatively mild weather by pointing to "the dress of the inhabitants, their houses etc.," which proved "this part of America is situated in a temperate climate, fit for cultivation."[25] These

observations can only have reinforced hopes about the prospects for a
Northwest Passage at higher latitudes since midsummer was still three
months off.

⚓

After the ships put to sea for the discovery phase of the expedition, the
number and specificity of Cook's latitude and longitude readings and
directional bearings increased, approximating his pattern from the first
two voyages. Jabbing at French cartographers, he steered "NW the direction I supposed the Coast to take." Crossing the fiftieth parallel, Cook was
intrigued about getting a glimpse of land because he was passing "where
Geographers have placed the pretended Strait of Admiral de Fonte." He
lent "no credet to such vague and improbable stories, that carry their own
confutation along with them[,] nevertheless I was very desirous of keeping
the Coast aboard in order to clear up this point beyond dispute." Cook
never got close enough to demolish this story, which was so preposterous
that it strains credulity to think that some learned men in the age of supposed Enlightenment put stock in it. He allowed that "it would have been
highly imprudent in me to have ingaged with the land in such exceeding
tempestious weather, or to have lost the advantage of a fair wind by waiting
for better weather." This is the mission-centric Cook of the third voyage
at work, a man who refused to lose time chasing exotica, pursuant to his
instructions.[26] With favourable winds pushing him toward 65° N, he played
well offshore outside the Queen Charlotte Islands – now known as Haida
Gwaii – but in so doing he inadvertently helped perpetuate the Fonte
legend for another decade and a half.

   Angling to once again encounter the continent, Cook traced the 135th
meridian northward. On May 1, at 55° 20' N, the coastline returned to
view for the first time in six days and six degrees of latitude. At this rate,
Cook would have hit 65° N by mid-May, nicely ahead of schedule. The
ships slipped by a passel of small islands at 57° N that guarded the approach
to "a large bay" (Sitka Sound) adjacent to which Cook saw "a round elevated Mountain." Mount Edgecumbe was the first place name he lent to
the geography of Alaska, honouring a British admiral. Only 3,201 feet
high, this extinct volcano's singular placement on a coastal promontory
makes it a landmark at sea that can be seen from many miles distant. John
Gore was captivated by the mountain's presentation, calling it "the most
Rimarkable" he had seen on America's Pacific slope. He called it "Mount
Beautiful." Cook observed in more restrained tones that the coast range
was "of a considerable height" but that Edgecumbe "far out tops all the

rest."[27] Scrutinizing the shore, Cook correctly inferred that the alignment of bays filled with islets and adjacent channels formed the western outline of large islands. He detected Cross Sound (a name again derived from the liturgical calendar), which is the oceanic gateway to Icy Strait, Glacier Bay, Lynn Canal, and other inlets that form Alaska's Inside Passage.

On May 3, a high "Peaked Mountain" came into the northeast view. Vitus Bering had recorded an eminence in this vicinity, thus Cook had finally reached the southern edge of the voyage's discovery zone. He named it "Mount Fair weather," the originating term for the Fairweather Range, which hosts La Perouse and other glaciers on its Pacific face. Cook saw terrain "wholy covered with Snow from the highest summit down to the Sea Coast," but he failed to make the connection between these glaciers and those he had seen rimming South Georgia and the South Sandwich Islands.[28] (He was viewing this boreal shore from a greater distance, plus the glaciers that calve icebergs originate on the Glacier Bay side of the range, not the ocean-facing one.) The salubrious climate that the name "Mount Fairweather" denotes starkly contrasted with the bone-chilling circumstances Cook had experienced near the icy South Atlantic islands virtually antipodal to his current position.

The good weather held. On May 4, Cook spied "the summit of an elevated Mountain which we supposed to be Mt. St. Elias" bearing north. Far removed from the tracks of Tasman, Bougainville, and Wallis, Cook was now clearly following in the wake of Bering, the only explorer who had preceded him here and the mountain's namer. Advancing slowly

▼ Matthew Paul, *The Peaked Hill as Seen on the 7th of May, 1778.* When Cook entered the waters adjoining Southeast Alaska, the discovery phase of the voyage began in earnest. Cook named the mountains in this view the Fairweather Range, reflecting the surprising but nonetheless salubrious climate found seasonally in these latitudes.

because of light wind (the start of a recurring problem), *Resolution* and *Discovery* reached the fifty-ninth parallel on May 6. Just east of the 140th meridian, Cook spotted a bay to starboard. He and King poured over Russian charts trying to reckon what they could see with what had been reported previously. They theorized that the inlet (Yakutat Bay) was where "Commodore Behring" had anchored. Historians now believe Bering's harbourage was Controller Bay, closer to the Copper River delta.[29]

Ominously, Cook found "the Coast to trend very much to the west inclining hardly any thing to the North." Compounding matters, what little wind availed was an opposing westerly. Contrary to what French cartographers and Staehlin had speculated, the continental shore was not providing a direct route to 65° N and an easy turn east. Instead, it was taking Cook farther away from Baffin Bay. In this respect, the coastline more closely resembled Müller's projection, which preceded Staehlin's. Massive Mount Saint Elias was a particular impediment because, Cook reported, it was attached "to a ridge of exceeding high Mountains" extending far to the west. Laying down the directional trend of the Northwest Coast would be one of Cook's many accomplishments on this voyage, but in the moment everyone was discouraged by this literal turn of events. Samwell was representative when he observed that the trajectory of the continent "reduces our Hopes of finding a Northwest passage to a low ebb."[30]

When a fair wind returned, Cook soon raised land "to the Southward of west." This was not propitious. The continental hemline did not run north, nor even west, but began to curve back toward the equator. Near the entrance to what was later named Prince William Sound, Cook tried to make sense of the situation. The coast's southwesterly inclination was "so contrary to the Modern Charts, founded upon the late Russian discoveries, that we had reason to expect that by the inlet before us we

should find a passage to the North, and that the land to the West and SW was nothing but a group of islands."[31] This was a candid, overly optimistic, and ultimately mistaken forecast about the short-term possibility of turning the continent's western flank.

Cook was at 60° N, proximate to where he had been directed to start looking for the Northwest Passage. The term "Modern Charts" was an explicit reference to Staehlin's cartographic image, whose reliability Cook was beginning to test, much like he had Tasman's and Vaugondy's on his first voyage and Bouvet's and Dalrymple's on his second. Cook was inclined toward the opening to the north because the only alternative was to sail to the southwest. He laid out this intention in his journal, and although subsequent events would reveal this decision to be unproductive, to his credit he made no attempt to edit or back sight his entries. We get a sense of the tactical dynamic on deck from King. He describes having Staehlin's map "constantly in our hands, expecting every opening to the N[orth]ward will afford us an opportunity to seperate the Continent, and enable us to reach the 65° of Lat[itude] when we understand we are to examine accurately every inlet." Here, Cook's emphasis on the mission's directive is palpable. King concludes: "We are kept in a constant suspense, every new point of land rising to the S[outhward] damps our hopes till they are reviv'd by some fresh openings to the N[orth]ward."[32]

⚓

Cook steered into Prince William Sound on May 12, 1778. Samwell noted: "Our Hopes of a Passage were somewhat revived, especially as the entrance here is wide and it had at first a very promising appearance." Similarly, Ledyard stated that the ships pursued a "course up the inlet not without hopes of the dear Passage, which was now the only theme."[33] That may have been how the corporal saw it, but for Cook it had always been the theme.

As the ships nosed into the inlet, a brief and friendly encounter occurred with some Chugach people, a subset of the Sugpiaq/Alutiiq, whose territory encompassed the Gulf of Alaska. Cook had crossed a distinct ethnic boundary between the cultures of the Northwest Coast and the southeastern extension of the Inuit world. The meeting surprised the Englishmen, and probably the Chugach too. According to Samwell, because the adjoining landscape was "almost entirely covered with Snow we did not expect to find it inhabited." Also intriguing were the blue and green beads the Chugach had in their possession. "How they got them," Samwell wrote, "was the Subject about which many Conjectures were formed." It did not seem likely "that they had got them from the Southward along the Coast" because the expedition had not seen beads at Nootka. Instead, this "afforded a plausibility of a Passage by which they might have come from the other side of the Continent." Samwell admitted this "corresponded more with our wishes than our Hopes, as there was still another Way for them to come and that the most likely, which was from the Russians of Kamtschatka who we had reason to suppose traded with the Indians along the American Coast for their Furs." This, of course, was the true source. Some sea-otter skins were acquired, part of the stock later sold for a tidy profit when the expedition reached Asia. Following Nootka, this was the second cross-cultural encounter in North America where nothing untoward happened.[34]

Cook then engaged in a problematic search for a protected anchorage to repair *Resolution,* which was leaking. According to Clerke, foggy and hazy weather rendered attempts "to delineate matters here, almost abortive." Land to the north had "the appearance of Islands," a promising turn since an insular pattern by definition would allow for a route "to get through to the N[orth]ward." The alternative, Clerke pondered, was that "if we find ourselves under the Necessity to go as far to the W[est]ward and S[outh]ward again ... it will be a most unhappy Retardation of that business we are now so anxious to forward." It was only mid-May, but Clerke's text suggests the seasonal timeline was beginning to make the expedition leadership anxious. Formerly well ahead of schedule, the concern now was that the expedition might soon be running late, especially since the ships were still four degrees shy of where they had expected to meet the Northwest Passage.[35]

On May 13, Cook found what he called Snug Corner Bay, now known as Snug Corner Cove, on the east side of Prince William Sound, south of modern Valdez, Alaska. Heeling *Resolution* to the port side, the carpenters were able to take off the copper sheathing and make repairs to the outer planking. The lull gave Cook the opportunity to study the Chugach and their material culture. His library contained David Crantz's *The History of Greenland.* Cook had regularly cited the book during the Antarctic voyage in regard to sea-ice formation, and it was now open before him again. From its illustrations, Cook determined that the canoes abounding in Prince William Sound "were built and constructed in the same manner" as Indigenous craft at comparable latitudes on the Atlantic side of the continent. He also used this tome to make comparisons on dress, body adornment, and weaponry, which, he concluded, had been "very accurately discribed by Crantz."[36]

A View of Snug Corner Cove in Prince William's Sound. Engraving by W. Ellis, after a painting by John Webber, 1783–84. Prince William Sound was perceived as the first opportunity to explore a route to a possible Northwest Passage. As part of an eight-day survey of this inlet, Cook moved the ships into Snug Corner Cove (in today's Port Fidalgo) for repairs. Small-boat parties ventured into the sound's many arms, all of which were enclosed by the Chugach Mountains, precluding the possibility of an exit leading toward Baffin Bay.

Cook had previously differentiated between Polynesian and Melanesian peoples in the South Pacific and now, as a proto-anthropologist, he noted that the inhabitants of Prince William Sound were "not of the same Nation as those who Inhabit [Nootka] Sound, both their language and features are wide[e]ly different." Indeed, to Cook, this northern people seemed to bear "some affinity" to Crantz's description of "the Greenlander," and he was intrigued by the similarities he saw in print. Cook reasonably concluded that he would have more encounters and could "reserve the discussion of this point to some other time."[37]

Cook's instructions, drawing on published accounts of Russian experiences in this zone, told him that he was likely to encounter "the same Race of People ... as the Esquimaux" inhabiting Hudson Bay, but Cook nonetheless verified the existence of a great Indigenous nation in the North equivalent to the one spread across Oceania.[38] Their existence also seemed to imply, in theory, an ability to travel from one side of North America to the other, presumably by boats of some kind, through the same passage he was seeking. This was an illusion, at least until the modern Anthropocene era, but it is vintage Cook: still studious, thoughtful, and open to ideas.

By May 17, the leak in *Resolution* was fixed, and Cook steered northwest, thinking if there "was any passage to the North through this inlet it would be in that direction." His course ran toward Valdez Arm. Tides were detected. To Cook, "this did not make wholly against a passage, it was however nothing in its favour." The next day, Cook, who by this point could read a landscape from a seaborne vantage as well as anyone ever had or would, came to a preliminary verdict. With a "distinct sight of all the land round us, particularly to the Northward ... the land seemed to close and left us with little hopes of finding any passage that way, or indeed in any other direction." This implied a remote chance he was wrong, so to "form a better judgement" and to remove all doubt from his expectant colleagues, Cook sent William Bligh and Gore, his most optimistic officer, in small boats to explore the upper reaches of the Valdez and Fidalgo arms.[39] (Years later, Vancouver would name a feature in the former Bligh shoal the bank that was the site of the infamous *Exxon Valdez* oil-tanker wreck in 1989.)

When Gore and Bligh returned, they disagreed on whether there was a passage north. True to form, Cook reported that Gore believed an opening extended "a long way to the NE by which a passage might probably be found." Gore was contradicted by Henry Roberts, *Resolution*'s master's mate and a member of the subexpedition, who told King that "the inlet was so narrow as to leave no doubt of its ending in a river or small cove." Bligh concluded that the arm he had explored merely "communicated with the one we last came from." Cook had seen and heard enough and cut his losses: "I resolved to spend no more time in searching for a passage in a place that promised so little success." King recorded him saying: "We shoud soon come to more Passages that woud be less equivocal than any we had seen in this Sound."[40]

Cook's strategic calculus was accentuated by the relative lateness of the exploration season. With an eye on his instructions, which advised reaching 65° N by June (now less than a fortnight away), he could not afford to follow Gore into exhausting all the recourses that Prince William Sound availed. There had to be a more expeditious route. Cook reasoned that if the land to the west seen before entering the sound formed a part of the "late Russian descoveries" (code for Staehlin's "New Northern Archipelago"), "we could not fail of geting far enough to the north and that in good time, provided it was not spent in searching places where a passage was not only doubtfull but improbable." The fingers of Prince William Sound fit this definition. Still several degrees shy of his presumed eastward turn at 65° N, Cook determined he was "upwards of 520 leagues [1,560 miles] to the Westward of any part of Baffins or Hudsons bays." He reasoned that any passage "or some part of it must lie to the North of 72° latitude" because "who could expect to find a passage or strait of such extent?"[41] This truly was inconceivable other than in the fantasies of French geographers or Arthur Dobbs. Clearly drawing on his understanding of Samuel Hearne's supposed northern terminus at 72° N, Cook dismissed the idea of a strait running over a thousand miles from Prince William Sound to the Atlantic outlets. (Hudson Bay was actually 1,700 miles distant, and Baffin Bay was 2,000 miles away.) Such a strait would have been *sui generis,* one of the wonders of the world.

. The Prince William Sound foray served several important functions in Cook's third voyage. First, it showed that the geography of *terra borealis* was far more intricate than any map had conveyed before. The inlet was a disappointment since it did not have an exit to the north, but the experience taught Cook a valuable lesson, and he would put that knowledge into effect when the opening named for him appeared shortly thereafter. Cook entered Prince William Sound because his ship needed repairs and because the inlet seemed to provide an opening northward. At the next inlet, he was immediately skeptical about its prospects. The only point of navigational import on this voyage that would be more disappointing than Prince William Sound would be the Arctic ice pack when the expedition hit it later that summer.

By May 20, 1778, *Resolution* and *Discovery* were back in the Gulf of Alaska. The earlier projection of the continental shore bore out. Cook found the coast trending west by south "as far as the eye could reach." What we know as the Kenai Peninsula filled the starboard view, but the next day only a distant promontory was discerned at its southwestern point beyond which no land could be seen. Once again, Cook wrote: "We were in hopes it was the western extremity of the Coast," meaning the tip of North America. This promise was short-lived because he soon saw land bearing "WSW."[42] Cook was detecting the base of the Alaska Peninsula, which, with its Aleutian Island extension, stretches toward Kamchatka. From a distance, it seemed to be connected to Kenai, but before he could sail in that direction to make that determination, a gale blew the ships offshore. After it abated, he stood back for Kenai's tip and from there ran a track across the opening to Cook Inlet.

Before entering, Cook saw still more land "extending to the southward" (Afognak and Kodiak Islands, though they had not yet been distinguished as insular in nature). This terrain was more worthy of Cook's attention because "on it was seen a ridge of Mountains covered with Snow." That was troubling enough, but Cook's immediate difficulty was reconciling what he could see from the deck versus "the account of Behrings Voyage and the Chart that accompanies it in the English eddition." This was a reference to the translation of Staehlin's publication by Royal Society secretary Matthew Maty. Cook derided the chart as "extremly inaccurate." King said he found it tiresome checking Russian cartography "with the realities" of the North Pacific. Regardless, the ships penetrated the apparent opening. Framing his outlook, Cook hoped for "a passage Northward without being obliged to proceed any fa[r]ther to the South."[43]

Elaborating on his thoughts about this northerly course, so soon after exiting Prince William Sound, King noted "there is yet a great space to the NW where we have seen no land." From that circumstance sprang the enchanting prospect "of breaking thro this (as we hope) imaginary continent, although the high Snowy land to the S[outh]ward [i.e., Kodiak Island] damps one's expectations." Taking his cue, the enthusiastic Gore called this gateway Hopes Return. Clerke reflected Cook's attitude, fearing this "spacious opening" might prove yet another "extensive Sound, and after searching its various crooks and corners, find ourselves under the necessity of returning, from whence we came."[44] This proved prophetic.

⚓

On May 25, Cook took his ships into the arm of the sea later named after him. When he passed by the sentry-like Barren Islands at its gate, Cook was two degrees south of his most northern latitude (61° N) in Valdez Arm. He saw the summits of "two exceeding high Mountains" to the west. He named this elevated promontory Cape Douglas in honour of the editor of his second voyage (and later the third). This cape forms the eastern anchor of the Alaska Peninsula. The next day, Cook sighted, off the port side, Mount Augustine, named after another saint's feast day. Gore, a romantic, titled this eminence "Mount Welcome."[45]

Was this the way around the "continent"? King wondered. Ledyard thought this new opening "gave us hopes again of a Passage." Alas, the high peaks of the Alaska Range to the west, formerly thought to be islands, were soon found to be the summits of a linked chain of mountains. Cook reported that they were "connected by lower land, which the haziness of the horizon had prevented us from seeing at a greater distance." Now at the zenith of his power to discern land and water, Cook knew for certain what Clerke had previously suspected. With the terrain west of the inlet "every where covered with Snow from the summits of the hills down to the very sea beach," Cook concluded it "had every other appearance of being part of a great Continent."[46] This opening would not take his ships around the western edge of North America.

Thus, as early as May 26, his second day in the inlet, Cook deduced it would not advance the mission. Not everyone agreed. His journal captures the shipboard dynamic: "I was fully persuaided that we should find no passage by this inlet and my persevering in it was more to satisfy other people than to confirm my own opinion."[47] The retrospective nature of writing travel literature provided all explorers with the option to always appear knowledgeable if not omniscient. Still, there is no reason to doubt Cook's truthfulness about giving some rein to his officers. Only two days earlier, he had made the decision to sail in when the only alternative was running farther southwest. Within the framework of orthodox interpretations of the voyage, including Cook's supposed arbitrary engagement with subordinates, his decision to spend a week probing this inlet, thereby allowing members of his staff to prove him wrong, shows his collegiality.

As *Resolution* and *Discovery* proceeded northeast, the inlet narrowed, as Cook would have expected. For a moment, no land was seen to the north, but whatever promise that held was overturned by a strong tide running counter to the ships' course, evidence of an enclosed body of water. Cook's suspicions were confirmed by sightings of seaweed and driftwood in water that was now "thick like that in Rivers," that is, infused with sediment. Gore was confused, stating that they were heading into "a Gulf, River or Streight" and hoping for the latter. Cook was not falling for it. Each side of the opening had "a ridge of Mountains rising one behind a nother." There were no gaps. The ships were headed into one of the "corners" Clerke had written about.[48]

On May 30, the expedition reached a narrowing between Kustatan, Alaska, on the mainland, and Kenai. Here, "a prodigious tide" ran. Indeed, Cook Inlet has some of the world's most formidable tidal action. Cook could not tell whether the water's "frightfull" agitation "was occasioned by the strength of the stream or the breaking of the waves against rocks or sands." The ships were anchored for testing. Thus far, the water had been "nearly as salt as that in the ocean," but analysis showed it was now "considerably fresher than any we had taken up lower down."[49] Cook was verifying objectively what his eyes told him subjectively – there was no Northwest Passage here.

Over the week, Cook had recorded his doubts in his journal, but he was now "convinced that we were in a large River and Not a Strait that would communicate with the Northern Seas." No other officer was holding out much hope either. Nevertheless, to remove all doubt, Cook pressed the ships still farther north with the next flood tide. This manoeuvre was, in microcosm, a replay of the grand strategy from his second voyage, when he criss-crossed the Southern Hemisphere while assuring himself there was no continent even if every other circumstance – currents, winds, and swells – already told him so. Having "proceeded so far I was desireous of having stronger proofs," Cook wrote.[50]

Reaching the vicinity of modern Anchorage, Alaska, Cook saw that the inlet split into two arms. This complication prompted him to anchor (thus the place name). On May 31, two boats were hoisted out for a reconnaissance led by the ship's master, Bligh. After Bligh departed, Cook tested the water again, finding that up to a foot below the surface it was "perfectly fresh." Deeper than that, it was quite saline. This was another among many "evident proofs of being in a great River; such as low shores, very thick and Muddy water, large trees and all manner of dirt and rubbish floating up and down with the tide." Cook never deigned to name this river; he hedged his bets, as explorers were wont to do when uncertain about what a particular geographic feature portended. He simply left a blank space in his journal after the word "river" to be filled in later.[51] (After the voyage, British officials named it Cook's River. Vancouver properly recast it as Cook Inlet in 1794 during his survey of the continental coastline.) Cook Inlet is fed by several small rivers and other streams, but none separately or together represent the "great" river Cook imagined. Melt from the adjoining glaciers infuses the inlet with most of its freshwater quotient, which is what misled Cook.

At 2 a.m. on June 1, Bligh returned in the daylight of the boreal summer. He reported having ventured far into what Cook termed the "main channel," Knik Arm, all the way to Anchorage's port district. From there, Bligh could see the mountains from which two small rivers flowed. After learning this, Cook noted glibly: "All hopes of a passage was now given up." However, the ebb tide was nearly spent. Not wanting to exit against a flood, he took the opportunity "to get a nearer view of" the upper inlet's "Eastern branch." Cook weighed anchor at slack and drifted in with the tide, but then a breeze from the northeast, opposite his course, settled in. Cook despaired "reaching the entrance of the River we were plying up to before high-water," so King was lowered in command of two boats "to examine the tides, and make such other observations as might give

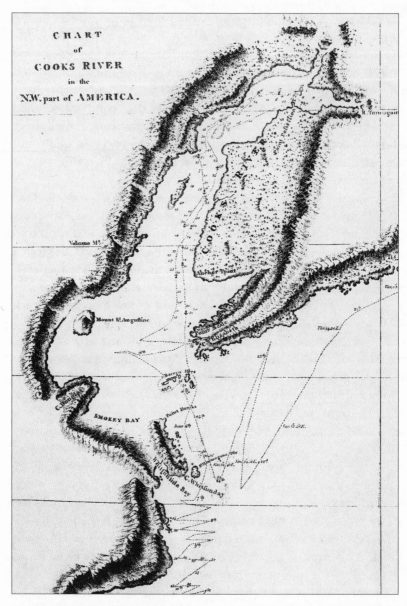

Chart of Cook's River in the N.W. Part of America. Prepared by Henry Roberts, 1783–84. Cook Inlet was, initially, another promising prospect for a Northwest Passage. As in Prince William Sound, boat parties were sent out to establish the head of navigation in each of the inlet's two arms. Cook, who was a great seafarer but not as capable as a river man, mistakenly concluded that the inlet was fed by a sizable stream, which, after the expedition, was named for him. After accounts of Cook's final voyage appeared in print, Canadian fur traders spent the next fifteen years fruitlessly looking for the headwaters of "Cook's River."

us some insight into the nature of the River." Cook thought this reconnaissance could be conducted expeditiously and thus at no further cost
to the timeline. But when the tide turned, the ebb became "too strong
for the boats to make head[way] against it." Accordingly, he signalled for
them to return "before they had got half way to the entrance of the River
they were sent to examine."[52]

The signal to King ended exploration of Cook Inlet. With the ebb, *Resolution* and *Discovery* turned to exit the waterway, a manoeuvre that prompted
the naming of Turnagain River, now Turnagain Arm. Cook thought it was
reasonable to suppose that both channels were "Navigable much farther
than we examined them." This line proved to be one of the more important in the history of the Northwest Passage as a cartographic image. After
the expedition, it would be argued that Cook had not exhausted the
possibility that this "river" flowed from deep into the continent. But at
the time, and unlike Prince William Sound, where Gore held out to the
end, everyone, according to King, was satisfied the inlet did not offer "a
passage to an open sea."[53]

Cook's decision to end the investigation was dictated by the calendar.
It was now June, the month he was supposed to reach 65° N and formally
commence searching for the passage. Cook was still short of that desid-
eratum (Anchorage is at 61° N) even presuming he had a clear path to
the target parallel, which at the time he did not. Thus, when the ebb tide
foreshortened King's exploration of Turnagain Arm, Cook wanted to catch
it so he could quickly exit from an inlet that his instincts had told him
early on was another dead end. Still, he left himself open to a thread of
criticism that ranks second only to his refusal to explore Fiji and Samoa.
Not having seen the head of either arm of the inlet, while concurrently
planting the mistaken idea that a "great river" emanated from the American interior, Cook inadvertently resuscitated the kind of geographical
hallucination that appealed to those who found the Fuca and Fonte stories
compelling. George Vancouver from the Pacific and Peter Pond and
Alexander Mackenzie overland from Montreal would spend most of the
next fifteen years looking for this nonexistent river. Within the context
of his mission, Cook's characterizations of the inlet were reasonable, but
he could not have anticipated the extrapolations others would draw from
his survey once the concept of the Northwest Passage evolved into new
modes of cartographic projection.

When Vancouver revisited the inlet in 1794, he opined that if Cook
had "dedicated one more day to its further examination, he would have
spared the theoretical navigators, who have followed him in their closets,

the task of ingeniously ascribing to this arm of the ocean a channel, through which a north-west passage existing according to their doctrines, might ultimately be discovered." Doing his best to defend "the great and first discoverer of it, whose name it bears," Vancouver unthinkingly contributed to the view that Cook was not his usual thorough self on the third voyage. Though Cook has been blamed for not foreclosing the later excesses of "theoretical navigators," surely he felt he had gone as far as necessary to satisfy his mission. Inexplicably, Frank McLynn states that "Cook claimed it was moral certainty that one of the two branches would lead into the sea at Hudson's Bay"[54] to justify spending so much time investigating a river. Others on board may have believed that, but the captain's journal never comes remotely close to suggesting any such thing.

Cook could not anticipate the specific interpretations others would lend to his exploration of the inlet, but he was not oblivious to the effort's cost:

> If the discovery of this River should prove of use, either to the present or future ages, the time spent in exploring it ought to be the less regreted, but to us who had a much greater object in View it was an essential loss; the season was advancing apace, we knew not how far we might have to proceed to the South and we were now convinced that the Continent extended farther to the west than from Modern [i.e., Staehlin's] Charts we had reason to expect and made a passage into Baffin or Hudson bays far less probable, or at least made it of greater extent.

This was an astute judgment that shows Cook on top of his game as an explorer, making solid risk-and-probability assessments. Cook also anticipated his role within the long history of the Northwest Passage, evidenced by his statement that if he "had not examined this place it would have been concluded, nay asserted that it communicated with the Sea to the North, or with one of these bays to the East."[55] Ironically, this happened anyway.

This whole sequence of events gnawed at Cook, as if he knew he had left himself open to future criticism. Now that it was clear that the ships had entered a cul-de-sac, he rued wasting the favourable wind that had blown the ships off their southwestern course near the Barren Islands on May 21–22. Had he persisted along the coast, that wind "would probably have carried us to [land's] extremity in that direction." Now he had to transect it anyway. In his log he regretted being induced, very much against my own opinion and judgment, to pursue the Course I did, as it was the opinion of some of the Officers that we should certainly find a passage to the North, and the late pretended Discoveries of the Russians tended to confirm it. Had we succeeded, a good deal of time would certainly have been saved but as we did not, nothing but a triffling point in Geography has been determined, and a River discovered that probably opens a very extensive communication with the Inland parts, and the climate seemed to be as favorable for a settlement as any part of the world under the same degree of latitude.[56]

Historians have found much to analyze in this paragraph. Notwithstanding Cook's refusal to pursue Fuca and Fonte, Glyn Williams scores him for a "credulous reliance" on Russian maps. Compounding matters, Cook was caught between this errant cartography and "the unshakeable optimism of Gore and others, and his own realization that the season was passing fast." McLynn also criticizes Cook for being swayed by "the enthusiasm of his officers," which he found explainable only by the allure of a £20,000 prize.[57] Whereas Cook has been deemed erratic for not having explored Fiji and Samoa when others urged it during the second voyage, in this case he is considered gullible for attending to similar wishes.

Beaglehole was surprised Cook considered any point of geography trifling, but the great navigator was deprecating the inlet's utility relative to the mission – the Northwest Passage – not its inherent qualities. This is borne out by his unprecedented and perspicacious encomium about the habitability of this zone. Today, Anchorage is by far the largest city in North America proximate to the sixty-first parallel, serving as an analogue to Oslo, Stockholm, and St. Petersburg. The sovereignty ritual Cook made King conduct at Possession Point across the bay from Anchorage was intended to preserve the potential for the region becoming a British possession, much like the one he had conducted in Australia near the end of *Endeavour*'s discovery phase. In the earlier instance, Cook's gambit was aimed at Holland's explorers, who had touched that continent's west coast, and Bougainville, who may have sailed nearby. In the North Pacific, for well over a month Cook had been in waters first charted, however sketchily, by Bering. If he had any intention of improving Britain's discovery claim to Drake's New Albion, then Nootka Sound might have been a more logical location for the rite. The fact that Cook thought he was at the mouth of a large river that drained an extensive portion of North America probably dictated his action. Not coincidentally, the chart of

"Cook's River" and the adjoining lands (see page 236) is as competently and minutely delineated as any in the Cook oeuvre – not bad for a navigator who had supposedly lost interest in exploration.[58]

As was the case with his encounters at Nootka and Prince William Sound, Cook's engagement with the Dena'ina, an Athabascan people inhabiting the inlet, was peaceably nondescript. The Dena'ina possessed iron, the Pacific basin's most precious commodity, which they made into weapons. Glass beads were another item "not of their own Manufacture." Cook inferred, correctly, that this currency was passed along by neighbouring peoples to the west who were in regular contact with traders. He ventured that "the Russians were never amongst these people, nor carry on any commerce with them, for if they did they would hardly be cloathed in such valuable skins as those of the Sea [otter]; the Russians would find some means or other to get them all from them."[59]

At this juncture, Cook had a minor epiphany. Seeing the otters that abounded, he thought "a very benificial fur trade might be carried on with the Inhabitants of this vast coast." We cannot be sure whether by "coast" Cook meant just the Gulf of Alaska or the Pacific slope as far south as Nootka. In any event, historians have read into this observation an intimation of the expedition's subsequent experience in China, imputing that he founded this commerce. But this is wrong because Cook quickly added that such a trade could only prove practicable if "a northern passage is found"; otherwise, "it seems rather too remote for Great Britain to receive any emolument from it."[60] Cook never envisaged the triangular trade between Europe, the Northwest Coast, and the south China ports that King would outline in the voyage account. Without a shortcut from Europe to the North Pacific, a Russian monopoly seemed inevitable to Cook, and in large measure this is how things turned out in Alaskan waters during the ensuing decades.

⚓

From June 2 to June 6, 1778, *Resolution* and *Discovery* backtracked out to sea. Little of any consequence occurred as they exited the inlet except for the sighting of Mount Redoubt, a volcano in an eruptive phase. This leg closed out a challenging four weeks for Cook, going back to May 12, when the opening to Prince William Sound appeared. This period is not held in high regard by most historians. Vanessa Collingridge is representative. She sees evidence of "indecision of whether to stay and explore, or just sail on." Though Cook offered his own sensible defence, King did him better. As the ships left the inlet, he wrote that it may have been "strange not to have tried all large Openings" and, further, he was certain it would be said "that every one we did leave unsearchd was a Passage." Indeed, Cook could have abandoned Staehlin's insular cartography and simply proceeded to Kamchatka and then north from there, but that "was not the Plan" laid out by the Admiralty. Besides, King continued, the American coast had been found "to have a tolerable regular direction to the NW to 60°." To have gone that far only to leave "unexamind all large openings would have been a very Extraordinary proceeding." For King, it would have been strange to have done other than what Cook did. The reason for his persevering up the inlet was "owing to some Officers not being so soon convinced, as I think the Captn himself was, of its being a river, and after all our pains to have left it a doubtful matter would have been Vexatious indeed."[61]

Nicholas Thomas offers a modern supplement to King's explanation. Cook was learning a new practicum of exploration in the North Pacific. To this point, the physical conditions on the third voyage had been less difficult than those experienced in the high southern latitudes on the second expedition, but the long-standing "logic" of Cook's discovery style was now inverted: "For a decade, he had honed his feeling for signs of land and signs of its absence. But now he looked not for land but for passages through land, which he had no concerted experience in seeking out."[62] In this sense, Cook, deep into his third voyage, was learning new skills along a complex and largely uncharted coast, alternately shrouded in fog or buffeted by gales that necessitated a course far from shore. That Cook would eventually find his way around the western edge of North America without missing any viable prospects for a Northwest Passage in the Pacific's high latitudes is testament to the vigour and ability he brought to his final expedition.

ELEVEN

Blink

By early June 1778, James Cook had reached the pivotal stage of the third voyage, a fact barely discernible in the Cook literature. Most authors treat his cruise along the Northwest Coast and into the Arctic Ocean as an afterthought. Even the most thorough treatments skimp on the mission-centric aspect of his final expedition. This is the palm-tree paradigm at work.[1]

As the ships left Cook Inlet, David Samwell wrote of being "disappointed a second time in our Hopes of finding a Passage." Cook was mute on this score, but rare is the explorer who trumpets his lack of success. (Meriwether Lewis, realizing he had not been on the main stem of the Columbia River after crossing the Continental Divide, stopped writing altogether.) James King also addressed the psychological import of what had just transpired, confessing "we have lost (to us) a deal of very precious time." Worse, he was afraid of ranging "along the Coast, which we are Apprehensive will stretch along to the SW." The practical consequence was "that by the time we get to the 65° Latitude, where I understand our Particular search for a Passage is to commence, the Season will be so far gone as to oblige us to leave a farther examination to another Year, in which case we shall have lost two Seasons; however it will be soon enough to lament when that happens." Few passages in journals kept during Cook's voyages are as forthright as this, or as prescient.[2]

⚓

On June 7, Cook was back in the Gulf of Alaska, heading southwest and looking for that elusive opening to the north and the sixty-fifth parallel. "The Weather was gloomy," he noted, in a passage that perhaps reflected his mood as well. Cook pondered whether the land to starboard was a contiguous extension of North America or a dense set of islands. Either way, it delayed their arrival in the discovery zone. Mist and haze often obscured the coast and, for the first time in the high northern latitudes, Cook noted "raw and cold" temperatures in his journal. The liturgical calendar inspired the naming of both Kodiak Island's Cape St. Barnabas on June 12 and Cape Trinity at the island's southern tip two days later. There was only one encouraging aspect reflected in Cook's prosaic descriptions of land and water – the hills were "but little incumbered with snow."[3]

Cook's predicament, as he searched the murk for an opening through or around "the continent," was summed up by Charles Clerke on June 16: "I hope and trust Providence will favour us with a little clear Weather: never had a set of fellows more need of it, here's such a Labrynth of rocks and Islés, that without a tolerable distinct vision, they will puzzle our accounts, confoundedly." Unlike Clerke's description, Cook's navigational text is dry and uninteresting. Running along the Shumagin chain, Cook saw so many islands that he left them "without names."[4]

At this juncture, *Discovery*'s cannon signalled *Resolution* to come to. Cook feared the consort had sprung a leak since no other exigency was evident. A small boat brought Clerke abreast. He reported that several Alutiiq canoes had trailed *Discovery*, rowing vigorously from off the Shumagins. They had induced his sailors to drop a rope so a message could be attached. Cook reported in his journal that one of these ambassadors had "fastened a small wood case or box" and then "droped astern and left." Little attention was paid to this box until "it was accidentally opened and found to contain a piece of paper." It was inscribed "in the Russian language, as was supposed, with the date 1778 prefixed to it." Another numeral seemed to refer to 1776, but that was "all that could be understood of it, as no one on board either ship could read it."[5]

Nothing like this had happened at any point in Cook's ten-year exploration of the Pacific. Clerke theorized that the note had been written by shipwrecked Russian sailors. Cook dismissed that idea. He reasoned that marooned seamen would have boarded the canoes themselves; they would not trust their rescue to intermediaries. Cook thought the note had been "left by some Russian trader who had lat[e]ly been in these parts, to be deliverd to the next that came, and that the Indians seeing us pass by and supposeing us to be Russians were determined to give us the note." (This was not far off. The forty or more Russian joint-stock trading companies operating in the Aleutians and points eastward between 1743 and 1800 regularly issued receipts to Alutiiq hunters indicating the number of pelts collected. When the expedition reached Kamchatka in 1779, after Cook had been killed, Lieutenant James Burney had this note translated and learned that it was "a receipt for Skins given as tribute." This circumstance surprised Russian officials there because they were unaware of operations "so far to the Eastward.") Cook did not want to be bothered. It was getting late in the exploration season, or was perceived to be, so he simply stood away "westward along the coast, or islands." This passage betrayed his uncertainty about whether the adjoining land "was continent" or not. *Discovery*'s Heinrich Zimmerman, a seaman who would rush an illicit account of the voyage into print prior to the Admiralty's edition, thought it was "amiss" that Cook did not tarry. John Ledyard also grumbled a bit. Their petty though public criticism of Cook contributed to the thesis that he was an unfeeling tyrant.[6]

On June 20, near the southwestern tip of the Alaska Peninsula and subcontinental mainland, Cook steered south to avoid the numerous rocks and breakers extending far offshore. This put the ships out of view of the opening to Ikatan Bay, near False Pass, Alaska, which could have led into the Bering Sea. Cook regretted being far from land, but the "thick weather" gave him no choice. Outside of the Sanak Islands, he saw "some hills whose elevated summits were seen towering above the clouds." Cook reckoned these heights were a part of what he called "the Main" in his shorthand lexicon. In fact, he was looking at Unimak Island, the largest unit in the Aleutians at the northeastern end of an archipelago that runs a thousand miles toward Asia. One eminence "continually threw out a vast column of black smoke." Known today as Shishaldin, this landmark was called "Smoaker" by John Gore. Cook, in his more laconic style, termed it simply "the volcano." Nevertheless, Cook appreciated that its ash rose to a great height before "spreading before the wind into a tail of vast length [making] a very picturesque" scene. In these waters, Cook's men also discovered Alaska's rich halibut fishery, which proved very "refreshing." The appearance of an Alutiiq hunter dressed in green cloth pants and a blue jacket was interpreted as evidence that the ships had entered the Russian zone.[7]

The "thick rainy weather" and highly variable winds stymied nautical discernment and made Cook apprehensive of "falling in with the land." On June 25, he detected a miles-wide opening at the southwestern end of Unimak Island. Cook headed toward the pass, but in a rare misreading he took it for an inconsequential inlet. Cook steered away, missing the first good and in Beaglehole's estimation "the best passage" through the eastern Aleutians. Following Beaglehole's lead, Frank McLynn suggests that this was further proof that "Cook had lost his touch" because the route was "the only way north."[8] In truth, with the recurrent fog and volcanic haze that lowered visibility, it would have been remarkable for Cook to have run a coast over ten degrees of longitude – a coast peppered with shoals, rocks, and islands – to then find the first sizable gap between the mainland and the Aleutian chain.

On June 26, Cook stated: "Day light availed us little as the Weather was so thick that we could not see a hundred yards before us." Since the wind was moderate, he proceeded on but soon "the Sound of breakers" was heard. Cook "immideately brought the ship to" in twenty-five fathoms of water, signalling *Discovery* to do the same. After waiting a few hours for the fog to burn off, Cook realized "we had [e]scaped very emminant danger." Off Sedanka Island, Cook wrote that "providence had conducted us through between these rocks where I should not have ventured in a clear day, and to such an anchoring place that I could not have chosen a better." This was a forthright admission of good fortune. He had avoided any penalty for the kind of navigational temerity of which he famously wrote after *Endeavour*'s brush with danger on the Great Barrier Reef. Other journal keepers also remarked on this miraculous escape. Clerke joked it was "very nice pilotage, considering our perfect Ignorance of our situation." Thomas Edgar, *Discovery*'s master, opined that had they continued on their course "a few Minutes longer both Ships must have been Ashore, however we had steer'd clear of all Danger and Anchor'd in such a good situation ... that we could not have done better had it been Day Light."[9]

Was Cook's zeal to find his way around land's end irrational, as has been charged? James Trevenen, who was not shy about criticizing him, addressed the question in his memoir:

The coolness and conciseness with which Captain Cook passes over the relation of dangers, the bare recollection of which makes every one else shudder with horror, is very remarkable. These imminent dangers and hairbreadth escapes in other hands would have afforded subjects for many labored and dreadful descriptions, and would even have justified them. The want of such may make him lose the credit of having avoided or surmounted them by his skill and prudence with readers who reflect not on causes and effects, or on natural and unavoidable consequences; but he who once revolves in his mind the immense extent of the coast that Captain Cook has in this voyage surveyed ... [and] the intricacies of the coast, the inlets, rocks, and shoals that would make, when well known, the boldest pilot tremble to venture on it.[10]

The near miss was providential in another way. When the fog cleared, Cook saw an inlet landward and nosed into it looking for greens that could be used in "either soups or as a Sallad." Entering, he found "land in every direction" and no hint of an exit. But when he proceeded around the northeast side of Unalaska Island, he saw a narrow opening heading northwest. It was Unalga Pass, which separates Unalaska from Unalga Island

and is one and three-quarter miles wide at its neck. Cook steered for this channel hoping he had found "a passage ... out to Sea," meaning a route to Bering Strait. With palpable relief, he recorded the supposition that the ships were "amongst islands and not an inlet of the Continent." The course from the Cook Inlet apex (at 61° N) had taken him to 54° N (farther south than modern Prince Rupert, British Columbia), but the Bering Sea beckoned at last.[11]

On June 28, the ships transected the gap between the islands. In what Cook termed a rapid tide, "Resolution got through before the Ebb made, but the Discovery did not, she was carried back [and] got into the race and had some trouble to get clear of it." According to William Bayly, this surge carried *Discovery* "round and round several times." Clerke told of being "confoundedly tumbled about for an hour or two." Once both ships passed through the narrows, Cook noticed that by looking northeast, as if toward Baffin Bay, "the Continent here took a remarkable turn in our favour." King, viewing this prospect, said everyone was "very sanguine in our hopes of meeting no more of Beerings Continent," meaning the northwestern extent of America. For the moment, Cook stood back into a harbour the ships had passed when they ran the strait. The Russians later named this haven English Bay; it was twelve miles east-northeast of the better-known Dutch Harbor.[12]

After anchoring, Cook was presented with another note written in Russian. It came from one of the neighbouring Unangax̂, the Indigenous people of the Aleutians whom he complimented for their friendly mien. As before, Cook could make no sense of this missive, so he "returned it to the bearer" along with a note identifying the ships' names, "their Commanders, and date, written in English and Latin." (As it turned out, this unproductive encounter provided the Admiralty with its first hint of Cook's whereabouts.) "Thick fogs and a Contrary Wind" detained Cook until July 2, 1778. When the weather cleared, the ships were "put to sea and steered to the North meeting with nothing to obstruct us in this course."[13] It was just ten days shy of two years since Cook had departed from England, but with the crux of his voyage finally at hand, all that had transpired up to this point was mere prologue to the prospect now before him.

⚓

After securing fresh water (but no firewood because Unalaska is barren), Cook left English Bay and headed into the Bering Sea. In the boreal summer daylight, there was no land ahead, but to starboard he observed Akutan Peak "covered with everlasting Snow." This mountain's insular character prompted acknowledgment that he had missed an earlier opening to the north. Nevertheless, Cook saw "that the Coast of the Continent took a NE direction and I ventured to steer the same Course." Cheeringly, Clerke stated: "We may find our way to the N[orth]ward, without any capital Impediment, and if we are fortunate in a mild Season, may still have time to look well about us."[14]

The ships had been driven so far southwest that this new track north of the Aleutians and the Alaska Peninsula essentially represented a reverse course. Drawing on previous experience from his continent-hunting voyages, Cook noted: "A great hollow swell from the WSW assured us there was no main land in that direction." It had been nearly a month since the expedition had gained a degree of latitude, but the voyage was finally on a productive course. Then land was seen joining from the northwest. Cook wrote: "Thus the fine prospect we had of geting to the North vanished in a moment." During the second voyage, Cook had sailed around the Southern Hemisphere at will. North America refused to cooperate. Cook was in an arm of what he would later name Bristol Bay.[15]

On July 10, Cook reversed course again, heading northwest with the Alaskan mainland to starboard. This was shallow, shoal-infested water, what Clerke termed a set of "ill-omen'd" discoveries. Compounding this difficulty, the ships were now again in murky weather. King reported that "we could not see the Discovery, yet we could hear her motion thro the water." Cook wrote of plying the coast of "the continent" in "so thick a fog that we could not see the length of the Ship." This gloomy business was broken only by jumping salmon and an occasional rumble of thunder. Cook tried "to get forward by anchoring when the tide was against us, yet we hardly advanced." On July 15, Clerke bemoaned a "lame 24 Hours [of] business, in respect to the purpose of our Voyage." In this predicament, Cook employed a tactic common to terrestrial explorers. Lieutenant John Williamson was sent ashore to climb "an elevated promontory" to "see what direction the coast took." It bore to the north, which was promising. Williamson had stood on a headland that Cook, in a goodwill gesture, said the lieutenant could name after anyone he wished. Williamson named it Cape Newenham after an Irish nobleman, while Cook named the body of water Bristol Bay to honour an admiral.[16]

When the ships turned Cape Newenham, they found little wind. Accordingly, Cook reported that they "got but slowly to the Northward." After creeping into the shoals of Kuskokwim Bay, the expedition was forced to back out, literally, as Cook could not find a channel other than the one he had entered. With summer waning, he began to ponder the wisdom of tracing the Alaskan coastline instead of heading out into deeper water to expedite travel northward (like he had run the Northwest Coast in April). This tactic was "attended with vast risks and if we should not suceed much time would be spent which we could ill spare." Cook feared missing an opening that pointed toward Baffin Bay. On the other hand, he found it encouraging that the ships had nearly returned to 60° N and the inland heights were still "free from Snow."[17] This was a far cry from the icy mansions of South Georgia and Southern Thule that he had found in comparable southern latitudes.

Before laboriously exiting from Kuskokwim Bay, a quick trade was conducted with Yupik people. Unlike the Unangax on Unalaska Island, Cook remarked that they appeared "wholy unacquainted with people like us." Since Prince William Sound his ethnographic observations had been sparse, and this was no exception. Detailed studies required a long stay at anchor. Referencing Yupik garb, Cook made a telling environmental point: their dress suggested "that they some times go naked, even in this high latitude."[18] This was the anthropology of hope, analogous to the more common geography of hope that informed so much of the Northwest Passage's history.

King called the forays into Bristol and Kuskokwim Bays the "third disappointment in our attempts to get to the N[orth]ward," following Prince William Sound and "Cook's River." Accordingly, Cook executed the strategy previously considered: he directed the ships 250 miles to the northwest into the heart of the Bering Sea and, perforce, out of sight of the North American continent. This sequence precipitated another discussion about the reliability of Russian cartography. King stated: "We can no longer frame any supposition in order to make our Charts agree with" Müller's map; he added that "Dr Mattys map is equally faulty." Matthew Maty was secretary of the Royal Society, and it was under his imprint that Staehlin's infamously inaccurate map had been republished before the voyage began. King concluded that "never was a Map so unlike what it ought to be."[19]

Fog beset the ships again on July 30 near St. Matthew Island at 60° N. Cook was unruffled; his journal entries are a dry recitation of bearings and minor observations about the weather and sea conditions. Clerke, however, called the situation "miserably ill calculated for our exploring business." He described the two ships foraging their way north, enveloped in grey mist, colourfully: "The Resolution just by us: we've been firing Guns, beating Drums, and ringing of Bells, a great part of these 24 hours, to understand each others situation." That was the bad news. But a northerly swell was "a fortunate omen," possibly signifying an open ocean ahead.

Driftwood and other flotsam suggested the ships were "in the Neighbour-hood of some River" or "Woodland Country, which wou'd now be a new Prospect to us." Beaglehole derided Clerke for overreading the situation since he was far offshore, which was true but beside the point.[20] The ex-pedition was now west of the Yukon River delta, and detecting the con-tinent's influences meant the expedition had not lost contact with it. Wooded terrain was auspicious because it portended the continuation of a temperate climate.

The mood was dampened on August 3, when William Anderson, *Resolution*'s surgeon and de facto ethnographer, died of consumption. Cook rarely invoked higher powers in his journal, but recognizing Anderson's congenial disposition and valuable skills, he wished "that it had pleased God to have spar'd his life [which] might have been usefull in the Course of the Voyage." When Bering's St. Lawrence Island came into view, Cook attempted to attach Anderson's name to it, but the honour did not endure.[21]

⚓

As Cook approached the Arctic Circle, his notes on course and distance became more detailed. Off the Seward Peninsula, he perceived land that he "supposed to be the Continent of America." This tentativeness reflected the navigational cost of abandoning the running survey of the coastline after escaping Kuskokwim Bay – four degrees of latitude without visual contact with the shore. Cook's decision in favour of an expedited route north had saved time, but the "considerable height" of the land to the northeast portended more delay in reaching a point where the turn toward Baffin Bay might commence. The silver lining was that these hills were still "free of snow," this at 63° N.[22]

On August 5, Cook landed at what he named Sledge Island. It was five miles off the mainland near present-day Nome, Alaska, and merely half a degree south of the long-wished-for 65° N. This was a high-latitude first for Cook, as he had never disembarked in the Southern Hemisphere poleward of 60° S. The island drew its name from the discovery of a sledge on shore. After consulting his library, Cook concluded that its design was similar to what "the Russians at Kamtschatka" had used to convey goods or people over ice and snow. Cook did not encounter anyone, but he sur-mised the sledge was of Indigenous, not Russian, design. He withheld comment on the climatic conditions that placed it five miles from the mainland. The island's inhabitants, the Iñuipiaq, obviously had dogs, like the people at Nootka, but unlike those living in the Aleutians. To the studious lieutenant, this offered "some proof" that the land opposite the island was "the Continent" (i.e., North America). The discovery of the sledge "also serves to prove that the Sea between this small Isl[an]d and the Main ... is frozen over in the Winter."[23]

Bearing northwest, Cook's ships approached the narrows dividing the continents of Asia and North America, though he was not yet aware of that fact. On August 9, the weather cleared, providing a better view of his surroundings, and to the east Cook spied "some high land" he supposed was a "continuation of the continent." He named it Cape Prince of Wales and estimated its location at 65° 46' N near the 191st meridian east of Greenwich. Its actual position is 65° 37' N and 168° 06' W (translated to 191° 54' east of Greenwich), a small error for a calculation at sea. Cook discerned the cape's geographic significance, noting it "is the more remarkable by being the Western extremity of all America hitherto known."[24]

A sudden wind carried the ships west, and in the evening light land was seen in that direction. According to Samwell, Cook had "run along the western Coast of America till he fell in with the Coast of Asia." The following morning, Clerke wrote of having "risen in a new World," *Resolution* and *Discovery* having emerged from a "succession of hazy, wet, disagreeable Weather" into the "clear Atmosphere and enlivening rays of the Sun." Pulling into St. Lawrence Bay on the Siberian coast, Cook landed with some marines and a few officers to encounter "40 or 50" Indigenous men armed with spears. He comprehended them as the Chukchi, a people Bering had previously described. They "were so polite as to take of[f] their Caps and make us a low bow: we returned the Compliment." Cook considered this a friendly encounter, and we know from impressions conveyed to Russian officials in Kamchatka (and in abbreviated form from there to St. Petersburg and London) that the feeling was mutual. In the Russian account, the Englishmen "addressed themselves politely" to the Chukchi, treated them to folding knives and snuff, and shared a meal of whale meat and onions. Cook, as proto-anthropologist, recognized their differences from "the Americans," meaning the Sugpiaq, Unangax, and Yupik encountered earlier in the summer. The conventional understanding of Cook's last voyage posits he was not up to his former standard, yet his 1,400-word description of Chukchi physiognomy, behaviour, and material culture compares well with anything he wrote while overlooking a sandy beach during his first two voyages.[25]

Cook had landed on the Asian mainland, but he was still under the influence of "Mr Staehlins Map," so it took him a few weeks to rule out the possibility that he had landed on "a part of the island of *Alaschka*."

John Webber, *Captain Cook Meeting with the Chukchi at St. Lawrence Bay*, 1778. This Webber image is an excellent example of the expedition's artist capturing (as in a photograph) a key event described concurrently in journal text. Ironically, Cook's encounter with the Chukchi in the summer of 1778, after being reported to St. Petersburg, provided his countrymen back in England in November 1779 with the first hint of his whereabouts since departing Cape Town in December 1776.

Staehlin and Maty had placed a large island by that name between Asia and North America proper, with spacious straits to either side accessing the Arctic Ocean. But Cook was reluctant "at first sight" to conclude he had made Asia because that meant Staehlin's cartography was irredeemably wrong or worse – "a mere fiction, a Sentance I had no right to pass upon ... without farther proof." Beaglehole did not know whether to admire Cook's geographic restraint or score his credulity since Staehlin's reliability had been questioned since May. The immediate problem, per Beaglehole, was that Staehlin's theory of "Alaschka's" insularity required two straits, and Cook had yet only spied one. In the end, Cook got it right. His log refers to the Chukchi landing site on "Alaschka," but the journal cites the "more probable" scenario of adjoining continents.[26]

On August 11, Cook steered east "in order to get nearer the America Coast." From this point, his intention was to keep North America to starboard until he reached Baffin Bay. Cook's reserved text for August 12–13, as the ships transited Bering Strait, obscures the enthusiasm building in others. King's entry for August 13 was expansive. Tiredly dismissing the

Staehlin-Maty confusion as a matter of "conjecture ... we cannot determine," he said the men were "in high spirits." The North American shore trended far to the northeast and the Asian coast to the northwest. To King, this spoke to the prospect of "an open sea to the N[orth]ward free of land, and we hope of Ice, as we have hitherto seen no signs of any."[27]

This was truly auspicious. Doubling America's most western point and entering the Pacific gateway to the Arctic, the expedition was headed northeast and, it was imagined, over the top of the continent toward Baffin Bay. As Samwell phrased it: "We are now in Bering's Straits which divide Asia from America as high as these two Continents are known." A passage lay ahead "round the Northern Extremity of America."[28] The ice-free sea contrasted with Cook's Antarctic voyage, where floes were routinely encountered at comparable latitudes. The excitement on deck must have been palpable, although one would not know this from reading Cook's laconic journal. He even failed to make mention of crossing the Arctic Circle (66° 33' N) on August 14. When heading toward the South Pole, he had celebrated his first transect of the Antarctic Circle, but polar zones were ordinary business for him now.

☍

The few days of August 14–16, 1778, were truly intoxicating aboard *Resolution* and *Discovery*. On August 14, in clearing weather and rapidly approaching the latitudes of Hearne's northern reach, Cook noticed land ten miles distant, Alaska's Cape Lisburne. The hills were still "free from snow." On August 15, as Cook kept the continent to starboard, "the sun shone out," and the astronomers "by observation" established their latitude at 68° 18' N. Only once before had he reached a higher latitude – during the second voyage's extensive run above the Antarctic Circle southwest of Tierra del Fuego. Now in the extreme North, Cook was sailing freely without any icebergs or loose ice floating about. On August 16, bearing northeast unimpeded, a euphoric King wrote: "All our Sanguine hopes begin to revive, and we already begin to compute the distance of our Situation from known parts of Baffins bay."[29]

Foggy weather occasionally hid the shore, but birds indicated that land was near. Cook then made a note that would foreshadow future events – his first passing mention of seeing "seahorses" or walruses. On the morning of August 17, he reported: "The weather was now tolerable clear in every direction, except to the Eastward, where lay a fog bank, which was the reason of our not seeing the land." He veered northeast looking for deeper water. Toward noon, "both the Sun and Moon shone out at intervals," which allowed the astronomers to conduct "flying observations" that yielded his latitude at 70° 33' N. Passing the seventieth parallel was a noteworthy accomplishment but was otherwise unremarked on. Instead, Cook noted in his journal this portentous observation, clearly retrospective in nature: "Some time before Noon we percieved a brigh[t]ness in the Northern horizon like that reflected from ice, commonly called the blink; it was little noticed from a supposition that it was improbable we should meet with ice *so soon.*" This explained "the sharpness of the air" for the preceding two or three days, which had been masked by the enchanting work of measuring the distance to Baffin Bay.[30]

Cook continued: "At 1 PM the sight of a large field of ice left us in no longer doubt about the cause of the brightness of the Horizon we had observed," meaning the blink. The generally fair weather of their months-long Alaskan cruise, and more recently the snowless hills above the Arctic Circle, had lulled Cook and his men into complacency. The officers seem to have expected open water all the way to Hearne's 72° N or, since Hudson in 1607 and Phipps in 1773 had reached 80° N, perhaps higher. How else to explain being surprised by meeting ice "so soon"? Based on his Antarctic voyage or his Newfoundland career, Cook might have expected icebergs or slushy field ice as harbingers of the main pack, but here there were none. This Arctic ice, from its first appearance, "was quite impenetrable and extend[ed] from WBS to EBN as far as the eye could reach." In tactical response, Cook temporized, plying about the edge of ice, keeping an eye on the walruses.[31]

Cook tried to stand south, but in the face of contrary winds the ships "gained nothing." At noon on August 18, his latitude was 70° 44' N, eleven minutes farther north than the previous day before the "blink" was observed. This proved to be the expedition's highest reach, but in the moment, Cook was more concerned about the ice. It was "as compact as a Wall and seemed to be ten or twelve feet high at least, but farther North it appeared much higher, its surface was extremely rugged." (At the North Pole today, the ice is only six feet thick.) As Cook intimated, flat, smooth ice is rare in the Arctic except in sections proximate to land. Pressure ridges and hummocks abound, formed after a lead opens thermodynamically or mechanically by wind and current, before refreezing causes compression. Unlike Antarctica, in the land-bound Arctic Basin, sea ice has nowhere to go when it is stressed by atmospheric conditions and piles on top of itself. In this situation, Cook saw "low land" to the southeast "3 or 4 miles distant." This promontory "was much incumbered with ice" and, accordingly, he named it "Icey Cape."[32] Having lent hundreds of

John Webber, *The* Resolution *Beating through the Ice with the* Discovery *in the Most Eminent Danger in the Distance,* 1778. Webber sketched Cook's ships at the edge of the Arctic ice pack in August 1778. Cook's encounter with the ice blink and the impervious pack was, short of his death in Hawaii, the most dramatic moment of the third voyage. The walruses huddled at the edge of the ice guided Cook with their incessant roaring.

names to places in the Southern Hemisphere and a few on the west coast of North America, this was his only Arctic toponym.

Icy Cape is above 70° N, a degree shy of the better-known Barrow, Alaska, 140 miles to the northeast. Cook was certain this was "a continuation of the Amirica cont[i]nent," but the larger issue was the four-way juxtaposition of the shore, his ships, the weather, and the ice – a situation that was "more and more critical." He explained: "We were in shoaled water upon a lee shore and the main body of the ice in sight to windward driving down upon us. It was evident, if we remained much longer between it and the land it would force us ashore unless it should happen to take the ground before us; ... the only direction that was open was to the SW."[33]

⚓

Had the series of manoeuvres Cook made over the next eleven days happened on either the first or second voyage, it would be the stuff of nautical legend, but since this occurred during the fatal third voyage, the sequence is little known. First, the ships backed away from both ice and land to avoid becoming marooned. However, the wind-driven ice was also moving, and the next day, August 19, Cook was again "close in with the edge of the main ice." Its consistency was now closer to what he had customarily seen at the Antarctic ice edge, for it was not as "compact as that which we had seen more to the Northward." Nevertheless, it was still "too close" to manoeuvre around, and the floes were too large "to force the ships through it."[34]

With the immediate prospect of finding a passage forestalled, Cook abruptly transitioned to making reflections in his journal on the prodigious number of walruses populating the ice edge. He had been relying on their barking to warn him about the dangerous proximity of ice, a charming aspect of the expedition. But his use of them to help feed the crew is, in standard renditions of the voyage, oft-cited as an example of an out-of-touch commander with no regard for his men. Vanessa Collingridge, for example, allows that Cook's recourse to walrus meat eventuated in a "near-mutiny."[35]

Cook had always believed in procuring local provender. On the second voyage, near Cape Horn, he had fur seals and sea lions slaughtered for their blubber and meat. They were found unpalatable, so penguins and cormorants went into the pots instead. Even that grew tiresome, and the men reverted to salted beef and pork. As for the walruses, Cook stated the seamen had been "feasting their eyes for some days" at the prospect of a novel repast, adding: "Nor would they have been disappointed now, or known the difference, if we had not happened to have one or two on board who had been in Greenland." These men declared no one there ate them. In Cook's estimation, there were "few" who did not prefer walrus to salted meat. Clerke agreed: walrus was "pleasant and good eating; and ... infinitely more nutritive and salutary than any salt Provision." *Discovery*'s commander reported having to "keep peace in the Galley" because many seamen were eager to commence "mastication" before it had been cooked.[36]

Walrus meat can be a delicacy, if cured properly, but the critics were more voluble than the defenders, and therein lay Cook's problem. Ledyard called walrus "an ill reward" for the labour that had brought them so far, and it was eaten by some "through mere vexation" to "consume it the sooner." George Gilbert called it "disgustfull" and said it had only been ingested "thro extreme hunger." James Trevenen, after noting that Cook's sense of taste was "the coarsest that ever mortal was endued with," admitted that the men had eaten the meat eagerly at first, "but that was only because they were rapaciously hungry, having been fed on nought but Salt Meat for several months past." They only continued to consume it because Cook "would let us have nothing else to eat, having put a stop to the usual allowance of Ships provisions." It was only after the discontent "rose to such complaints and murmurings," Trevenen tells us, that Cook "restored the salt meat."[37]

Scholars later seized on these critical comments to help frame the argument that Cook had lost his grip. But this interpretation is based on a selective parsing of what the journal keepers had to say about the episode. Ledyard said some "tars" would eat anything Cook did, echoing a view Cook himself had famously formulated on *Endeavour.* King extolled the brilliance of Cook's dietary plan, which directed that the crew "live upon the Produce of the parts he visit'd, whether it afforded only Grass or berries, or its shores Seals or Gulls, and to which he steadily adher'd and set himself an Example." Constancy was a necessity in this regard to countervail the crew's "prejudice and caprice." Cook's difficulty was the dietary custom of those sailors who had worked off Greenland, where walruses were considered "bad food." The resistance induced Cook to restore a "½ allowance of salt meat," and thereafter he "left it to the free choice to eat [walrus] or not." King was adamant that if Cook "at his table, and we at ours had nothing else, the whole would have been very happy at eating what was far preferable to any salt meat." Trevenen concurred. Walruses were "wholesome food, because tho' coarse and disagreeable they abounded with the nutricious juices totally wanting in the salt meat." This was Cook's motive, and Trevenen admitted he was "right" to "prevail on his crew to eat it," but disapproving historians have found the midshipman's modestly critical comments more conducive to their thesis.[38]

Whatever might be said about Cook's use of walrus meat for the crew's diet, he ably described them in their native state. His thoughtful reflections, complete with measurements and citation of published sources, are thorough and enchanting to read. He depicts them lying about "in herds of many hundred upon the ice, huddling one over the other like swine, and roar and bray very loud, so that in the night or foggy weather they gave us notice of the ice long before we could see it." There were always some "upon the watch" so that, "on the approach of the boat," the guards "would wake those next to them and these the others, so that the whole herd would be awake presently." Since the walruses were unfamiliar with gunfire, "they were seldom in a hurry to get away till after they had been once fire[d] at," whereupon they tumbled "one over the other into the

SEA HORSES.

Sea Horses. Engraving by J. Heath, engraving by E. Scott, after a painting by John Webber, 1783–84. Among his other contributions to navigational practice, Cook pioneered the use of local food resources, in this case walruses. Webber was probably in one of the boats sent to kill some of the animals. Walrus meat, however, evoked a debate on the quality of the fare being served to the crew. Walruses were also the subject of affective natural history descriptions by some of the men on the voyage.

sea in the utmost confusion." Cook argued that walrus were fiercer in appearance than temperament. "Vast numbers of them would follow and come close up to the boats" in their natural curiosity, and they were quick learners too. The "flash of a Musket in the pan, or even pointing one at them would send them down [into the water] in an instant." Female walruses "defend the young one to the very last ... whether in the Water or on the ice; nor will the young quit the dam though she be dead so that if you kill the one you are sure of the other. The Dam when in the Water holds the young one between her fore fins."[39]

On August 20–21, Cook stood southwest along the edge of the "main ice" in a thick fog "partly directed by the roaring" walruses. When the weather cleared, he could see "the Continent of America" about fifteen miles off his port side. Even though Cook had moved toward Cape Lisburne, a degree both south and west from where the blink had been detected, the ice pack "was at no great distance" from the ships. It seemed as if he were being chased by the ice because the pack "now covered a part of the sea which but a few days before was clear." Cook was not surprised by this phenomenon. Drawing on his experience with the mobility of the Antarctic ice pack, he explained: "It must not be understood I supposed any part of the ice we had seen fixed, on the contrary I am well assured that the whole was a moveable Mass." The inquisitive Yorkshire-man, now into his tenth year of exploration, conducted an experiment. When the wind stilled, William Bligh was dispatched in a small boat to see if he could detect any current but "found none." Cook did not expand on this test until late September, by which time he had returned to English Bay at Unalaska Island. There, he concluded that a land connection be-tween Asia and North America or its functional equivalent in the form of a contiguous ice pack running from one side of the Artic Basin to the other lay still farther north. Either would preclude the appearance of surface currents and dampen wave action. Modern studies of the Arctic have demonstrated that the ice shifts principally because of wind.[40]

On August 23, after the fog lifted, Cook "hauled to the Westward." Now for the first time, he recorded his thinking relative to the strategic import of running into the ice, as he thought of it, prematurely: "finding I could not get to the North near the [American] Coast for the ice, I re-solved to try what could be done at a distance from it."[41] This gambit was right out of his Antarctic playbook. Contrary to the common (mis)under-standing about Cook's supposedly flagging vigour during this voyage, what ensued was the boldest navigation of his career. A truly exhausted explorer, tired of discovery, would have simply turned for home after striking the ice. But not Cook, now more than two years into his third global exped-ition. Since he could not head directly east, he sailed west hoping to turn the flank of the pack which, according to reigning theory, was presumed to be fastened to the continent of North America. Get far enough from land, lose the ice.

King expanded on the theory behind this stratagem: "The higher we could get, the greater would be the Space, and possibly free'r of Ice, and if the Ice we have now seen should be said Yearly to be meltd away, and renew'd by the breaking up of the Ice in the Siberian rivers, and what we will allow in the Corresponding rivers on America – that on such a supposition by being early, before these rivers broke up, we might get much higher, and therefore in a more open and a more enlarg'd Sea." This was a compilation of several age-old notions: saltwater did not freeze; rivers were the major source of ice found in the Arctic; polar ice was an annual not a perennial phenomenon; landfast was the dominant factor in its location; and, above all, the sea might be open at the pole. King's ruminations were the last full expression of classic polar hydrology in the journals of Cook's voyages. Cook himself no longer fully subscribed to all of these conceptions, and his doubts would grow over the summer of 1778.

Indeed, in the postexpeditionary account, King stated: "Captain Cook, whose opinion respecting the formation of ice had formerly coincided with that of the theorists we are now controverting, found abundant reason, in the present voyage, for changing his sentiments."[42]

Resolution and *Discovery* moved steadily west, usually in fog or other adverse circumstances in which the ice was unseen and so, too, the other ship. Still, they never lost contact with each other. This was a remarkable navigational achievement and, in the estimation of historian Barry Gough, yet another display of Cook's "uncanny sense" for sailing through danger, a practical skill honed through many years' experience "surveying the Newfoundland shore ... in fog." Over the next three days, attempting to turn the ice's flank, Cook left the North American zone and sailed above eastern Siberia. Trevenen, describing Cook navigating amidst the ice in a constant fog, stated that one "must wonder at and admire the boldness of daring and skill in executing projects big with every danger; but as his mind was more impressed with the thought of duty, and the general consequences of his undertaking, no danger or difficulty had the power of turning away his thoughts from this object." Martin Dugard is the only modern historian to recognize Cook's accomplishment in these waters. He writes of his "swashbuckling confidence" sailing "back and forth between the North American and Asian landmasses, trying to find a gap in the ice." Two centuries apart, Trevenen and Dugard both argue against the now common view, as recently stated by Andrew Lambert, that Cook's greatest accomplishment was detecting tropical shorelines, "charting, making accessible the new land, and preparing the way for others."[43] No one would follow Cook and Clerke to these ice fields for another fifty years.

Cook's flanking manoeuvre was foiled because the Arctic ice pack is not principally a landfast phenomenon; it freezes on its own with sufficient cold and is mobile besides. On August 26, Cook reported dryly: "We made the ice," that is, it moved into the latitude (69° 36' N) he was running westward, almost a full parallel south of where he had first struck it off Icy Cape. After two hours of probing for an opening, he noted that the ice "extended from NW to EBN and appeared to be thick and compact." This surely was a daunting moment. Cook confided to his journal that "we had no better prospect of getting to the North here than nearer the shore," meaning the North American coast. A lesser man facing this obstacle would have terminated the effort. After all, he had run fourteen degrees of longitude westward after sighting the blink, trying get around the end of the ice. During his second voyage, he had never sailed laterally at the ice edge for so long a distance; he chose instead the tactic of quick

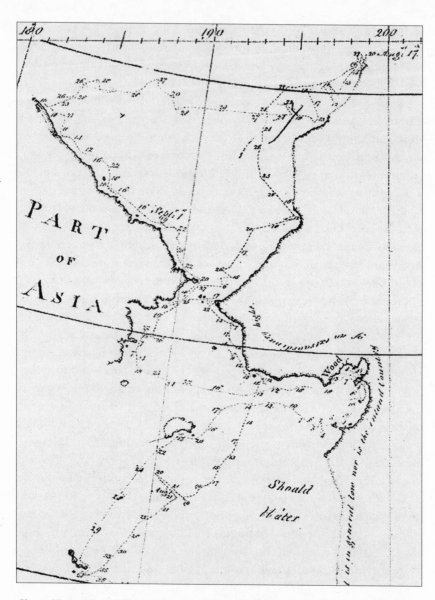

Chart of Part of the NW Coast of America Explored by Capt. J. Cook in 1778 (detail). At Unalaska in October 1778, Cook dispatched this chart (and related correspondence) to London, through Russian intermediaries, to give the Admiralty a sense of his whereabouts and accomplishments to date. This image accurately recorded the juxtaposition of Asia and North America for the first time, but it has the added value of showing the extent of Cook's vigorous search for a way through or around the Arctic ice pack. In contrast to his practice from the second voyage, here the great navigator sailed for many degrees of longitude along the edge of the ice.

needle-nose runs up the ladder of parallels toward the South Pole and then, after encountering the pack or exhausting his crew, heading down the register toward the temperate zone before trying again. But in the face of his Arctic foe, Cook stated: "I continued to stand to the Westward."[44]

Vanessa Collingridge alleges that, "to everyone except Cook, it was clear that their journey north was over." Rather than admitting failure "in the face of an old and hated enemy," the ice, and turning for home, Cook, with the obsessive "fixation of a madman," persisted. Frank McLynn also assails Cook's "unconscionable stubbornness" for not conceding that ice had beaten him. Modern critics have the benefit of normative understandings of the Earth's geography and the dynamics of polar hydrology. The irony, of course, is that this knowledge was secured through the efforts of an explorer, Cook, who refused to give up. Neither Collingridge nor McLynn volunteer just how much time should have been spent testing the ice edge. McLynn also describes the captain's "extreme disappointment" at this turn of events, yet Cook's journal clearly indicates that, if circumstances had allowed, he would have continued westward atop Russia, as his instructions countenanced, "in further search of a North East ... Passage, from the Pacific Ocean into the Atlantic Ocean, or the North Sea."[45] (In 1878–79, Finnish explorer Adolf Erik Nordensköld made the first confirmed passage of this route, having overwintered north of Bering Strait. The first in a single season occurred in 1932.)

After sailing a few more hours to the west, Cook discerned that the ships "were in a manner embayed by the ice, which appeared high and very close in the NW and NE quarters, with a great deal of heavy loose ice about the edge of the Main field." His log notes that the view from the masthead (implying he climbed there himself) revealed "it was a matter of doubt whether a ship could get through it or no, and I had no motive for making the experiment." This was vintage Cook: always pressing to the limit, always probing, always looking, but never willing to engage in undue risk. The situation might not have been as stark as the prospect of being run aground by the pack at Icy Cape, but the light, baffling breeze gave him little room to manoeuvre. Accordingly, Cook "stretched to the Eastward, being the only direction in which the sea was clear of ice."[46]

But Cook was not done. Early the next morning, August 27, he had the ships stand westward again. Having turned the previous day only to keep from getting embayed, he was still exploring for a way around the ice or perhaps over Russia to the Atlantic. But soon the ice, "which lay

ENE and WSW as far each way as the eye could reach," blocked his way again. With "little wind" in the sails, he "went with the boats to examine the state of the ice."[47] The spirit of Enlightenment inquiry still burned in Cook, who was at this point three hundred miles west of Bering Strait's centre line.

Here, he reopened his internalized dialogue about sea-ice formation, the last great passion of his life. Cook's reflections on Arctic ice are on par with his deliberations covering polar hydrology from the second voyage. The "loose pieces" on the perimeter of the pack, he wrote, were "so close together that I could hardly enter the outer edge with a boat." For the ships, the ice was impossible. After looking at ice in the way a navigator would, he adopted the outlook of the incipient ice scientist that he was. He noticed that "it was all pure transparent ice, except the upper surface which was a little porous." We know from modern science that Cook was examining desalinized multiyear ice, but the explanation he offered was a reapplication of his faulty Antarctic theory: "It appeared to be intirely composed of frozen Snow and had been all formed at sea, for setting [a]side the improbability or rather impossibility of such masses floating out of River[s] in which there is hardly Water for a boat, none of the production of the land was found incorporated, or fixed in it, which must unavoidably have be[en] the case had it been formed in Rivers either great or small."[48]

Cook's Arctic ice analysis was consistent with his major conclusions from the voyage toward the South Pole. There, he had concluded that (except for a theoretical Antarctica near the pole, comparatively small and perpetually frozen) the high southern latitudes were vacant of land (and therefore rivers) to convey ice to the sea. In the Arctic, as was the case in the Southern Sea, the ice was the result of serial accumulation from repeated snow falls.

Unlike Antarctica, however, the continents rimming the North Pole were largely a known quantity, especially the rivers flowing toward the Arctic from Eurasia. King observed that the thickness of the ice did not match "the Shallowness of the Siberian rivers." The Arctic's ice was "far too deep, even the detachd pieces" to have come from any river in Russia. Besides, King added, "we should have seen some signs of Earth, or some produce of the banks of the rivers upon some of the bodies." Instead, "it seem'd as if it had been froze, and upon its frozen surface great bodies of Snow had fallen, which also became froze and added to the Mass." King's journal reads like the byproduct of a deck-side tutorial from Cook, up to and including the captain's mistaken notion that ice found in salt-water was an accretion of frozen snow. Cook and King were wrong about this. Snowfall in the Arctic is comparatively light, usually around four to six inches a year, and it is dry, serving mostly as an insulating layer – but through scientific observation, the Englishmen were moving theory in the right direction.[49]

The largest floes that Cook saw were forty to fifty yards wide and had a base that "reach'd 30 feet or more under the surface of the Water." This remarkable sight forced Cook into the kind of prolonged reflection that routinely populated his account of the voyage toward Antarctica:

> It appeared to me, very improbable that this ice could be the produce of the preceding Winter alone, but rather that of a great many, or that the little that remained of the summer would distroy the tenth part of what now remained, sence the Sun had already exerted the full influence [of] his rays. But I am of opinion that the Sun contributes but little towards reducing these great Masses, for altho he is a considerable time above the horizon he seldom shines out for more than a few hours at a time, and some times is not seen for several days together. It is the Wind, or rather Waves raised by the wind, that brings down these great Masses by grinding one piece against another, by undermining and washing away that which lies exposed to the surge of the sea. This was evedent by the upper surface of many pieces being partly washed away while the base or under part remained firm for several fathoms round that which appeared above water exactly like a shoaled round an elevated rock. We measured the depth of Water upon one and found it to be 15 feet, so that the ships might have sailed over it. If I had not measured the depth of Water, I would not have beleived that there was a sufficient weight of ice above the surface to have sunk the other so much below it.

Cook may have been describing some landfast ice that had calved from the Eurasian shelf or perhaps a large floe that had capsized, forming a kind of iceberg, but the more significant point is that he took the time to measure it. He drew a sketch of the underwater platform that projected out from the floe to "ilustrate what I have said on the subject."[50] Regrettably, this image has not survived.

The ice Cook described may have eroded, but nothing in this text suggests his skills had. He underestimated the solar effect, but his main observation is now settled science: erosion plays a large role in the annual diminution of the polar ice caps. A little more than five months before he would die, his mind remained inquisitive and far from exhausted, at least in regard to a subject that still interested him. Indeed, Cook was as feisty as ever. On his first two voyages, the geographic myths that Alexander Dalrymple had propagated were skewered regularly. Here, deep into the third expedition, Cook cut loose with another cannonade: "Thus it may happen that more ice is distroyed in one Stormy Season, than is formed in several Winter[s] and an endless accumulation prevented, but that there is always a remaining store, none who has been upon the spot will deny and none but Closet studdying Philosiphers will dispute."[51]

This salvo was probably aimed at Daines Barrington, the armchair geographer and Northwest Passage promoter. It has been argued that Cook "was now certain" that Barrington's premise about an open polar sea was faulty but that he stopped short of conceding the impossibility of a Northwest Passage.[52] Indeed, Cook would never have enunciated a plan to return for a second season in the North if he believed his chances for success were nil. On his southern voyage, Cook and Forster had noted the mobility of ice over many degrees of latitude. Accordingly, it was conceivable that a follow-up exploration would reveal that the pack had receded from its envelopment of Icy Cape, allowing an opening toward Baffin Bay. Cook did not live to make that attempt, and he never developed an entirely correct glaciological theory. Nevertheless, he was without question the central figure in Enlightenment science's dawning awareness as to the naïveté of the old ice-formation thesis. The essential experimental observation was his: ice packs at both poles were more extensive than previously believed possible and unquestionably the product of many years' worth of cold temperatures and not a terrestrial emanation.

⚓

After reboarding *Resolution,* Cook "spent the night Standing off and on amongst the drift ice." The next day, August 28, the fog cleared, and boats were dispatched to hunt walruses. Cook reported that "by this time our people began to relish them," and they were eaten, presumably without complaint. A trick of the northern trade was applied at this juncture. Lieutenant James Burney on *Discovery* noted that if walruses were bled immediately after they were killed this "cured them of their rank fishy taste." This meat might not have supplanted fresh beef, but many seamen now thought it "preferable to the Ships salt meat."[53] Burney's commendation is not as well known in the interpretation of this voyage as are some crewmembers' negative comments about walrus meat from the week before.

The final reckoning for Cook's northern ice-edge survey came on August 29. The ice pack ran from the north toward "land bearing SWBW," only three miles away. *Resolution* and *Discovery* were being hemmed in by the fastening of the ice to the Asian mainland near Cape Shmidta, repeating the Icy Cape scenario. When the drizzling rain ceased, Cook stated that he had "a pretty good View of the Coast," which strongly resembled "that of America," meaning Alaska's Arctic coast: "Low land next the Sea with elevated land farther back." The landscape "was perfectly distitude of Wood and even Snow, but was probably covered with a Mossy substance that gave it a brownish cast," an apt description of the northern tundra. With an accuracy that would astonish Beaglehole, Cook established the latitude and longitude of various benchmarks in this vicinity, including a still more westerly headland that King would later name Cape North. Cook deduced, correctly, that beyond that point the coast "must take a very westerly direction, as we could see no land to the Northward of it though the horizon was there pretty clear."[54]

The irredeemably curious Cook desired "seeing more of the Coast to the westward." Finding the northernmost point of Siberia intrigued him. He tacked hoping to "weather" Cape North, "but finding we could not, the Wind freshening, a thick fog coming on with much snow, and being fearfull of the ice coming down on us, I gave up the design I had formed of plying to the Westward." First the path to Baffin Bay had been cut off and now, too, the incipient voyage to the North Sea. In his journal, Cook reflected: "The season was now so very far advanced and the time when the frost is expected to set in so near at hand, that I did not think it consistant with prudence to make any farther attempts to find a passage this year in any direction so little was the prospect of succeeding."[55]

Cook was clearly open to a full exploration of a North*east* Passage above Russia. He judiciously concluded that the 1778 season had closed, but he determined on retrying the following year. So much for his exploratory fatigue. On all of his expeditions, others tired of the effort long before the great navigator did. When Cook announced his decision to turn away, King reported that the crew, which had been "in the dread of being enclosed" in the ice, erupted in "the general joy that this news gave."[56]

Cook's run from Icy Cape to Cape Shmidta covered nearly twenty degrees of longitude, a sail of over four hundred miles along the ice edge. He gave the Arctic Ocean every chance to provide an opening to the north. This cruise was diligent and vigorous, but that is not the way some historians see it. Vanessa Collingridge, having chided Cook for staying at the edge of the ice as long as he did, mocks him for conducting a two-year voyage that spent "just three weeks in the Arctic Ocean and a mere eight days in sight of the American continent."[57] By deprecating Cook's effort and failing to credit his commitment to returning to the North, she provides yet one more example of jaundiced interpretations of his final voyage.

Glyn Williams treats the question of the timing of Cook's arrival in the Arctic and his durability more equitably. He argues that Cook's instructions were severely misguided in two respects: first, they hinted that a strait or passage of some sort would quickly get him to the Coppermine River, and second, they minimized the risk of getting icebound above North America. Cook's comment about striking the ice "so soon" substantiates this last point. Williams argues that by conceding to the desires of his officers and giving "the Russian maps every chance to prove their accuracy ... he arrived at the polar sea too late."[58]

Was Cook tardy? During his second voyage, he had always reached the icy latitudes in December, the start of the austral summer, which corresponded to June in the Northern Hemisphere. Coordinating polar exploration with the long days surrounding the summer solstice was an eminently sensible model to follow, but as the experience of many adventurers had indicated and modern science has proved, the best time to explore polar waters is not the end of spring but at the end of summer. In that sense, Cook got to the Arctic too early because the sea-ice minimum is in September, when the cumulative effects of solar radiation and wave erosion have forced the ice to recede.

Glaciologist Harry Stern has explicated the problem of Cook's Arctic timeline through several counterfactual scenarios that trace the progress Cook might have made under variable intervening sea-ice states. Based on the records of explorers and whalers who came later, and assuming that summer sea-ice conditions in 1778 "were no different from those of the nineteenth and twentieth centuries," Stern posits Cook had a "fifty-fifty chance of arriving at Icy Cape in an August that would permit a sailing ship to reach Point Barrow." Between 1979 (the commencement of passive microwave telemetry indicating sea-ice extent) and 1995, the route to Point Barrow on Cook's schedule was open eight years and closed nine. Since 1996, Cook's route has been closed once. But if Cook had reached Point Barrow, then what? Given sea-ice conditions since 2002, Stern thinks Cook could have reached the mouth of the Mackenzie River by early September and then entered the labyrinth of the Canadian Arctic Archipelago. From there, Stern concludes, "he might have retreated, knowing that he was still far from Baffin Bay, to avoid becoming locked in the ice for the winter. He would have rounded Point Barrow heading west in early October, about two to three weeks before the coastal waters east of Point Barrow typically freeze nowadays, confident that he had found a navigable western approach to the Northwest Passage."

Stern then considers whether Cook was fortunate to have been halted when and where he was in 1778 and hypothesizes: "Cook hit a relatively heavy ice year, but one that was not outside the range of natural variability as we know it from the 1850s to the mid-1990s." In that sense, Cook was unlucky. On the other hand: "If the ice had retreated away from the coast and Cook had proceeded northeast, he would have exposed his ships to the very real danger of becoming trapped or crushed in the ice in the event of an unfavourable wind shift." Thus, Stern avows, Cook "was lucky that the ice did not retreat and tempt him to go farther." He echoes Beaglehole's view that in a good ice year Cook might have been able to make it "towards where Samuel Hearne had stood, in the direction of the veritable Passage, and been lost forever."[59]

Cook had to postpone his discovery agenda in August 1778, but there was still much to think about. He wrote: "My attention was now directed towards finding out some place where we could Wood and Water, and in the considering how I should spend the Winter, so as to make some improvement to Geography and Navigation and at the same time be in a condition to return to the North in further search of a Passage the ensuing summer." Cook was keeping his options open, but Beaglehole, advancing the negative cast he lent to this voyage, said of this moment: "The ice had beaten him."[60]

Would a beaten man be planning a "return to the North?" Cook's comment shows him firmly committed to the mission with a strong appetite for discovery. The intermittently critical Trevenen phrased it best when he wrote: "indefatigability was a leading feature of his Character." Trevenen also astutely observed that if Cook "failed in, or could no longer pursue, his first great object, he immediately began to consider how he might be most useful in prosecuting some inferior one." In a concluding observation that might well serve as an epigram for the entire voyage, he said of Cook: "Action was life to him and repose a sort of death."[61]

Having reached the Ultima Thule of this northern season, Cook bore "eastward along the Coast" of Asia. Validating his intuition, soon after turning for Bering Strait, "the wind blew fresh with a very heavy fall of snow," which necessitated sailing "with great causion." The ships were "brought to" on the night and early morning of August 29–30. At daylight, guided only by a lead tossed overboard because of the "thick" weather, Cook coasted the Siberian shore. He noted that in the days since bending away at Cape Shmidta the temperature had "been very little above the freezing point and often below it." Even "the Water Vessels on deck were frequently covered with a sheet of ice." King, speaking for all in language indirectly commendatory of Cook, stated: "We are very glad that our backs turn'd upon this forerunner of a miserably cold season." This was written on August 30, fourteen days after he had calculated the distance to Baffin Bay.[62]

Coping with gales from the northeast and "attended with Showers of Snow and sleet," Cook struggled to stay off the lee shore, but his geographic confidence was growing. He was "now well assured that this was the Coast of Tchuktschi, or the NE coast of Asia, and that thus far Captain Behring proceeded in 1728." Cook had reached a stretch of Siberia's coastline that had been mapped to such a degree of accuracy that what he perceived on deck actually matched a charted image. Müller's map depicting the Chukchi Peninsula was imperfect but was workable to a degree, something Staehlin's supposedly more modern map never was.[63]

On September 2, "fair weather and Sun shine" enabled a reading of the sun's altitude and yielded a latitude of 66° 37' N, almost exactly the Arctic Circle, which the ships had crossed northbound twenty days earlier. This was by far Cook's longest stint in a polar zone. (His three sojourns across the Antarctic Circle lasted one, four, and eight days.) That day, *Resolution* and *Discovery* passed Siberia's eastern extremity, Cape Dezhneva. On August 11, Cook had preliminarily reckoned this the "East point" of Staehlin's fanciful "island of Alsascha." Now fully informed by his own experience, which surpassed anything the Staehlin-Maty map depicted, he declared with emphasis that this was the "Eastern Promontory of Asia." Cook had been frustrated by Russian inaccuracy, but he understood the dialectic of discovery and in that sense could afford to be patient. Transiting Bering Strait, he forsook sorting out the inconsistencies and vowed that, "as I hope to visit these parts again I shall leave the descussion ... untill then." In that same moment, Ledyard delighted in "the pleasure to see both continents at once," perhaps seeding his idea of walking around the world.[64]

Looking for a source of wood to fend off the cold, Cook found an anchorage on Russia's Chukchi Peninsula. He also inspected Lavrentiya Bay because he was "very desirous of finding a harbour in those parts where I could come to in the Spring." He desired a more proximate jumping off point for a second season of northern exploration than Unalaska but found it wanting. As Clerke phrased it, the landscape was "wretchedly barren" and "as destitute of good Things, as when upon the Ocean."[65]

Resolution and *Discovery* headed south, enjoying what Clerke termed "a fine Breeze and pleasant Weather, which render tracing a Coast an agreeable Amusement." The ships passed the opening to the Gulf of Anadyr, "into which," Cook reported, "I had no inducement to go." There were geographic enigmas in East Asia that Gore and King would attempt to investigate the following year, after both Cook and Clerke had died, but at this point Cook was more intrigued by unfinished business on the American side. Leaving the Siberian shore, he praised a fellow explorer: "In justice to *Behrings Memory*, I must say he has deleneated this Coast very well and fixed the latitude and longitude of the points better than could be expected from the Methods he had to go by." A churlish John Ledyard would later accuse Cook of being envious of Bering and claiming credit for Russian discoveries, but these allegations simply reveal the American's anti-British animus and call his own reliability as an objective chronicler into question.[66]

For the Bering assessment, Cook drew on John Harris's *Collection of Voyages and Travels*, which was published in London in the 1740s and contains Bering's chart of Asia's eastern extremity, an image that preceded Müller's map (1761) and Staehlin-Maty's (1774). This evolution puzzled Cook. Enlightenment-era cartography was presumed to be progressively more accurate in its depictions, but in this part of the world that did not hold true. Cook allowed that Harris's map was more accurate in its depiction of the Siberian coastline than Müller's, but Staehlin's was completely confounding. "The more I was convinced of being up the *Asia* Coast," Cook confided to his journal, "the more I was at a loss to reconcile Mr *Staehlens* Map of the New Northern Archipelago with my observations." He "had no way to account for the great difference" other than the possibility that he had mistaken some part of America for Staehlin's "Alaschka" and missed "the Channel that separates them." For Cook, it was "a matter of some Consequence to clear up this point this Season, that I might have but one object in View the next."[67]

Beaglehole posed Cook's tactical predicament thusly: had he "sailed in his ignorance past the west coast, and not the east coast, of Alaschka?" Cook was content to have to put up with the dubieties of Müller and Staehlin-Maty for one year, but he was not about to suffer such diversions during a second. If Cook had somehow missed what might be called "Staehlin Strait," which separated "Alaschka" and the North American mainland, better to find out while time still allowed in the moderating weather of early September 1778 than take part of another exploring season to do so. But the ships were still in need of wood, and for this combination of reasons Cook steered *Resolution* and *Discovery* "over for the America coast" into a bay he would later name Norton Sound.[68] This set the stage for the last phase of James Cook's northern explorations.

TWELVE

Northern Interlude

James Cook's 1778 discovery season had been remarkable. His geographic accomplishments that year alone compared well with any previous voyage in its entirety. Starting with the discovery of Hawaii, Cook established, in succession, the trend of America's Northwest Coast, the shape of the Alaskan subcontinent, the separation of Asia from North America at Bering Strait, and the baseline of the Chukchi Sea ice pack. J.C. Beaglehole argued that Cook was occasionally stymied by ice during the second voyage but at no cost to that expedition's success, yet in the North similarly impenetrable masses "preordained" failure. This was no more true than Cook's being unable to find another nonexistent geography – Terra Australis Incognita – during his celebrated second voyage.[1]

Cook's standard for every voyage was to make incremental additions to cartography and navigation, and his final expedition was no exception. This work was never done, and after leaving Bering Strait, it led him back to North America. His probe of Norton Sound, generally regarded as a Cook backwater, demonstrates the gritty diligence that led to his extraordinary success. His immediate goal was to certify whether the coast from Cape Prince of Wales to Icy Cape was the west side of Staehlin's "Alaschka" or, as he surmised, the continental mainland. Put another way, after backing out of Kuskokwim Bay but before striking Sledge Island on his outbound course had he missed a putative "Staehlin Strait" east of Bering's? As James King conceived it, Cook now had more time and better weather than that which had availed when they had travelled north a month earlier and could assure himself that "nothing might be left undetermin'd on this Coast, which otherwise would naturally interfere with the plan of Next season, and which indeed could not be well laid, till we were sure of the Continent." Here was a geographic concern central to the mission that, unlike Tonga and Fiji, had to be fastened down.[2]

⚓

Sledge Island and the American coast returned to view on September 6, 1778. Cook wrote: "If any part of what I had supossed to be [the] America coast was the island of *Alaschka* it was that now before us and that I had missed ... by steering West instead of the East after we first fell in with it; so that I was not at a loss where to go to clear up these doubts." The skies were auspicious. Clear with a "fine breeze," Charles Clerke reported. He added that the ships sailed eastward, "with all Chearfullness and Satisfaction," along the northern shore of what would be named Norton Sound. Late on the night or early morning of September 7–8, Cook noted that "the water shoald pretty fast." This was a harbinger. Anchors were dropped to await dawn. Taking up, the ships came abreast of Cape Darby, the southernmost projection of the Seward Peninsula. Rounding it, Cook could see that the coastline took a "northerly direction," but he was in only thirteen fathoms of water. In fading light, anchors were again dropped "over a muddy bottom" three miles from shore. The "drean of a current" flowed to the southwest, but the tide did not rise or fall markedly – mixed signals as to whether this opening had an outlet.[3]

At daybreak on September 9, the ships weighed anchor and sailed north. Due east, a view of land was raised that, in the morning haze, was surmised to be two islands. On the near port side, the adjoining shore was timbered. This was an agreeable sight because it had been over two months since the expedition had seen any land other than barren tundra. To Clerke, the wooded landscape was "a most pleasing and satisfactory Innovation in the face of the Country, as we begin to be much in want of that most essential Article." Equally auspicious, the sea remained open to the north. Clerke was effervescent about this new prospect for a "passage between the Island of Alaschka and America." He remarked: "I wish with all my Heart we may find it so, but I shou[l]d like it much better if we did not swim quite so near the bottom." The water in some places was only three fathoms deep.[4]

King showed the same enthusiasm as Clerke. He recorded that the earlier sighting of what were presumed to be islands to starboard after rounding Cape Darby "gave us fresh spirit, and we hoped ... we should find the main Coast of America far to the East." It was as if, on September 9, the thrilling week of August 9–16 had magically returned. To King, it seemed "too late this Year to push for a Passage into Baffin's Bay," but for the next there was "well grounded hope of Succeeding ... which we cannot now have if we meet with no other rout of pushing to the N[orth]ward but by the Channel we have just left," meaning Bering Strait. This new prospect was soon dimmed. In the midday sun, Cook saw "high land" behind the "islands" seen to the east in the morning haze. These were the Nulato Hills, which rim the Yukon River as it bends toward its delta at the southwest corner of Norton Sound. Cook's theoretical prospect of a course between the continent and "the island of Alaschka" became more doubtful when the water shoaled as he proceeded north. Two boats were lowered to find a deeper channel. The ships inched north behind them, occasionally bringing mud up from the bottom. He anchored when the water shallowed at eighteen feet.[5]

Cook was determined to get to the head of the inlet to avoid replaying the Prince William Sound scenario, where his decision to call off exploration had precipitated some grousing on deck. When water sufficient to float the ships "was not to be found in any part of the Channell," Cook retreated. With a southern breeze at his back, he dreaded the "wind increasing and raising the Sea into [waves] so as to put the Ships in danger of striking" the shallows. James Trevenen doubted "whether any other Man in the world would have ventured so far as he had already done; and indeed whether it was not highly imprudent" to have done so.[6] Trevenen's comment reveals the predicament Cook occasionally found himself in during his voyages and in the estimation of historians. On the one hand, when he failed to survey some islands in the South Pacific and exhaustively explored Prince William Sound, his decisions led to grumbling. On the other hand, the Forsters found his constant faceoffs with the Antarctic ice pack obsessive. Historians have tended to see Cook's consistently strenuous efforts as a kind of mania, not fully appreciating that what made him a great navigator was not inborn talent but rather his constant investment of time and energy toward mastering tasks at hand. It was not Cook's enthusiasm that distinguished his career; Daines Barrington and Alexander Dalrymple had plenty of passion too. It was Cook's endurance – the ability to persevere – that was the rarest of his skills.

After plying in the shallows overnight, two boats were dispatched, with Cook aboard, to the west shore of the bay "to see what the Country produced and to look for a convenient place to get wood and Water." Berries, always of interest because of their rejuvenating qualities, were found in profusion, and there was "no want of fresh water" either. But to secure firewood, Cook "stretched over" to the eastern shore of Norton Sound and anchored off a headland he called Cape Denbigh. He observed several Iñupiaq on the peninsula, one of whom "came off in small Canoe." Cook gave this man "a knife and a few Beads." Seeing he was pleased with the gift, Cook built on this positive start by signing "for him to bring us something to eat." After paddling to shore, the man returned with two

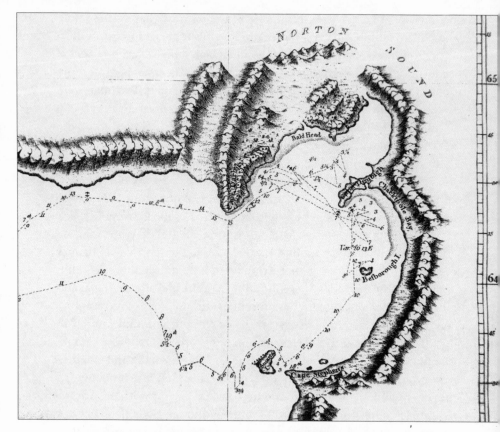

Chart of Norton Sound and of Bherings Strait, Made by the East Cape of Asia and the West Point of America (detail). Prepared by Henry Roberts and engraved after a missing survey by Cook, 1783–84. No subexploration of the North American coast demonstrates the thoroughness that Cook brought to his third-voyage mission more than the survey of Norton Sound. After proving to himself that he had not missed a more southerly route toward Baffin Bay, Cook turned for Unalaska and then Hawaii.

dried salmon. This friendly encounter took a curious turn when the Iñupiaq man said he would give the salmon to no one but Cook, whom, it was thought, he had asked for by the name "Capitane." Cook thought this had to be a mistake in euphony because he could not conceive how the man knew that title. He left the matter there, proceeding to other trades – fish for knives, tobacco, and "such trifles as they could get."[7]

Wood was plentiful around Cape Denbigh, but the shallow water made loading difficult because the ships had to maintain a distance from shore to avoid grounding. "Finding it would require much time and trouble to

Wood here," Cook "stood back" to the opposite side of the inlet off Cape Darby near where he had landed two days earlier. Driftwood abounded on the beach. Clerke stated this was "singularly fortunate" because "its qualities are far superior to those of any we could possibly cut, for we find plenty among the great abundance of it, by no means water-soaked, but that burns well and makes a strong and durable fire."[8]

Cook went ashore at Cape Darby, again to inspect the landscape. His account of this trek shows how vigorous and observant a man he still was. His walk "into the Country" alternated between woodland, heath and "other plants," which were producing "berries in abundance." Cook noted that "it was bad traveling in the woods," which consisted of "birch, willows, Alder and other shrubbery," mixed in with scrawny spruce trees, a distribution of species, Cook astutely observed, that was in contrast to the fir driftwood on the beach.[9]

This jaunt was also followed by the most poignant human encounter Cook had on any of his voyages. While wood was being loaded, four people (likely Yupik) approached. This party consisted of what Cook took for a husband and wife, their small child, plus "a nother man who bore the human shape and that was all, for he was the most deformed cripple I either ever saw or heard of, so much so that I could not even bear to look at him." The husband "was almost blind," and his wife was not as robust as most Indigenous northerners Cook had seen. The group had some blue beads, hinting at indirect Russian trading influences, "but iron was their beloved Metal." Cook's armourer fashioned four knives "out of an old iron hoop" in exchange for an astounding four hundred pounds of fish that he believed had been caught in the previous thirty-six hours.[10]

The next day, Cook sent a party ashore to cut switches for brooms and spruce-tree branches for brewing beer. This last intention, we know from Lieutenant John Rickman, caused "great murmuring" among the seamen. It also anticipated a better-known incident that would occur off Hawaii three months later. Beer did not flow over the palate in colder climates as well as liquor, Rickman explained, "especially when they were given to understand that their grog was to be stopped, and this beer substituted in the room of it."[11] Cook makes no mention of this complaint, but Rickman notes that Cook responded by alternating spruce beer and grog each day. By this point, the expedition had been at sea for just over twenty-six months; it had crossed the two-year threshold of nostalgia and discord that had arisen with Cook's crews at the same juncture during his first two voyages.

Fully provisioned, Cook was eager to get back to the work of geography. Having been frustrated by shoals on September 9, when he first tested the hypothesis that he might have missed "Staehlin Strait" on the outbound leg, he recrossed to the east side of the bay off Cape Denbigh on September 15 for a follow-up assessment. King was dispatched in command of two launches to establish, with finality, whether, as Cook phrased it, the "western land" (i.e., Seward Peninsula) "was really an Island or part of the Continent of America." This mission was expected to take a few days, but as a precaution King took provisions that would last a week. Cook directed King to the deepest recess of the bay northward, where he was to go ashore "and endeavour from the heights to discover whether the land you are then upon, supposd to be the Island of Alatska, is really an Isl[an]d or joins to the Land on the East supposd to be the Continent of America."[12]

⚓

After King departed, some men landed at Cape Denbigh for another berry-picking expedition. Cook later landed there and found "a good deal of very good grass." At every turn, the terrain "produced some vegetable or another." He took other extensive notes, including this comment on the geomorphology of Cape Denbigh: "It appeared to me that this Peninsula must have been the Isthmus and it was even kept out now by a bank of Sand, Stones and Wood thrown up by the waves. By this bank it was very evident that the land was incroaching upon the sea and one could see the progress it had made from time to time."[13]

Cook's voyages occurred within what scientists call the Little Ice Age, generally considered to have lasted from 1300 to 1850, with the first half of the eighteenth century being consistently the coldest. As the climate cooled over this half-millennium, more of the Earth's water was taken up into the form of ice, lowering the sea level and presenting incrementally more land. (This is the opposite effect from what we are experiencing with modern climate change, which has meant warmer temperatures that diminish the polar icecaps, raise the sea level, and inundate land.[14])

Cook's remark about land "incroaching upon the sea" is consistent with what his contemporaries in North American exploration had discovered concerning the continent's changing climate. Samuel Hearne noticed that the woods on the margin of the Barrens were ringed with old juniper stumps and windfall. "Those blasted trees are found in some parts to extend to the distance of twenty miles from the living woods," Hearne wrote, proving "that the cold has been increasing in those parts for some ages." He recounted the story of older Indigenous people, who told him their grandfathers "remembered the greatest part of those places where the trees are now blasted and dead, in a flourishing state; and that they were remarkable for abounding with deer." On July 6, 1771, just before Hearne reached the mouth of the Coppermine River, he referred to the "uncommon badness of the weather." Many of his Indigenous companions deserted him because the "constant sleet and rain made it so wet" that it became impossible to start a fire. Worse, "a very sudden and heavy gale of wind came on from the North West, attended with so great a fall of snow, that the oldest Indian in company said, he never saw it exceeded at any time of the year, much less in the middle of Summer." The snowflakes "were so large as to surpass all credibility, and fell in such vast quantities, that though the shower only lasted nine hours, we were in danger of being smothered in our caves."[15]

Both Alexander Mackenzie and Charles Duncan, fur-trade explorers who would investigate the Canadian north a little more than a decade after Cook's final voyage, recorded climatologically consistent findings. On Mackenzie's expedition, which took him to Pacific tidewater in 1793, by which time the Little Ice Age had begun to loosen its grip, he heard a compelling story along the Peace River in present-day northern Alberta. He reported: "An Indian in some measure explained his age to me, by relating that he remembered the opposite hills and plains, now interspersed with groves of poplars, when they were covered with moss, and without any animal habitation but the rein-deer. By degrees, he said, the face of the country changed to its present appearance, when the elk came from the East, and was followed by the buffalo; the rein-deer then retired to the long range of high lands that, at a considerable distance, run parallel with this river." W. Kaye Lamb, the editor of Mackenzie's journals, asserts that "so great a change could not have taken place in a single lifetime, and the Indian must have been recounting a tribal legend rather than his own experience." But, as Brian Fagan relates, prior to the prolonged warming trend line of modern times, the climate pattern often zigzagged dramatically in quarter-century increments. Thus, today's science allows us to consider the story Mackenzie related a credible one.[16]

In 1791, maritime fur-trade explorer Charles Duncan, near Hearne's old post at Churchill on Hudson Bay, was told by Indigenous people that "canoes have passed within their remembrance over rocks, where the highest tides do not flow at present by several feet."[17] Duncan was at a loss to explain this but, again, modern climate science can verify the Indigenous account. Duncan's informants were describing the effects of isostatic uplift, the gradual rebound of the Earth after the immense weight of the Laurentian Ice Sheet was lifted when it melted at the end of the last full Ice Age, thousands of years ago. This process is still ongoing, amounting to sixty centimetres of rebound every one hundred years.

⚓

King returned from his reconnaissance of Norton Sound's northern extremity far sooner than expected. Prepared for a week, it was a one-day excursion. He concluded, grimly, "our business was finish'd." Conveying his findings to Cook, King related that the captain "was pretty certain of the Event" – that is, he was not surprised. In his journal, Cook recounted that King had proceeded up to twelve miles farther than he was able to get with the ships. After ascending a height of land, King saw that the bay's previously opposing "coasts" met at "a small River or Creek, before which were banks of Sand or Mud and every where Shoald Water." To the north (the Koyuk River basin), the terrain "swelled into hills and united those which are on the East and West sides of the inlet."[18] It was, then, September 17 – not August 17–18, off Icy Cape, as standard accounts hold – that the theory of a northwestern shortcut to Baffin Bay was dashed.

With King's reconnaissance in hand, and with it the certain knowledge that there was no Staehlin Strait, Cook named the body of water Norton Sound after a British barrister and member of Parliament. He was also "now fully satisfied ... that Mr Staehlin's Map must be erroneous and not mine." Accordingly, Cook determined "it was high time to think of leaving these Northern parts, and to retire to some place to spend the Winter where I could procure refreshments for the people and a small supply of Provisions."[19]

Cook was entering the portentous nexus that would define the balance of this expedition, if not his entire career. His instructions had highlighted a possible recourse to Petropavlovsk in Kamchatka, but they also allowed him to choose any place he judged "more proper." He determined the Russian harbour "did not appear to me a place" where he could reprovision "so large a number of men." Cook had never been to Kamchatka, nor anyone with him, but he had seen its neighbouring coast south of Bering Strait, and he had access to publications describing the place, such as Harris's *Collection of Voyages and Travels*. His instinct on the unattractiveness of Petropavlovsk was sound (as the expedition's experience there the following year under Clerke would prove), but the dispositive factor was not the climate per se. Petropavlovsk was only ten degrees south of Norton Sound, and that fact framed Cook's logic. The reason to avoid Kamchatka, Cook professed, "was the great dislike I had to lay inactive for Six or Seven Months, which must have been the case had I wintered in any of these Northern parts." The only place in the Northern Hemisphere that offered the "necessary articles" of refreshment and productive discovery work was Hawaii. Cook wrote: "To these islands, therefore, I intended to proceed."[20] This was a popular decision with the crew. For Cook, it would prove fatal.

In his parting assessment, Cook called Norton Sound "an indifferent place," a grumpy characterization heavily informed by the fact it did not have a suitable harbour. Nonetheless, its "remarkable fine Clear Weather" availed seventy-seven sets of astronomical observations over eleven days. The result is visible on the "General Chart" published after the expedition returned to England. Norton Sound is the only indent on the west side of the Alaskan subcontinent delineated in any meaningful fashion. Cook had still not fully replenished the ships' water supply, so he "resolved to search the America coast for a harbour" while proceeding south. This examination could also fill in some gaps on his chart. In his earlier haste to get north, Cook had sailed far off the mainland, so this was an opportunity to investigate the west coast of Alaska, a shore that was "intirely unexplored, and is probably inaccessible to every thing but boats or very small Vessels." If such a tour failed to yield a harbour or proved difficult to navigate, Cook intended to sail directly to Unalaska, a place that would also double as a rendezvous should the ships become separated. This was a concession to the infamous fogs that were a summertime routine in the Bering Sea.[21]

On September 19, the ships started exiting Norton Sound. Because they were trying to keep the coast in view for cartographic purposes, the ships were in very shallow water, as Cook had predicted. He noted that "the Water was very much discoloured Muddy, and considerably fresher than at any of the places we had lately Anchored at." Applying his long experience as a geographer, Cook recorded that he "supposed that there is a considerable River in this unknown[n] part." Cook's intuition did not fail him. Although he did not see it directly, he was off the Yukon River delta. In a telling expression, he called this stretch of Alaska coastline (the estuary and Pastol Bay, which adjoins it) "the Chasm." The name represented both an immediate hole in his chart and the introduction of uncertainty regarding his projected return the following spring, when, inevitably, he would need "wood and water." Cook had not yet found "in all these Northern parts ... such a thing as a harbour, or even a well sheltered bay."²² Far from being tired of exploration, late into the 1778 exploring season Cook was actively planning for the next year's.

Concluding it impractical to explore the coast in launches, which navigating around shoals mandated, Cook turned his back to the delta and headed west into the deeper water of the Bering Sea. King, now the leading naturalist on the expedition after William Anderson's demise, offered some parting reflections on Norton Sound. He admired its "very extraordinary Weather that would have been deem'd fine at any time of the Year, but which in the latter end of September was thought remarkable." The adjoining hills were still "clear of snow." Such "novelties" as "very excellent Raspberries" led King to conclude that Norton Sound should be regarded "a very pleasant part of the World, and considering its Latitude it is so." The salubrious late summer weather prompted King to ponder whether the Arctic Ocean was "much clearer" of ice in mid- to late September than when the expedition had struck it in mid-August. Subsequent explorers would prove this surmise correct but, as King pointed out, attempting a northern passage that late (instead of in August or even earlier, as the Admiralty instructed) would involve "a very great Alteration" to the plan and considerable risk in pushing "we know not where so late." Furthermore, perhaps the fair weather in Norton Sound had been an isolated local phenomenon that could not be extended to the adjoining Arctic Basin since the bay was sheltered "by a body of high land to the NW from the Icy Sea," a reference to the mountain chains occupying the Seward Peninsula.²³ Cook would not live to act on any of the potential ramifications of the prospects King described so astutely. A sick and inherently more cautious Charles Clerke abided strictly with Admiralty guidance on the timing of the return to the Arctic in June 1779. Nevertheless, King's discourse provides insight into the range of conversation engaged by Cook and his officers as they looked ahead to their next year in the icy latitudes.

Sailing into the heart of Bering's sea, the expedition came into contact with the islands Bering named St. Lawrence and St. Matthew. According to Beaglehole, Cook here suffered "an odd lapse of perception," not realizing that, outbound, he had named the former Anderson's Island. Cook had recognized it as Bering's St. Lawrence when he crossed over to Norton Sound on September 4, but when he reached a different part of the same large island from yet another direction on September 20, he named it after Clerke. Beaglehole grants that Cook's first view of St. Lawrence or Anderson Island was a distant one (amounting to a mere bearing) and also submits that the issue could be explained away by the distractive quality of Staehlin's "shamble of islands." But in the end, Beaglehole reverts to the trope of the diminished captain: "Assuredly the Cook of the second voyage would have disentangled the truth."[24]

But Cook's confusion had another source. When the ships had left Chukotka for the American shore to test "Staehlin Strait," they had been struck by "showers of Sleet and Snow." While Cook was in Norton Sound, he enjoyed mild conditions; however, as King recorded, the moment they left it, the Bering Sea's "customary unsettled bad weather" prevailed. These notoriously difficult conditions were an insufficient excuse for Beaglehole, who believed the younger Cook "would surely" have seen St. Lawrence clearly and distinctly "in spite of the cloudy and snowy weather."[25] Recall that in equally tempestuous weather Cook never "disentangled" the nature of Sandwich Land either.

As was so often the case in the development of his master narrative for Cook's final expedition, Beaglehole took his cue from a stray third-party comment. His critique was derived from William Bligh's snide note on a copy of *Resolution*'s charts, which in form resembled George Gilbert's gratuitous comment that had informed Beaglehole's appraisal of Cook's failure to explore Fiji and Samoa. Thus, generally speaking, Beaglehole's fault-finding with this voyage revolved around some minor proposition relative to the scope of insular aggregations and dimensions. In so doing, he miscast what Cook and his patrons had deemed important about the mission. Cook had long since moved past a pointillist delineation of the Pacific's 12,933 habitable islands in favour of depicting the globe on a grand continental and oceanic scale. This outlook was the hegemonic impulse driving British interest in the Northwest Passage and any related commercial opportunities that could be derived from it.

⚓

From St. Lawrence and past neighbouring barren St. Matthew Island, the ships proceeded toward the Aleutians, what Clerke termed the "good Archipelago." *Resolution* had sprung a leak, so when the ships anchored in Unalaska's English Bay on October 3, the first order of business was a repair of its timbered hull and new sheathing. Once the overhaul started, Cook searched for fresh vegetables. This effort came to naught, but he noted that "this loss was more than made up by the great quantity of berries every where found a Shore." Though Cook has generally received less recognition for his dietary regimen on this voyage than the first two, the one constant throughout them all was his concern for the men's health.

The crew – "one third of the people by turns" – had shore leave to pick berries. Cook stated confidently that if there were "seeds" of scurvy "in either Ship, these berries and Spruce beer which they had to drink every other day, effectually removed it." He also secured fresh fish from the islanders, including some dried salmon in "high perfection." Yet, after netting a single halibut weighing 254 pounds, there were few who did not prefer it over salmon.[26]

Cook's attention was drawn to the Russians after receiving notes "written in a language none of us could read." They were delivered by Indigenous intermediaries and covered a gift of seasoned salmon loaf. Cook dispatched his own envoy, John Ledyard, to an inlet on the other side of the island (near Dutch Harbor), from which the emissaries had come, laden with "a few bottles of Rum, Wine and Porter." Cook considered Ledyard "an inteligent man" whose mission was to gather information. Should he meet "any Russians, or others," he was "to endeavour to make them understand that we were English, Friends and Allies."[27] These instructions indicated he understood the underlying diplomatic relationship between the two countries.

Ledyard was gone three days and returned with three seamen engaged in the Russian fur trade. This was, per King, an emotional occasion. Seeing "these three Sailors rais'd a peculiar sensation in the breast of every one from the Captain downwards which was visible enough in our countenances, and in our behaviour towards them." Finding Europeans "in so strange a part of the World ... was such a novelty, and pleasure, and gave such a turn to our Ideas and feelings as may be very easily imagined." There was, however, "a sensible chagrin" about the lack of a "common Language" to facilitate understanding. Clerke's and Cook's outbound receipt of a similarly unintelligible missive was merely regrettable. Now, with Russian emissaries standing before them, the situation was embarrassing. Historian Glyn Williams sees the absence of an interpreter as part of a pattern that lends credibility to the notion that Cook was acting "out of character" on this voyage. And maybe this was an oversight. (This episode was instructive for La Pérouse, who took aboard a Russian translator when he sailed in Cook's wake.) On the other hand, to paraphrase Beaglehole's famous characterization, perhaps scholars have come to demand perfection when it comes to Cook, to expect him to anticipate every contingency. As Evguenia Anichtchenko points out, "had the expedition succeeded in its main goal of finding the Northwest Passage," there would have been no need for a translator. Like the failure to strengthen the ships against prospective encounters with ice, the lack of a Russian interpreter reflected the Admiralty's overly optimistic voyage planning.[28]

Some things were learned despite the communication barrier. They inferred that the leader of the Russian delegation was the sailing master of the sloop that had transported trade goods between the islands. Here was an opportunity to advance geographic comprehension. Whenever Cook had been in the company of other European sailors in well-travelled Cape Town or Batavia, he had driven the discussion into the cartographic realm. He now had the same prospect in the Aleutian Islands, of all places. While the Russian master visited with Cook, King recalled that "we in the Gunroom entertaind the two rough Sailors." Cook concluded that his guest had not "the least idea of what part of the World Mr Staehlins Map refered to." He was probably not surprised by this, and from the construction of his journal text appeared to take some comfort from it. The Enlightenment's empirical method dictated the testing of working hypotheses, so Cook said he next "laid before them my Chart." The Russians, he wrote, "were strangers to every part of the America Coast except what lies opposite to them," meaning across Bering Strait.[29]

Cook secured a promise for a return visit and their "chart of the islands lying between this place and Kamtschatka." Two days later, he met "the principal person amongst the Russians in this and the neighbouring islands." "Erasim Gregorioff Sin Ismyloff," as recorded in Cook's journal, had travelled twelve miles from Iliuliuk (now Unalaska Harbor, southwest of Dutch Harbor), the principal Russian establishment in the Aleutians. Gerasim Izmailov was accompanied by an extensive entourage, probably to impress Cook. He was an important character in the extension of the Russian fur trade eastward and eventually to the Alaskan mainland, and he later served under Aleksandr Baranov. Izmailov was captain of *St. Paul*, from the merchant firm of Orekhov, Lapin and Shilov, now on its third voyage to the Aleutians. Izmailov had been conducting a "census" (probably his euphemism for collecting tribute from Alutiiq hunters) on the neighbouring islands when word reached him that two mysterious ships had arrived. He hastened to English Bay and, as he later recorded in a report to his superior in Kamchatka, went out of his way to be polite to these strangers "from the island of London, who call themselves Englishmen." Izmailov had been on the Sindt expedition that hoped to reach America. Sindt only made it to St. Lawrence and St. Matthew, but his expedition prompted an update of cartographic projection in the form Staehlin presented. Izmailov was clearly a well-travelled individual, a circumstance that informed King's opinion that he was "a much more intelligent person than we expected to have found in ... this Country." For this reason, Cook "felt no small Mortification in not being able to converse with him any other way then by signs assisted by figures and other Characters."[30]

Izmailov did not have the chart his sailors had promised earlier, but he did pour over Cook's copy of the Staehlin-Maty map, pointing out errors and inconsistencies. At this, King stated he and Cook felt vexed and chagrined "at the Publication of such a Map, under the title of an Accurate one, and the attention we had paid to it." Neither Izmailov nor anyone in his retinue knew anything about the American mainland. Staehlin's name for the continental mass east of his imaginary island of "Alaschka" – "Stachtan Nitada" – was, Cook said, "a name quite unknown to these people," both Indigenous and Russian.[31]

As for what King called the "grand object of our enquiries ... the state of the Sea to the N[orth]ward," Izmailov said he "knew of no expedition into the frozen Sea or to the N or Siberia" since Bering. Furthermore, he was of the opinion "that there never was nor could be a communication" with Arctic waters "because of the constant Ice upon these Northern shore[s]." Cook had just disproved that, but as King discerned, the Russians were only interested in practical geographic information. King declared that "the principal object of the Russian colonies amongst these Islands is ... procuring the Sea otter," a species not found in the Arctic. Indeed, as eager as Cook was to query Izmailov, he in turn was keen to know the number and extent of sea otters found in the waters the Englishmen had visited. The Russians, Cook reported, "have made several attempts to get a footing upon that part of the Continent which lies adjacent to the islands," referring to Kenai Peninsula and points east. These forays, Cook was told, had been "repulsed by the Natives" at great cost to Russian lives (an experience that would be replicated in the Baranov era).[32] More than three dozen joint-stock companies operated in the Aleutians between 1743 and 1799, when an imperial charter granted a monopoly to the new Russian-American Company formed from the merged entities.

Izmailov recounted, through sign language and sketches, "a voyage he himself had performed" in 1771, which was mistakenly interpreted by the Englishmen as a fur-trading venture that ran from Kamchatka past the Kurile Islands and Japan to Canton and from there "to France in a French Ship." According to Cook, this "ingaged our attention more than any other" subject. Interest centred on three aspects of Izmailov's story: the Kuriles, an island chain then poorly understood cartographically; Japanese antagonism to outside influences; and the dynamics of the fur trade. Izmailov's inability "to speak one Word of French made this story a little suspicious" to Cook, but when King later reached Kamchatka, he confirmed the general outline of what had been murkily related at Unalaska. In the event, Izmailov had been caught up in the Kamchatka insurrection of 1771, executed by east European exiles banished by the czarist regime. Izmailov was among the Russian seamen the rebels pressed into service to sail and row a small galley to Canton. He and several fellow sailors actually did sail to Europe from Canton on a French merchant ship and subsequently returned to the Russian Far East. Indeed, King met three of these seamen when he reached Petropavlovsk in May 1779. But in addition to Izmailov's remarkable saga, what also stands out in this story is the inquisitiveness of Cook's mind, merely four months before his death, by which time, as some historians suppose, the compromised nature of his energy, judgment, and sensibilities was made manifest.[33]

Izmailov returned to English Bay three days later, this time with the promised charts, which he allowed Cook to copy. To Cook, these manuscripts "had every mark of being Authentick," and who would know better? One detailed the Pacific Coast of Asia from Kamchatka to Japan, one of the few parts of the Pacific basin that Cook knew nothing about from personal inspection. But it was the second map that Cook concentrated on "as it comprehended all the discoveries made by the Russians to the Eastward of Kamtschatka towards America." Cook finally had access to an original chart depicting Russian exploration in the North Pacific, not a projection from St. Petersburg as reissued in London. He studied it intently, analyzing the document as only he could for divergences in latitude and longitude between it and published sources such as Müller and Staehlin.[34]

The most intriguing variance between Izmailov's manuscript chart and previously published Russian maps was the extension of the Aleutian chain toward Asia. In his journal, Cook vividly describes Izmailov standing at his side while the Russian "struck out" one-third of the Aleutian Islands on Müller's map, assuring him "they had no existence." The location of other islands was altered "considerably, which he said was necessary from his own observations, and there was no reason to doubt it." This was how Enlightenment map-making in the Republic of Letters was supposed to work. As for the origin of these mistakes, Cook explained: "These islands

ly all nearly under the same parallel," and a succession of Russian navigators had been misled by their "different reckonings," mistaking one island,
or group of islands, for another, leading them to think "they had made a
new discovery when they had not."[35] With that, he also explained his St.
Lawrence Island mistake, only he was more forgiving of the Russians than
Bligh or Beaglehole were of him.

Cook amended his charts to include islands Russian traders had visited
even though he had not. He did this because (using a telling phrase from
the Enlightenment ethos) there was "some authority" for doing so;
Izmailov. His journal incorporated a description of the Aleutian chain from
Kamchatka toward North America, since that time more fully discerned
as numbering 150 islands running 900 miles laterally across the North
Pacific.[36] This fastening down of mission-centric information is the *real
Cook* that Beaglehole, and the many historians who have subscribed to his
interpretation, cannot find on the third voyage.

Cook learned a lot about the North Pacific from Izmailov: names of
islands (notably, Kodiak); the location of passes through the Aleutian
screen that he had missed outbound; the eastern extent of the Russian
trade network; and what supplies might be procured in Kamchatka. Cook
was in his element with Izmailov, talking with a fellow explorer of sorts,
spreading out maps in the great cabin, trying to solve geographic puzzles.
Never effusive in emotional sentiment, Cook described his personal and
intellectual encounter with Izmailov with a depth that shows he was as
fully engaged in his work as at any time in his career. With the expedition's
mission centremost in his mind, Cook sought assurance "over and over
again that they k[n]ew of no other islands but what were laid down on
this chart and that no Russian had ever seen any part of the Continent to
the northward, excepting that part lying opposite the Country of the
Tchuktschis."[37]

With his own findings buttressed by Izmailov's authoritative information, Cook let loose with a rhetorical barrage aimed at his *bête noire:*
Jacob von Staehlin. This vituperative passage is worth quoting in full because it speaks to the Enlightenment soul that burned within Cook: "If
Mr Staehlin was not greatly imposed upon what could induce him to
publish so erroneous a Map? in which many of these islands are jumbled
in regular confusion, without the least regard to truth and yet he is
pleased to call it a very accurate little Map? A Map that the most illiterate
of his illiterate Sea-faring men would have been ashamed to put his name
to." The last sentence was an attempt at sarcasm because Staehlin's text

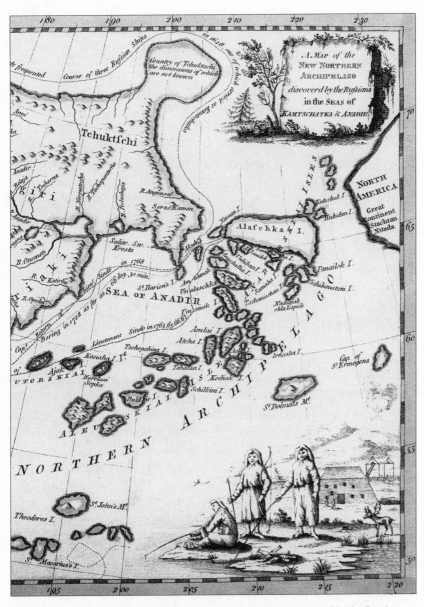

Jacob von Staehlin, *A Map of the New Northern Archipelago, Discovered by the Russians in the Seas of Kamschatka and Anadir, 1773* (detail). Perhaps no aspect of Cook's third-voyage conduct is more poorly understood than his engagement with this map. Cook and his officers tried, without success, to reconcile this fanciful depiction of Alaska (as an island) and the Aleutian chain with what they could see from the deck. Consistent with his practice on previous voyages, Cook proved the dubiousness of this projection.

(translated by Maty) refers contemptuously to the "illiterate accounts of our sea-faring men."[38]

Cook's indictment of Staehlin is one of the most misunderstood aspects of his career. It is usually interpreted as a temper tantrum or a form of self-flagellation for having invested in its imaginary depiction. John Norris characterizes Cook's third voyage as "little more than an endeavour to test Stählin's speculations." I.S. MacLaren cleverly coined the phrase "Gullible's travels" to suggest Cook and his contemporaries naively believed whatever they saw in print. Perhaps, but Barbara Belyea has defended the role of speculative geography by arguing that "scientists have always worked from hypotheses." In every discipline, "theory always precedes observation: the hypothesis is formulated as a problem or a tentative explanation, which observed phenomena confirm or refute." Citing Thomas Kuhn, Belyea asserts that increases in knowledge come "not as factual accumulation but as theoretical confrontation."[39]

Cook was hardly the first navigator to be misled. Sven Waxell, Bering's second in command on *St. Peter,* was highly critical of the instructions Joseph-Nicholas Delisle had written for the Dane's second voyage. Bering searched the North Pacific fruitlessly for Gama Land, a pursuit of fantasy that limited his opportunities to explore North America. Waxell stated that Delisle and his associates had "obtained all their knowledge from visions ... my blood still boils whenever I think of the scandalous deception of which we were the victims." Sketchy maps were an occupational hazard for explorers. Cook himself had complained about the reliability of maps depicting the Southern Hemisphere, as did Bougainville. Both were harried by charts composed with faulty scales and full of imaginary places. One hundred years *after* Cook, George DeLong, captain of the USS *Jeannette,* sailing north of Siberia in hopes of reaching an ice-free North Pole, could not reconcile his Russian charts with what he saw from the deck, resulting in a far more disastrous outcome than Cook's third voyage. Most of DeLong's crew expired.[40]

The key to understanding Cook's vehement denunciation of Staehlin is his ascription of reliable authority to Izmailov's charts. "Authority" within the realm of Enlightenment science was not derived from power or position (as in governance or command structures of either a secular or religious nature) but by a competent comprehension of objective reality. The scientific method in disciplines such as physics or chemistry, or geographic discovery, required that facts be recorded so that they could be scrutinized, cross-checked, and replicated. What historians have characterized as Cook's intemperance during the third voyage was actually an attempt to enforce scholarly discipline, the hallmark of an intellectual culture that was competitive but that also prized accuracy. The Republic of Letters operated on the presumption that the published findings of their peers were reliable. Staehlin, as secretary of the Imperial Academy of Sciences in St. Petersburg and a fellow of the Royal Society of London, was ostensibly a member of this learned peerage. Furthermore, the version of Staehlin's account that Cook had before him had been translated from the original German by no less a figure than Matthew Maty, secretary to the Royal Society, and published under its imprint. King often referred to it as "Maty's map."

Given modernity's normative understanding of world geography, we now know that Staehlin's chart was fanciful. But Cook could not have known that; in fact, he sensibly presumed the opposite on the basis of credentialed authorities – Staehlin and Maty. Their northern projection was equivalent to those Cook took on his first two voyages, which were previously thought to accurately depict the Southern Hemisphere. He tested the contents of those maps, too, elements of which were invalidated by experiment. His destruction of the notion of the unknown southern continent, propagated by Dalrymple and Bouvet, forms much of his deserved record of accomplishment for the second voyage. Inexplicably, in modern-day analyses of the third voyage, Cook is scorned for testing equally fraudulent cartography whose dubiety he proved. In any event, with this final judgment, Cook dismissed Staehlin from his journal. He then applied the term that Staehlin had popularized for a nonexistent island – "Alaska" – to the long peninsula southwest of Anchorage on his chart. He observed poetically that "the America Continent is here called by the Russians as well as the Indians *Alaska,* which is the proper Indian name for it ... and know very well that it is a great land."[41] In this manner, the word "Alaska," from the Aleut word "Alaxsxaq" (the mainland toward which the sea is directed), appeared in its now standard form for the first time. It was later applied to the entire subcontinent.

How or why did Staehlin garble the region's geography so badly? Conceivably, his strait east of "Alaschka" was a misunderstanding about Cook Inlet's deep cut into the continent, which even fooled some of Cook's men when they first saw it. Evguenia Anichtchenko suggests that a map had been developed circa 1711 by Russian colonial officials in Siberia depicting an island called the "Great Land" east of Chukotka, the likely origin of Staehlin's island. But there is a more devious explanation:

purposeful deception to thwart imperial rivals, Spain and Great Britain. Russia's commercial presence in the North Pacific was, in Cook's time, a relatively recent development. Claudio Saunt explains: "To safeguard its hard-won geographic intelligence, akin to today's atomic secrets, Russia concealed its discoveries and released falsified maps of the Aleutian archipelago and Alaskan coast." Ledyard's chronicler, James Zug, shares the view that Russian maps of the North Pacific were "willfully inaccurate."[42]

Regardless of whether he trusted sources, Cook always checked projections from other sources against what he could see with his own eyes. This was in keeping with the motto of the Royal Society – *Nullius in verba* – Latin for "take no one's word for it." Cook was particularly leery about dogmatic formulations from armchair geographers, a skepticism that in part was a function of his pride in disproving the theories of men who had more formal education than him. On the other hand, Cook tended to lend hypothetical credence to explorers who had visited waters before he had. This was the case, for example, with Dutch navigators who had depicted New Guinea and the islands constituting modern Indonesia with what Cook deemed tolerable precision. As is evident in his discourse with Izmailov, he made allowances for mistakes that inadvertently crept into the cartographic record. Staehlin's map, however, even if theoretically based on Sindt's truncated voyage of discovery, was malfeasance, not misfeasance, and it was this violation of trust that Cook found so offensive.

Izmailov took leave from Cook on October 21, 1778. "To his care," Cook wrote, "I instructed a letter to the Admiralty in which was inclosed a chart of all the Northern coasts I had Visited." Izmailov assured him this missive would arrive in Kamchatka the ensuing spring so that it might get to St. Petersburg (and thence to London) the following winter, which proved to be the case though in a roundabout way. Izmailov was himself unavailable for this duty, but a skipper from a rival firm operating out of nearby Umnak Island, Yakov Ivanovich Sapoznikov, was lined up. Cook was promised that his packet would reach Petropavlovsk the ensuing May and then pass to the governor of the Russian Far East, Magnus von Behm, headquartered at Bolsheretsk on the west coast of Kamchatka. In the event, through Behm's intercession, a copy of Cook's communique arrived in London in January 1780 (the original, with the chart of the "Northern coasts," was delivered in March). Through several twists of fate, Cook's packet from Unalaska, via Sapoznikov, arrived in Kamchatka about the same time Clerke himself reached Petropavlovsk, by way of Hawaii, in April 1779. There, after meeting with Behm face to face, Clerke added Cook's Unalaska material to a now larger collection of letters, journals, and other expedition documentation that he allocated into separate parcels. Behm, having informed the Englishmen that his business schedule would not allow him to reach St. Petersburg until February or March, convinced Clerke to send his more pressing correspondence, including a copy of Cook's Unalaska letter, by separate express. Behm was personally entrusted with a copy of Cook's journal and his northern chart plus Clerke's log to that point in time.[43]

Cook's last letter to the Admiralty and news of his death via Clerke's correspondence arrived in London at the same time. Cook's missive has a message-in-a-bottle quality and is compelling because it serves as an immediate, first-hand, unfiltered interpretation of the voyage. In it, Cook recapitulates the mission-centric portion of the expedition and notes that, after leaving Nootka Sound, he had been on "a Coast where every step was to be considered, where no information could be had from Maps, either modern or ancient: confiding too much in the former we were frequently misled to our no small hinderence." The Staehlin-Maty image, which Cook alludes to here, resulted in an exploratory effort "on an extensive coast altogether unknown." After "many obstacles," Cook continues, he managed to get "through the narrow Strait that divides Asia from America, where the coast of the latter takes a NE. direction: I followed it, flattered with the hopes of having at last overcome all difficulties, when on the 17th of August in the Latitude 70° 45', Longitude 198° East, we were stopped by an impenetrable body of ice."[44]

Cook then writes that "the same obstacle every where presented ... over to the coast of Asia." When "frost and snow, the forerunners of winter began to set in, it was thought too late in the season to make a farther attempt for a passage this year in any direction." With this text, Cook reconfirms that had circumstances availed he was prepared to run to the North Sea over Russia. Since that option was also foreclosed, he "steered to the SE along the coast of Asia, passed the Strait above mentioned, and then stood over for the America coast to clear up some doubts," referring to the Norton Sound survey. Looking ahead, Cook states that he intends to visit Hawaii and "after refreshing there, return to the North by way of Kamtchatka and the ensuing summer make another and final attempt to find a Northern passage." This imprecise language left open the prospect of going over the top of America or Russia. As for his chances, he writes: "I must confess I have little hopes of succeeding; Ice though an obstacle not easily surmounted, is perhaps not the only one in the way. The coast of the two Continents is flat for some distance off; and even in the middle between the two the depth of water is inconsiderable: this, and some other circumstances, all tending to prove that there is more land in the frozen sea than as yet we know of, where the ice has its source, and that the Polar part is far from being an open Sea."[45] Here, Cook restates an idea from his journal that a bridge of land might connect the continents north of Bering Strait, and from that idea he implies an allied consequence – the great quantity of ice found in the Arctic Ocean was anchored to a terrestrial surface. There is no such land bridge, but Cook's instincts did not fail him entirely. North of his farthest point lay the Canadian Arctic Archipelago, the world's second largest island chain.

In closing, Cook cites "the great dislike I have to lay inactive for six or eight months, while so large a part of the Northern Pacific ocean remains unexplored." He may have tired of the South Pacific, but its northern counterpart still challenged him. The ships had "stores and provisions ... sufficient for twelve months," a supply that would make it possible "to remain in these seas, but whatever time we do remain shall be spent in the improvement of Geography and Navigation." Cook's commitment was total. In this letter, it is difficult to see Beaglehole's "tired man," for whom the "continued responsibility for his own men, continued wrestling with geographical, nautical and human emergencies ... made him go limp." Characteristic of the age for which he was the emblematic figure, Cook's letter included a return address – the latitude and longitude of Unalaska "East from Greenwich."[46]

When Cook handed this missive to Izmailov, the trader gave him in return letters of introduction to Russian officials in Kamchatka: one for Behm at the regional capital in Bolsheretsk and another for the local commander at Petropavlovsk. Some of the contents were garbled and others misinterpreted, as the Englishmen would learn when they reached Kamchatka. But with this exchange, Cook's only encounter with a fellow explorer within a discovery zone closed. Izmailov, a fur-trade pilot, was assuredly not in the same class as Cook, but the great navigator generously allowed that the Russian was entitled "to a higher station in life than that in which he was employed." He found Izmailov "well versed in Astronomy and other necessary parts of the Mathematicks," the essential skills of scientific discovery. Cook gifted Izmailov a Hadley octant: "altho it was the first he had perhaps ever seen, yet he made himself acquainted with most of the uses that Instrument is capable of in a very short time."[47] Cook probably saw a bit of himself in Izmailov, for this tableau is reminiscent of the young Cook being tutored on Cape Breton Island two decades earlier.

Izmailov wanted to reciprocate by presenting Cook with a sea otter pelt. Cook's sense of propriety dictated that he decline this gift, though he accepted in its place some dried fish and Indigenous basketry. These transactions probably prompted Sapozhnikov, the courier, to visit the next day. Having seen Izmailov with his fancy new octant, Sapozhnikov was desirous of securing "some token" from Cook, who gratified him with "a small spy glass." More importantly, if Cook needed any further inducement to deviate from his instructions about overwintering in Kamchatka, on this occasion he secured from Sapozhnikov acute insights into what in modern parlance would be described as the cost of living in the Russian Far East. At Petropavlovsk, Cook recorded, "every thing we should want was there very scarce and bore a high price."[48]

Cook's journal entries from English Bay describe Russian, Indigenous, and creole customs plus the operation of the northern maritime fur trade. His natural history of the region runs thousands of words. In its depth and breadth, this narrative is on par with anything Cook wrote on *Endeavour* or the first *Resolution* voyage. His report focuses on animal life in the sea and, to a lesser extent, on the land. It is full of interesting detail heavily informed by comparisons to previous observations from the Southern Hemisphere's icy latitudes. For example, "to the North of 60°," Cook states, "the sea is in a manner quite barren of small fish of every kind, but then Whales are more numerous." He also asserts that "seals and that whole tribe of Sea Animals" are not as "numerous as they are in many other seas." But this, he continues, should not be "thought strange sence there is hardly any part of the Coast on either Continent or island that is not inhabited by Indians, who hunt these animals for food and Cloathing." Walruses, he remarks, helpful for both sustenance and navigational assistance, were extant "in prodigious numbers about the ice." Cook also comments on the ubiquity of the sea otter, a species, he allows, that could be "no w[h]ere found but in this Sea." Conversely, he considers it "a little extraordinary that Penguins which are common in many parts of the world" were not encountered in the Arctic.[49] As was often the case when he could not explain something, Cook was here content to point out the issue for the study of other naturalists.

⚓

Cook's departing notes on Alaska include a customary section on Indigenous culture. Except for Nootka Sound, his most extensive contact with the Indigenous peoples of North America occurred at English Bay.

He admired the Unangax for their adaptation to a hostile environment and their craftsmanship, especially the women's. The expedition's encounter with them was benign, but Cook was not unaware of the profound influence of Russian colonialism, evidenced by their unnatural obsequiousness around Europeans such as himself. A few months later, Cook would be dead within a context some scholars have argued was a consequence of his chronically contemptuous attitude toward Indigenous peoples. Yet his parting summation about the islanders from Unalaska stands against that interpretation. He called them a "remarkably cheerful and friendly" people who, "amongst each other" and around his crew, "always behaved with great civility."[50]

Earlier, Cook had alluded to the similarities between the "Esquemaus" he had met in Prince William Sound and people he called "Greenlanders," relying on information about the Inuit of Baffin Bay in David Crantz's study. In furtherance of this theory, King composed a table at English Bay, which would be published in the voyage's account, that shows linguistic affinities between the Indigenous languages spoken at Unalaska and Norton Sound with those spoken in Greenland. In this case, he drew on Crantz plus Cook's Newfoundland recollections. (This circumpolar culture was another anthropological breakthrough for Cook's third voyage.) Cook also noted the shared material culture between the Chukchi of northeast Asia and their neighbours "on the America Coast." This analysis included the design of watercraft, which, given the narrow constriction of Bering Strait, were capable of passing from the end of one continent to the other. Nevertheless, given their larger stature and facial characteristics, Cook concluded the Chukchi were a "different Nation" from the "Americans" he had encountered in Prince William Sound, Cook Inlet, the Alaskan subcontinent, and the Aleutians.[51]

Cook left the demographic implications of circumpolar migration there, but Ledyard offered a more expansive exposition in his account, where he lays out a theory of migration that attributes the peopling of America to a land bridge from Asia across the Bering Strait. This was not an entirely new idea. In 1759, the Franciscan chronicler José Torrubia, combining Aztec traditions with Chinese and Japanese geographic theories that the continents were narrowly separated, concluded that the Indigenous peoples of Mexico had migrated from Siberia. Georg Steller, who sailed with Bering, compared the Kamchadals and Chukchi in eastern Siberia with Indigenous peoples in Alaska and "found a clear indication that the Americans originated in Asia." Steller died in the Russian Far

East in 1746, and none of his work was published until 1781. Still, his views gained such currency that in the mid-nineteenth century Russia based its claim to Alaska, contesting Spain's and Great Britain's, on the premise that the country had been "peopled by Siberians."[52]

Today, the modern scientific consensus, based on gene studies and archaeological evidence, holds that the Americas were bereft of humans until the ancestors of today's Indigenous peoples crossed a temporary causeway that connected Siberia with Alaska in recurring waves. The first migrants from Eurasia and East Asia established a foothold on the Alaskan subcontinent thirty thousand years ago, before the full onset of the last Ice Age. In a demographic phase referred to as the Beringian Standstill, this population became isolated when giant glaciers created by global cooling blocked travel corridors to the south and east. Miles-deep lobes of the ice sheet extended almost halfway to the equator.

Deglaciation, commencing with a general warming phase fifteen thousand years ago, reopened migratory routes along the Alaskan littoral and inland for both the old population and successive waves of newcomers from Siberia. Genomic studies show that today's Indigenous peoples in North America share ancestry with ancient Siberians from the Lake Baikal region but that they are nonetheless distinct from the Indigenous peoples of northeastern Siberia or other East Asians. Over centuries, pioneering populations were constantly overwhelmed by new migrants from Asia, each with a different genetic script, in a process that repeated itself serially across the face of the Western Hemisphere to the very tip of Tierra del Fuego. The demographic result, geneticists believe, was not displacement, as in the post-Columbian era, but rather a continuous intermixing, during which the pre-existing gene pool became overwritten biologically. The Inuit, whose circumpolar migration Cook scoped, began moving from Alaska across Canada about fifteen hundred to two thousand years ago and after several centuries reached Greenland.[53]

More broadly, Felipe Fernández-Armesto trenchantly postulates an epoch of great divergence in human history: the antecedent serial migration of Homo sapiens from Africa to the ends of the Earth. This resulted in populations becoming greatly isolated from one another, thereby creating the racial and ethnic differences that are now seen as an axiomatic aspect of the human condition. He theorizes that divergence "has dominated most of the human past," but this "is not how most people see history, and it is certainly not how historians write it." Instead, we are more interested in the relatively recent period of convergence, when the many variegated and dispersed peoples of the world have, in a sense, re-encountered one another.[54]

Fernández-Armesto argues that the divergence is an untold story because of its distance in time and the evidentiary gap. We know the exploits of Cook, or Columbus before and Lewis after him, because a documentary record makes them accessible to understanding. We do not know who the great ancient explorers were by name, but the migrations they led across Polynesia and the Americas make the accomplishments of modern Euro-American explorers pale by comparison. This is especially true when one considers the juxtaposition of inherent risks and available technologies, such as compasses and chronometers.

Fernández-Armesto professes that "Europeans must be given some credit for meshing the native routes together" in a common understanding of human history. Cook, the Second Age of Discovery's pre-eminent figure, is the key transitional figure between these two great epochs. It was Cook who first comprehended the great Polynesian dispersion across the Pacific, now thought to be the culminating phase of an outmigration that had started in Africa thousands of generations earlier. In the Arctic, he became the most prominent figure to allude to the wide expanse of polar terrain covered by Inuit. Fernández-Armesto cautions that Western civilization's reconceptualization of the global village did not represent cultural superiority. Global route forging became "a European specialty" because Europe needed access to places whose peoples "were richer or more self-sufficient." To address "the relative poverty of their homelands," Europeans craved resources, and empire, with its elaborate trading routes, was the means to secure them.[55] Fernández-Armesto's hypothesis serves as the metanarrative that underpinned the impulse behind Cook's quests for both Terra Australis Incognita and the Northwest Passage.

⚓

On October 26, 1778, *Resolution* and *Discovery* stood to the west from English Bay. Cook was in no hurry to reach the tropics. His intention was to follow up on Izmailov's information about some of the islands reaching toward Kamchatka. Wind-whipped squalls of snow and hail soon dissuaded him from pursuing that track. When the ships regained Unalaska, King noted that the weather was the voyage's worst. Taking nature's hint, Cook concluded his northern interlude by repassing the strait between Unalaska and Unalga Islands that had first brought him into the Bering Sea. He then set a course toward Hawaii.[56]

Many scholars have not been kind in their assessment of Cook's boreal tour. Gananath Obeyesekere, for example, asserts that as the voyage wore on, "Cook withdrew more and more to himself, almost totally eschewing the company and confidence of his officers." That was not the contemporaneous view of King, Cook's principal adjutant, who recorded: "I cannot leave this Northern Sea, and depart for the Southern Islands ... without casting an Eye upon what has been done this Summer." He believed that the perseverance, skill, and judgment Cook displayed "were never surpass'd in such like services; and that during one Summer, an extent of Coast of above 1100 Leagues was examin'd and pretty accurately survey'd," a tabulation that did not include the extensive survey of Siberia's northeastern coast.[57]

And yet Cook was eager for more, perhaps singularly. Ledyard, who frequently damned Cook directly or with faint praise, offered a summary when the ships left Alaska. He noted that the expedition was destined for "the tropical islands where we were now going to wait the return of another season in order to make a second attempt for the Passage, though in fact we were well convinced already of its nonexistence. Cook alone seemed bent upon a second trial." Intended as a criticism of Cook's stubbornness, Ledyard's comment can also be read as a powerful antidote to the orthodox view that the captain was tired or exhausted.[58]

THIRTEEN

Intimations of
Mortality

The most commonly known fact about Cook is that he died in Hawaii. It could hardly be otherwise since most authors and documentary producers, in their coverage of Cook's third voyage, cannot wait to get to Kealakekua Bay, where he meets his seemingly predestined fate. For example, in Stephen Bown's estimation, Cook had been "slowly cracking under the strain of command," becoming "short of temper, unpredictable and losing his usually good judgment." Bown's argument parallels Frank McLynn's characterization of Cook's ultimate "breakdown" during the "crazy" sail off the Big Island of Hawaii that prefigures his demise a month later.[1]

Most discussions of Cook's third voyage focus on his death. Particular facets within the larger tableau hold a particular fascination, such as whether the Hawaiians perceived Cook as a god or why their early deference devolved into a fatal rage. Nicholas Thomas, in his path-breaking criticism of Cook orthodoxy, pointed to the possibility of new interpretations.[2] There is, indeed, another perspective that casts doubt on Cook's supposedly erractic deportment during his return to Hawaii, and the presumed inevitability of the end sequence.

In a way, the tired-explorer hypothesis was demolished long ago. Five years after Cook died, James King composed a much-neglected paragraph in the Admiralty's account in which he addresses the durability of Cook's geographic problem-solving skills. King stated that Cook's "eagerness and activity were never in the least abated." Cook resisted what King called the "incidental temptation," much to the annoyance of his crew who looked upon this pattern "with a certain impatience." For Cook, the goal was always "making further provision for the more effective prosecution of his designs."[3] This was never truer than during the most controversial portion of Cook's career – his return to Hawaii after the expedition's first season of exploration in the Arctic.

⚓

After sailing into the Gulf of Alaska in late October 1778, strong tailwinds carried *Resolution* and *Discovery* south quickly. On November 26, Maui was sighted. "The joy that we experienced on our arrival here," wrote midshipman George Gilbert, "is only to be conceived by ourselves or people under like circumstances; for after suffering excess of hunger and a number of other hardships most severly felt by us for the space of near ten months, we had now come into a delightfull climate w[h]ere we had almost every thing we could wish for." Gilbert suggests the men expected an immediate landing, a view made explicit by John Ledyard, who stated that Cook's first object was finding "a harbour, where our weather beaten ships might be repaired, and our fatigued crews receive the rewards due to their perseverance and toil through so great a piece of navigation as we had performed the last nine or ten months."[4] Cook's transect of the North Pacific and Arctic was his longest high-latitude tour between tropical harbours by a multiple of two. There is no doubt the crew was fatigued, but was Cook?

Cook's thinking at this time was heavily informed by extenuating circumstances. Approaching Maui from windward, he noted that "the sea broke in a dreadfull surf." The island was "well wooded and Watered," but the north shore did not present a harbour to access those vital materials, and the search was compromised by the large breakers. Continuing eastward, the Big Island of Hawaii suddenly appeared on December 1. Cook was surprised that its highest elevations (Mauna Kea and Mauna Loa) were snow-covered, which temporarily obscured their volcanic nature. No formal elevations of the peaks (each is over 13,000 feet high) were taken, but Cook did record a lunar eclipse on December 4, a phenomenon that he said "seldom happens, but when it does it ought not to be omited."[5] He was a cosmic voyager to the end.

The Big Island's triangular shape and the prevailing winds enveloping it shaped the ensuing events. The island's geometric form became evident as Cook circumnavigated it in a clockwise fashion, starting at the northern extremity, where he first approached. From there, the coast runs southeast to Cape Kumukahi, the island's eastern point. The shore then trends southwest toward Kalae, at the southern tip, then north back to the starting point. Kealakekua Bay is about halfway up the western side of the island. Cook conducted a supposedly dilatory or purposeless sail along the island's northeast and southeast coasts, during which he stood off and on for the better part of two months, refusing, so the story goes, to land. Beaglehole considered it poor judgment and evidence that Cook "was a little jaded and tired." Another historian, more dramatically, writes that Cook had "already cracked and scarcely knew what he was doing or why." Vanessa Collingridge says Cook, in the eyes of the men, had become "a vexed and prickly tyrant."[6]

At the time, however, everyone on board seems to have understood what Cook was doing. As Gilbert explained, Cook kept a distance from shore "till our stock was expended; and then went in and traided for more, as we had done off the other Island" – that is, Maui. Gilbert recognized that Cook "prefered this method of passing the time to going into a harbour as it was a great means of saving traid." George Vancouver also learned from Cook. When he conducted his own expedition to Hawaii more than a decade later, he likewise prohibited a "general traffic with the islanders ... until the more important business, of procuring the necessary supplies of refreshments, wood and water, be accomplished." Vancouver observed that if crews were allowed to trade freely, the cost of commodities increased 400 to 500 percent.[7]

For Cook, friction first developed over his innocent decision to brew beer, a customarily intermittent practice. As he plied the coast of the Big Island, he tried a new formula, flavouring the beer with sugar cane procured via trade transacted over the side of the ship. He deemed this "innovation" a "very palatable and wholesome beer which was esteemed by every man on board." (King recorded that the only test subjects were officers.) Using the modern connoisseur's terminology, Cook said it tasted "like a new Malt beer, and if the decoction is strong it hath a considerable body." After the test he ordered more brewed.[8]

Cook's tactical intention reflected a long-standing strategic goal – "saving our spirit[s] against a colder climate." He was simply replicating a practice from this and previous voyages. However, when he attempted a large-scale rollout, the product did not go over well with the seamen. The grumbling resembled what had happened when he had traded out grog for beer in Alaska a few months earlier. The situation escalated into the legendary sugar-cane beer mutiny. This characterization came from Cook's own text: "When the Cask came to be broached not one of my Mutinous crew would even so much as taste it," thinking it was injurious to their health. Cook probably had second thoughts about the term "mutinous" because he later uses the word "turbulent," and from his construction of a companion passage it is clear he regretted this kerfuffle.[9]

But Cook's journal is not expansive on the particulars. He writes that he was fine with the fact that the men did not want the beer. But then

Cook adds that, so he "might not be disapointed in my views I gave orders that no grog should be served in either Ship." In brief quasi-exculpatory remarks, he next alludes to the beer's value as an antiscorbutic; however, everyone, including Cook, knew there was little risk of a scurvy outbreak amidst isles of plenty. He reiterated that his principal motive had been husbanding grog for the crew's benefit during the next season of exploration in "a colder climate."[10]

Within the broad scope of the voyage's mission and the crew's needs, Cook's actions were hardly tyrannical. Frank McLynn argues that Cook's "fanaticism" warranted mutiny. Collingridge states that "the only predictable" thing about Cook at this point in the voyage "was his unpredictability." But titrating grog was consistent with long-established precedent. Spirits had been rationed when they left Tahiti a year earlier and in Alaska. On the Antarctic voyage, Cook had cut off the supply when the ships arrived at New Zealand after their voyage across the Indian Ocean. Lynne Withey plausibly posits that the crew interpreted the husbanding of grog as an unspoken but unmistakable reaffirmation of Cook's longer-term intentions. Resistance "indicated their displeasure at the thought of enduring another Arctic summer when the original plans had called for only one such stint."[11] Cook's instructions clearly anticipated a second attempt at a northern passage if the first was not successful (thus the possible layover at Petropavlovsk), but it is also true that the expedition had been a year late getting north of the equator. In any event, Cook said he and the officers continued to drink the beer "whenever we could get cane to make it; a few hops, of which we had on board, was a great addition."[12]

Others on board had more to add. Midshipman John Watts, for instance, made serial remarks on the incident from December 10 to 12, 1778. According to Watts, Cook did not use the term "mutinous" casually, and other interlocking events and fractiousness on the men's part tested Cook's patience. Watts remarked on the "substitute" beverage and the fact that people disliked it, but more importantly, he revealed that they expressed their displeasure in a letter. This missive's focus was "the scanty Allowance of Provisions serv'd them, which they thought might be increas'd where there was such Plenty and that bought for mere trifles." Thus, the beer was part of a larger complaint about provisioning, in writing no less, not Cook's refusal to land per se. This also explains why Cook had found the crew's distaste for the beer inexplicable. The test batch had been deemed palatable, and similarly strange brews had been concocted since Nootka. Only now were the men displeased with it. That did not trouble Cook; he had grown accustomed to such reactions. For instance, whenever he had added a new element to his regimen, it was first opposed by the seamen, most famously when he introduced sauerkraut on the *Endeavour* voyage. Off Hawaii, Cook affirmed that "few men have introduced into their Ships more novelties in the way of victuals and drink than I have done; indeed few men have had the same oppertunity or been driven to the same necessity."[13]

In response to the written complaint, Cook, according to Watts, "order'd the Hands aft, and told them, that it was the first time He had heard any thing relative to ye shortness of ye Allowance." Cook was under the impression "they had had the same Quantity usually serv'd them at the other Islands, that if they had not enough, they should have more and that had He known it sooner, it should have been rectified." Cook explained that the beer was not injurious, but allowed that if they chose not to drink it, it would be their loss. The kicker, according to Watts, was that the men would only get "grog every other day provided they drank ye Sugar cane" too. If that was not satisfactory, the brandy cask would be "struck" into the hold so the men "might content themselves with Water." The crew had "24 Hours to consider of it."[14] Thus, Cook was surprised at the state of provender but immediately rectified it. And his response was not unique. A similar episode had occurred on the Antarctic tour when the first mate approached Cook and told him the biscuits were rotten, in which instance the situation was remedied.

It is clear from Watts's journal that there was tension on December 10, the first day of this dispute, but Cook had yet to play the "mutiny" card. However, when the twenty-four-hour cooling-off period passed and the crew still refused to drink the beer, the midshipman's version of events, which had not been critical of Cook to this point, took on a more negative tone. Cook adhered to his threat, and the brandy was stowed away. The next day, as the ships zigzagged off the northeast side of the Big Island, events reached a crescendo. According to Watts, William Griffiths, a cooper, received "12 Lashes for starting ye Cask of Decoction" because, ostensibly, it was "sour." Watts recorded that Cook then addressed the "ships company" a second time, "telling them He look'd upon their Letter as a very mutinous Proceeding and that in future they might not expect the least indulgence from him."[15]

Griffiths, by "starting" the beer cask, had turned the tap so that its contents began to pour out on the deck. It was this act, not the letter per se that sent Cook over the edge. Griffiths had intended to provoke Cook with an undisguised act of disobedience, and he succeeded. The captain got angry, as would any eighteenth-century British naval officer. Still, this was not the first time Cook had faced a fractious crew. On *Endeavour*, when the ear of his clerk, Richard Orton, was cut off, Cook was perturbed by his inability to determine who committed the offence, and he considered the episode a threat to his authority and a betrayal of his good will. Additionally, many years after the *Endeavour* voyage, midshipman James Matra astonished an unaware Joseph Banks with an assertion that a mass desertion had been planned in Polynesia.[16]

There is risk in taking the sugar-cane beer incident out of context, not merely within the span of Cook's career but within British naval history and maritime tradition in general. In *The First Salute*, Barbara Tuchman argues that Royal Navy officers did not choose to be irascible, their office mandated it. She explains that navigating a ship "whose motor power was the inconstant wind not subject to human control, and whose action depended on instant and expert response by a rough crew to orders governing the delicate adjustment of sails through an infinity of ropes, hardly identifiable from another" required stern leadership. Tuchman concludes that "there is something about commanding a ship ... that brings out ill-temper." Stephen Bown stipulates that "Royal Navy ships in the late eighteenth century were definitely not floating republics guided by a code of human rights." They were "harsh and violent dictatorships, where the lash and other cruel punishments were used to keep men in line and to compel them to perform dangerous or unsavoury tasks." Fault-finding – or, in Cook's case, his nagging perfectionism – could make captains unpopular. According to historian Caroline Alexander, grousing about commanders was "simply an established fact of naval tradition."[17]

The conceit of Cook historiography is that the captain alone is responsible for shipboard friction during the third voyage. Tuchman argues that the American Revolution had an immediate effect on European sensibilities, and this incident may be evidence of it. An antiauthoritarian zeitgeist began to permeate Western civilization and became stronger as the eighteenth century wore on. No institution, including the Royal Navy, was immune. Compounding matters, as James King pointed out, Cook's crew on the third voyage was "very young," and most of them "had never before served in a ship of war." Johann Forster picked up on this theme, later asserting that "the young men that were with [Cook] must have been in some measure undisciplined and disorderly, otherwise he would not have lost his life." Of course, matters off Hawaii did not actually reach the state of mutiny, as they would on *Bounty* the following decade or at Nootka in 1788, when fur trader John Meares's crew grew impatient over the timing of their voyage to Hawaii and its implied enjoyments.[18]

We do not know what Cook meant by withholding the "least indulgence" from the men, or how Watts understood it, although Glyn Williams
evenly points out that, given the ominous repercussions of a mutiny,
withholding mere indulgences "was less frightening than some alterna-
tives." Other historians, based on a jaundiced comment from John
Ledyard, have interpreted it to mean holding out from the plenitude and
comforts of Hawaii for as long as he could to punish the men. Lynne
Withey, drawing on Beaglehole's seminal thesis, argues that this episode
reveals that Cook's "judgment, balance, and sense of fairness," for which
he had been "justly praised" during his first two voyages, had diminished.
His determination to persist in "unpopular" policies such as cutting off
the grog and remaining offshore was evidence of increasingly intemper-
ate behaviour. His characteristic "benevolence" was now giving way "to
anger and impatience," the implication being that this behaviour prefigured the circumstances surrounding his death. Frank McLynn trans-
mogrifies Beaglehole's "weariness" into the "seven crazed weeks" Cook
cruised around Hawaii.[19]

The problem with these viewpoints is they overlook similar tactics deployed by Cook during his first two expeditions. While transecting the
Arafura Sea on *Endeavour*, a tired, bored and cranky crew, nostalgic for
home, had wanted to land in New Guinea and on nearby islands to secure
some coconuts. Cook refused. Nicholas Thomas observes that if one is
looking for evidence of "Cook's diminishing empathy towards his men,"
there is evidence that it began during the second voyage "rather than
some point in the course of the third."[20]

⚓

It might have seemed like Cook had no reason to refuse to land while
circumnavigating the Big Island, but, as John Robson points out, "poor
weather and the lack of good harbours prevented their getting ashore."
Indeed, Cook's journal from late December 1778 into January 1779, and
those of King and Thomas Edgar, *Discovery*'s master, repeatedly mention
problems such as the wind "continually varying from one quarter to
a nother," heavy squalls, thunderstorms, and "dreadfull surf." The watch
word for these few weeks was "danger." In this tumultuous weather, the
ships became separated for an extended period on December 23. In the
official report, King states that a recurrently "heavy sea" was compounded
by the inability to take soundings. He also thought it "necessary to remark"
that the island's whole northeast coast running to its "southern extremity,
does not afford the smallest harbour or shelter for shipping." There is,

in fact, only one bay on the windward side of the island, where modern Hilo is located.[21]

Cook was naturally wary of being blown onto a lee shore by the northeast trade winds, but there were two additional factors that informed his decision to keep off. First, he wanted to determine just how far east the archipelago extended. Until that was determined, he had no interest in a harbour. Kenneth Munford is the rare historian to appreciate that "the delay in finding an anchorage in Hawaii seems largely to have stemmed from Cook's clear-cut ideas about a proper place for refitting and victualling and from his discoverer's instinct. He hated to stop when there was more to be seen and learned around the next point of land."[22]

Second, Cook wanted to experiment with the provisioning market. He had always been a micromanager when it came to this business because it impinged on the welfare of the men. Since the length and timing of a Pacific voyage was always uncertain, the stock in trade for procuring provisions, principally iron, had to be carefully conserved. At the previous stop, Unalaska, Cook's sailors had been so improvident with the tobacco he had provided for barter that it lost 90 percent of its purchasing power in two days. Frank McLynn argues that if Cook could not afford to pay retail prices for provisions in Hawaii, "he should have spent the winter in Kamchatka."[23] Now, had that happened, there might very well have been a mutiny. No one who saw Petropavlovsk the following spring would have thought it a good place for overwintering.

Cook could dare this undertaking because, unlike the Tahitians, the Hawaiians were prepared to conduct commerce offshore. They were knowledgeable merchants fully conscious of price and value propositions and the workings of a captive market. As Cook explained, Hawaiians brought items to his ships offshore "in great plenty, particularly pigs," and maintained price levels by taking produce back to their villages rather than selling below the asking price. This was a practice Cook admired, but not as much as another: "It is also remarkable that they have never once attempted to cheat us in exchanges or once to commit a thieft."[24] After the troubling aftermath from the previous winter, it is easy to see, from Cook's vantage, the attractiveness of an alternative mode of trade that obviated the necessity of landing. This was not vengeance against the men, it was a prudent attempt at keeping the peace. Cook was also concerned about his crew spreading venereal disease; the longer contact could be postponed the better. Minimizing landings also lowered the risk of desertion. In short, Cook's logic model for staying offshore was based on extensive experience, completely rational, and not the function of a deteriorated state of mind.

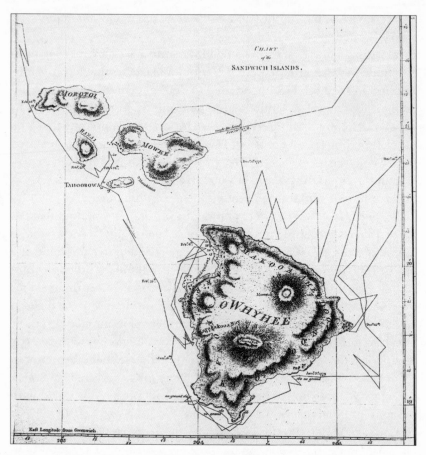

Chart of the Sandwich Islands (detail). Writing by W. Harrison and engraving after Henry Roberts, 1783–84. *Resolution* and *Discovery* arrived off the Big Island of Hawaii in December 1778. Cook spent all that month sailing off its eastern coast, much to the frustration of his men and the puzzlement of historians. Cook's track can be explained by poor weather, the lack of harbours, and a temporary separation of the ships. Finally, an acceptable anchorage – Kealakekua Bay – was found on the island's west side. This was the site of the most infamous encounter in the history of European exploration of the globe and Cook's resting place.

On Christmas Eve, *Resolution* finally rounded Cape Kumukahi, putting her on the equatorial side of the insular triangle. For Cook, this "accomplished what we had been so long aiming at": he had gotten into the island's wind shadow and shelter from the prevailing trades. *Discovery*, however, was "not yet to be seen." Cook had battled the blustery weather for three weeks but now had to continue laying at sea to make his ship visible to Clerke when *Discovery* rounded the point. He expected to stall for a day or two, but it ended up taking a fortnight. This kept his sex-starved men from their shore-side rendezvous still longer, to say nothing of the difficulty it created for Hawaiian merchants who conducted their floating stores far from shore. Cook regretted that "they had taken the trouble to come so far as we could not trade with them, our old stock not being yet consumed." He had learned that with an existing surplus "the hogs would not live, nor the roots keep many days."[25]

Cook's New Year "was ushered in with a very hard rain." The unsettled weather delayed him from meeting up with *Discovery*. The next three days were spent running down the southeast side of the island looking for his consort. During this leg, Cook realized that the foodstuffs being brought to the side of his ship were not of the same quality or quantity seen on the island's northeast coast. Looking at the adjoining countryside, he noticed it no longer seemed capable of yielding provender, the landscape having been devastated "by a Volcano, though as yet we had seen nothing like one upon the island." The destructive lava flows emanated from the mountains he had seen earlier, but he had been deceived by their snow-covered tops.[26]

After two weeks separation, *Discovery* rounded the south point of the island. Clerke came aboard *Resolution* and reported that he had been carried far from shore. The ships were brought together in the nick of time because, Gilbert noted, "we began to be in want of water." Cook dispatched William Bligh in a boat to sound the coast and look for a watering spot, but the sortie proved unsuccessful. He could not find bottom with a lead line nor fresh water once on shore. Bligh reported, in Cook's words, that the "surface of the Country was wholy composed of large slags and ashes."[27]

The ships proceeded up the west side of the island. At daybreak, on January 16, a cove appeared. A boat from each ship was sent in under Bligh's command for examination. According to Gilbert, the expedition had "very luckily found a small bay," for this "was the first we had seen the least appearance of." (This underappreciated line on its own warrants that Cook had not been purposefully standing offshore to torment the crew.) According to Cook: "Canoes now began to come off from all parts, so that before 10 oclock there were not less than a thousand about the two Ships." Bligh returned with favourable news: he had found "good anchorage and fresh water tolerabl[y] easy to come at." Cook wrote, fatefully: "Into this bay I resolved to go to refit the Ships and take in water."[28]

The Hawaiian welcome for *Resolution* and *Discovery* at Kealakekua Bay is one of the most famous episodes in the history of European exploration.

The frenzy that resulted, including extensive sexual liaising, was astonishing to all. Even Cook was amazed: "The Ships very much Crouded with Indians and surrounded by a multitude of Canoes. I have no where in this Sea seen such a number of people assembled at one place, besides those in the Canoes all the Shore of the bay was covered with people and hundreds were swimming around the Ships like shoals of fish." Though Cook would live for four more weeks, his voice as an explorer ends 183 words later.[29] He went ashore January 18 and entered a sequence of events that culminated in his death on February 14, 1779.

Cook's carefully written journal, which he intended to be the first draft of his book, actually ended on January 6. The entries in Beaglehole's annotation from January 7 to January 17 are fragmentary, taken from Cook's log, a more rudimentary record on which the polished and more reflective journal was based. This gap has raised eyebrows among later commentators. Beaglehole believed Cook must have written something after January 17 but that it was lost after the expedition returned to England. Scholarly opinion since has ranged across the spectrum, from Glyn Williams, who calls the gap "mystifying," to the more ominous view of Gananath Obeyesekere, who, with a self-confessed "suspicious mind," suggests that Cook's record for this period was suppressed because of its incriminating nature. Ian MacLaren professes that the termination of Cook's journal "before the ships moored in Kealakekua Bay sends up another red flag" but cautiously adds: "We will never know whether the disappearance of words resulted from plotting by friends or simple neglect."[30]

MacLaren's own exemplary deconstruction of exploration narratives points to an alternative and less mysterious or conspiratorial explanation for the supposedly missing text. He argues that travel writing is rarely subjected to systematic critical textual analysis. To address this problem, MacLaren formalized a progressive multistage typology of narrative composition, starting with the ship's log (or the terrestrial explorer's field notes), which serves as an *aide-memoire* for a more polished and contemplative journal. This document, in turn, is the basis for the draft manuscript that is passed to the printer. The process culminates in a polished, published text. MacLaren concludes that the reflective journals of explorers were necessarily composed in retrospect, commonly "at the end of a stage of a journey/voyage." Cook's journal is a second-stage narrative within this typology and, on the basis of its common use of the past tense, was clearly composed after the fact.[31] Historians have been tempted to assume that the abrupt end to Cook's record is exceptional. In one sense it is: Cook did not live long enough to inscribe events that occurred in the four weeks before his demise. But, given what we know about his writing process, the text describing events in mid-January could have been written on the night of February 13 or even the morning he died; the truncated text could be a narrative that was interrupted by the events that drew him to the shore where he was killed.

It is naive to assume that Cook's journals were ever up to date. Take an earlier Hawaiian episode, for example. When the islands had first come into view the previous winter, Cook sent Lieutenant John Williamson ashore on Kauai as the leader of a watering party. On January 20, 1778, Williamson shot and killed a Hawaiian, but in his journal entry describing the incident, Cook said he did not learn of it "till after we left the islands," a comment that validates MacLaren's typology. Cook, then, could not have recorded the Williamson fracas any earlier than February 2, for that was the Hawaiian departure date – thirteen days after Williamson's offence. Before Cook reached the proximity of North America on March 1 and became preoccupied with the discovery zone, he opened his journal to make three other brief inscriptions for February 7, 12, and 25. Theoretically, then, Cook could have inscribed the Williamson story on this last date – thirty-six days after the event, an interval that is eight days longer than the period of the supposed missing text from the weeks prior to his death.[32]

⚓

Although Clerke and then John Gore would command the expedition after Cook's death, it was King who unofficially became its primary narrator. His role was institutionalized after the voyage, when he was asked to author the balance of the Admiralty's account, post-Kealakekua. In his description of the exuberant welcome they received from the Hawaiians, King reports that they "express'd the greatest Joy and satisfaction ... of our coming to Anchor." This facilitated "a more intimate and regular connexion with them, nor was the Pleasure less on our side." On the surface, King seemed to lend some weight to criticisms of Cook when he then wrote: "We were jaded and very heartily tir'd, with Cruising off these Islands for near two months." But his ensuing observation exculpated Cook: "The Weather was often more boist'rous than we could have imagin'd in this Latitude, and almost a Constant and heavy Swell or Sea when off the NE side of the Island; our old ropes and Sails were daily giving way."[33]

In his journal, King also provides a backwards glance to the sugar-cane beer mutiny in a text that would highlight the incident for future historians. Acknowledging early December's tempestuous weather, he states: "Disappointment in not trying for a place of Anchorage had a bad effect on the Spirits of our Ships Company." This stands to reason, but as Richard Henry Dana Jr. writes in *Two Years before the Mast:* "To find fault with the seamanship of the captain is a crew's reserved store for grumbling." King also references the "mutinous letter" Cook had received about the beer, a "dicoction" he and other officers found pleasant enough. But he notes that what added to the crew's aggravation was the "scanty allowance of Vegetables and pork," a point of contention that Cook redressed "upon the first complaint." King also explains the captain's trading strategy – avoiding the vagaries of "either a glut or a Scarcity, particularly in respect to Vegetables," where "more would be brought to the Ship in one day than would serve a Month." According to King, if Cook bought everything presented, "the greatest part" of the produce would spoil. On the other hand, if he sent the Hawaiians away without buying anything, "they would not return again," injuring both parties. By staying offshore, King remarked, Cook could not only maintain suitable portions but also preserve "the Value of his Iron, which began to be a scarce article." In short, Cook did not want to run out of money. King thought this was a prudent strategy, but the hidden cost was "the great mortification of almost all in both Ships."[34]

King's assessment was consistent with what Cook had said. His narrative shows keen insights about the captain and demonstrates that he was Cook's confidant. In the Admiralty's account, King also added information not found in his journal about Cook's concern that the voyage was going to last at least a year longer than originally expected. According to King, Cook was determined to experiment with curing hog meat in a tropical setting, a process that had earlier been deemed impractical because of rapid putrefaction. This was a matter of great importance because, King wrote, Cook "was under the necessity of providing, by some such means, for the subsistence of the crews, or of relinquishing the further prosecution" of the voyage. According to King, "salting of hogs for sea-store was ... one of the principal objects of Captain Cook's attention."[35]

Cook had learned a lesson from the walrus-meat episode in the Arctic, and his mind was clearly still on the northern mission, not the overwintering that preoccupied the crew. This provisioning scheme had occurred to Cook through a combination of fortuitous circumstances. The size and plenitude of the hogs on the Big Island stunned him, but the meat could only be preserved by salting it. Serendipitously, he discovered that the Hawaiians, through the use of evaporation tarps, had perfected a means for producing a bright white salt "of a most excellent quality."[36] Pork cured in January 1779 was still edible at the voyage's end nearly two years later.

Because King often accompanied Cook on shore excursions, he became one of the principal chroniclers of the controversial events that transpired, including the remarkable homage the Hawaiians paid Cook by prostrating themselves when he walked past. It was King who divined that the honorific term "Lono," directed toward Cook, had supernatural implications. As for the circumstances leading to Cook's death, King noticed after two weeks in harbour that the Hawaiians began inquiring into "the time of our departing." They were pleased "that it was to be soon." He began to suspect the Englishmen's need for food was putting a strain on local resources and relationships. This explains not only the denouement at Kealakekua Bay but also the entire record of the third voyage's difficulties throughout Polynesia. The esteemed scholar Anne Salmond has cogently observed that Cook had only one ship (thus a smaller complement) on his first expedition, and that was also true of the second voyage after *Adventure* separated. But on the third, she notes, because "*Resolution* and *Discovery* were constantly together, their arrival in any Polynesian port was formidable." It was bad enough to be draining local supplies, but "as the sailors amused themselves on shore they created major disciplinary problems, both for Cook and for his Polynesian counterparts. These tensions eventually spiralled out of control."[37]

Despite the obvious plausibility of this hypothesis, most historians have followed Ledyard's contentious lead. He blamed frayed relations on the removal of a temple's fence for firewood, although, Nicholas Thomas argues, the actual grievance was that the "sailors carried off some carved gods along with the rest of the timber." King, a more reliable journal keeper than Ledyard, also discussed the incident. He was concerned enough to query sailors who told him "the Natives told them to do so and assist'd." King wisely confirmed this with Kao, a Hawaiian priest, who identified a single icon he wanted returned. King's intuition that the expedition was wearing out its welcome was amplified a few days later. Having described the encounter up to that point in generally favourable terms, he presciently noted that Hawaiian good will could "wear away from familiarity, or by length of intercourse, [and] their behaviour may change."[38] This proved prophetic.

King related that on departing Kealakekua on February 4, Cook intended to finish surveying the west coast of the island and then proceed to one of the neighbouring isles, where he hoped to find a more sheltered bay. The Hawaiians said the best prospect could be found at the southeast tip of Maui, what would become known as La Perouse Bay, named after the French explorer who sailed in Cook's wake. Curiously, they did not point to Pearl Harbor on Oahu. In any case, Cook's plan to vacate the increasingly contested shores of Kealakekua Bay was foiled. A few days out, a storm sprung one of *Resolution*'s masts. Much to his regret, the ships were forced back to Kealakekua for repair.

As King predicted, relationships were now much more strained. Ledyard's account meshed with King's theory: "Our return to this bay was as disagreeable to us as it was to the inhabitants, for we were reciprocally tired of each other. They had been oppressed and were weary of our prostituted alliance, and we were agrieved by the consideration of wanting the provisions and refreshments of the country, which we had every reason to suppose from their behavior antecedent to our departure would now be withheld from us or brought in such small quantities as to be worse than none." David Samwell ominously recorded that "iron daggers" had replaced adzes as the preferred trade item.[39]

Another contentious theft – one of the ships' four small boats – was the proximate cause of Cook's death on February 14, 1779. The attraction for the Hawaiians was almost certainly the iron in the craft's oarlocks and other fittings, not the boat's navigational capability. Frank McLynn argues persuasively that the prospective loss of this boat provoked Cook into an aggressive posture because in the Arctic it would be "dangerous folly to venture into the ice" without a cutter to guide the ships through the floes. In that sense, the theft of the boat was analogous to the pilfering of astronomical instruments in Tahiti prior to the transit, another incident where Cook sprang into action to recover mission-centric equipment. But the long-standing consensus is that Cook was so fatigued by many years of command that he showed poor judgment when he tried to take a Chief hostage to force a return of the launch. His actions led to a fracas in which he was stabbed and drowned. Beaglehole seized on a theory common in the immediate aftermath of the voyage suggesting that Cook's rashness or overconfidence had brought on his own death. But not every contemporary agreed with this premise. Samwell asserted that "nothing can be more ill-founded and unjust."[40] King provided the one explanation – the guests had worn out their welcome – that transcends time and cultures and best explains the circumstances leading to Cook's death.

⚓

Nervous speculation about Cook's whereabouts began to appear in the British press in November 1779. This concern was precipitated by the arrival of two letters that same month from Sir James Harris, the British ambassador in St. Petersburg, to Lord Weymouth, an assistant secretary of state. Harris relayed word from a Russian official who had received a "Report from the Commandant of Kamtschatka" that their traders in the Aleutians, "towards the Autumn 1778," had seen two ships peopled by a crew "dressed like Russians [i.e., Europeans], but talked a Language they did not understand." These visitors had "behaved with the greatest civility, and were received with the greatest hospitality." They had also been observed by the Chukchi from the "North East Extremity of Asia." Harris's informant indicated that these ships had then "sailed northward" and later "returned by the same Track they went." After touching again in the Aleutians, the craft "steered Southward." Harris told Weymouth that this "immediately suggested ... that it might be Captain Cook." In follow-up correspondence with the Home Office that same month, Harris asserted that these ships had "appeared in those distant Seas" in 1777, a year earlier than originally reported. The documents, which later turned up in Russia's national archives, say "1778," so Harris, or an informant in St. Petersburg, may have been engaging in some wishful thinking, concluding that it should have been "1777," the year Cook was originally scheduled to arrive in the Far North.[41]

The source of the information was Magnus von Behm, the governor of the Russian Far East based in Kamchatka. Given the vagaries of time, distance, and language, Harris's report was a remarkably accurate recapitulation of what had transpired, albeit without knowing the national identity of the visitors. News of these enigmatic ships in the Aleutians and off the eastern tip of Siberia came to Behm's attention from divergent sources. The shorter and vaguer report of strange ships in the Aleutians may have originated with Gerasim Izmailov (or one of his associates or a competitor), who would have learned of Cook's ships at Unalaska in July 1778 and encountered them during their return in October.

At the end of October 1778, when Cook had arranged through Izmailov for a letter and a chart of discoveries to be sent to London (via the same Kamchatka–to–St. Petersburg connection that informed Ambassador Harris), he had also received in return letters of introduction to Russian officials in Kamchatka. One of these letters reached Behm at his chancery in Bolsheretsk on May 1, 1779, a mere eleven days before Gore and King arrived there in person at Clerke's direction. Izmailov's introductory letter had been forwarded to Behm from Petropavlovsk on the other side of the peninsula by Russian officials who had just encountered King, who brandished it when the British ships first reached that harbour. In the letter,

Izmailov actually identified the visitors as Englishmen, provided the names of the ships and their command structure, and after allowing for some garbling of titular nomenclature, noted that Cook was the leader of the expedition. Consistent with what Cook was telling his own crew at the time, Izmailov's introduction conveyed the insight that the Englishmen intended to overwinter at an (otherwise undisclosed) location near 22° S and then arrive in Petropavlovsk the following May. Either the Englishmen had privileged knowledge about the Hawaiian Islands or the Russians had found an island group that far south uninteresting. Izmailov's letter explains why, when King and Gore appeared in Bolsheretsk, Behm could joke that he had been waiting for them. More to the point, the letter's specificity regarding the visitors' identity suggests that Behm had another less-informed source, perhaps the three sailors who initially met with Cook at English Bay or Izmailov himself before he met with Cook. Thus, Behm's vague report was probably dispatched to St. Petersburg before Izmailov's letter of introduction (or Cook's packet) arrived at his chancery.

The reference to the "North East Extremity of Asia" in Harris's story originated with another of Behm's field agents, Timofei Shmalev, commandant of the Russian fortress at Gizhiginsk, Chukotka. Shmalev was regularly in contact with the Chukchi on matters of colonial governance and trade. Two Chukchi emissaries visiting Shmalev apprised him of the mysterious ships they had seen in the summer of 1778. The sailors from these craft, Shmalev related, "addressed themselves politely" to the Chukchi by giving them presents and sharing a meal of whale calf meat, fish, and onions. Afterwards, the ships set sail and "entered the northern sea." They were spotted again "on their return, and this happened in the first days of September." These strangers were assumed by the Chukchi to be Russians on account of "their dress and language."[42]

These two sketchy and disparate reports, later conjoined in the Russian Far East and eventually reaching Harris's cognizance in St. Petersburg, beat the packet containing Cook's and Clerke's definitive reports to London by two months. Behm's initial hazy account was subject to wide-ranging interpretation in diplomatic and commercial circles. Empress Catherine II thought the unidentified ships might be American, but Ambassador Harris made the correct surmise: they were Cook's. All doubt was removed when Clerke's dispatch, sent from Kamchatka while the expedition was on its way from Hawaii to the Arctic, made it across Eurasia with the startling news that the great navigator was dead.

By fall 1779, when Harris secured his intelligence, Cook had been on expedition for more than three years. Had he kept to the idealized schedule in his instructions (and followed an imaginary ice-free path eastward), he might have passed through Baffin Bay by the fall of 1777. Under less optimistic scenarios, countenancing a delay here or there, he might have been expected back in England by late 1778. Now, another whole year had passed without report. What was truly alarming, London's *Public Advertiser* opined on November 5, 1779, was that in the presumptively failed attempt at making a passage through the North, Cook had not reached Canton or any port in South Asia on a more traditional route home. Mercantile ships that had left China in their traditional December 1778 time slot, arriving in London in November 1779 after an eleven-month voyage home, had no knowledge of Cook's whereabouts.

If something untoward had happened, the line of public speculation in London was that Cook had been waylaid by ships from a hostile foreign power. The *Public Advertiser* attempted to reassure its readers that these fears were unwarranted because the king of France had issued a mark of favour that allowed Cook to pursue discovery and "voyage home without Apprehension of Molestation." Even the nascent United States, through the office of Benjamin Franklin, its minister plenipotentiary in Paris, had issued a safe-conduct letter to the attention of "all Captains and Commanders of armed Ships acting by Commission from the Congress." In his March 1779 edict, Franklin stated that a "ship having been fitted out from England, before the Commencement of this War, to make Discoveries of new Countries in unknown Seas, under the Conduct of that most celebrated Navigator and Discoverer Captain Cook ... is now expected to be soon in the European Seas on her Return." (At the time of Franklin's pronouncement, Cook had just been killed.) Franklin's decree advised that, should Cook's ships be encountered, they were not to be considered enemy craft, "nor suffer any Plunder to be made of the Effects contain'd." American captains were instructed to "treat the said Captain Cook and his People with all Civility and Kindness, affording them as common Friends to Mankind all the Assistance in your Power which they may happen to stand in need of." These noble Enlightenment sentiments, crafted by a member in good standing with the Royal Society (and an avid reader of Cook's accounts) who had served as Pennsylvania's agent in London, were countermanded by the Continental Congress once it learned of them.[43]

News of Cook's death reached London (from Petropavlovsk via St. Petersburg) on January 10, 1780. Arriving simultaneously were Cook's

October 1778 letter from Unalaska and Clerke's June 1779 dispatch from Kamchatka. The Admiralty's first lord, Sandwich, immediately wrote a letter to Joseph Banks, which began: "What is uppermost in our mind allways must come out first, poor captain Cooke is no more, he was massacred with four of his people by the Natives of an Island where he had been treated if possible with more hospitality than at Otaheite." Armed with information from Cook and Clerke, Sandwich also conveyed an observation completely counter to the modern understanding: "The voyage has been in general very successfull." The letter concluded by noting that Clerke "means to make one more attempt for a northern passage."[44]

On January 11, the *London Gazette* broke "the melancholy account of the celebrated Captain Cook," whose death occurred on "one of a Group of new discovered Islands" in the Pacific. This report mentioned an event that has dominated discussions about Cook in Hawaii ever since: "The Captain and Crew were first treated as Deities, but upon their revisiting that Island, some proved inimical." The London newspapers were, as Arthur Kitson phrases it in his 1907 biography, a "chorus of regret and appreciation" about Cook's death and his accomplishments. The staggering news spread across Europe. Lord George Augustus Herbert learned of his death while travelling in Italy via correspondence from the Reverend William Coxe. In reply, Herbert called Cook's death "truly a great loss to the Universe."[45]

However, on January 17, the London *Morning Chronicle* published an anonymous letter questioning Cook's reputation. Impliedly written by someone who had travelled with Cook previously, the commentator, after conceding Cook's seafaring skills, stated that the Pacific's Indigenous peoples considered him a pirate. However, the day after this anonymous writer suggested that Cook had it coming, another unidentified writer came to Cook's defence. This rejoinder, by someone who seems to have also travelled with Cook (possibly William Wales or George Forster), included the salient observation that "no person who is capable of reflection ever thought it a remarkable circumstance that he should be killed, *but that he should so often have escaped.*" In his account published four years later, James King adopted the same theme as the *Morning Chronicle*'s contributor, admitting he was "always fearful" that Cook's confidence in handling such affrays "might, at some unlucky moment" cost him his life.[46]

Thus, the now centuries-long debate over Cook's deportment in the Pacific was engendered within one week of news of his death reaching London and a year *before* Gore brought the expedition home. The context of Cook's death inevitably flavoured perceptions of the voyage during which it took place, entirely to its detriment. During his expeditions, there were many episodes similar in portent to Kealakekua Bay where Cook might have met the fate that ultimately befell him. During *Endeavour*'s voyage, for example, he dodged missiles aimed at him by Māori at Poverty Bay, one of which nearly hit home. On the Antarctic expedition, at Nieu, a spear barely passed over his shoulder as he bent out of its way. And early in the third voyage, Cook allowed a landing party in New Zealand to be seriously outnumbered, as happened again in Kealakekua Bay. Simon Baker, evaluating Cook's career within a largely adulatory context, wondered if the great navigator's career was built on "a series of lucky escapes."[47]

King's post-mortem also implied that perhaps Cook's successful hostage-taking tactic had been employed one too many times. Historians tend to posit that Cook deviated from his normal practice at Kealakekua, a theory that is consistent with the idea that he had broken with his previous pattern of exemplary conduct. But, as Nicholas Thomas reminds us, there was nothing "anomalous in his behaviour on the morning of 14 February 1779." Like any person, Cook was subject to the laws of mortality, misfortune, and mere chance. George Gilbert reported that Cook had wanted to send Clerke ashore to confine the Hawaiian Chief "till the Boat was returned according to his usual custom in these cases," but Clerke declined for reasons of health. So "Capt Cook said no more about the matter; but went himself."[48]

The discourse in the *Morning Chronicle* reached a crescendo on January 22, 1780, when a writer with the nom de plume "Columbus" joined the fray. In his biography, Kitson infers that "Columbus" was Banks because the pseudonym was resonant with maritime history, Banks was familiar with Cook's procedures and had a direct claim of friendship with the deceased, and the letter included some idiosyncratic turns of phrase. Provoked by comments critical of Cook, "Columbus" determined in his letter to "immediately put pen to paper, to express my concern that any one should talk unfeelingly of that great man" and to "rescue him ... from the animadversions" of those who had not known him.[49]

The letter was a languid commentary that reads in equal measure as eulogy and epic poem: "His paternal courage was undaunted. His patience and perseverance not to be fatigued. His knowledge in the art of practical surveying inferior to no man's. His skill in mathematics and astronomy was compleat, as far as those sciences are necessary to a seaman." (Students of American exploration will note that the first of these tributes was later borrowed by Thomas Jefferson in his eulogy for Meriwether Lewis, which was later adapted by Stephen Ambrose for the title of his bestselling book.) Columbus capped his remarks by citing Cook's commitment to his crew's health and diligence in preparing his account of the second voyage, written as it was by a man in "want of education." He also commended Cook for his stewardship of resources and modest manner. In short, Cook was entitled to "the approbation of mankind" and "surely a tear is due to his untimely fate." A fortnight of remorse in the London press concluded with the publication of an item in the *Gazetteer* on January 24, which reported that King George III, "who had always the highest opinion of Captain Cook," had "shed tears when Lord Sandwich informed him of his death."[50]

⚓

Interpretive theories suggesting Cook had an unusually fractious relationship with his crew in the two months before his death contribute mightily to the view that the third voyage was a failure. The crew's reaction to his death belies this hypothesis. Following Cook's demise, Clerke's first step, after moving into the captain's cabin on the flagship and naming Gore as his successor in command of *Discovery*, was to secure *Resolution*'s new foremast, the circumstance that had dictated the expedition's return to Kealakekua in the first place. He was anxious about the prospect of another "unlucky accident" on shore, which, if it resulted in the failure to secure a new mast, would have "totally ruin'd ... another N[or]thern Campaigne, which is certainly now my principal object to forward." This comment is remarkable. Cook was dead, but he had inculcated such fidelity to mission that Clerke, who was himself in bad health, immediately recommitted to the goal of finding the Northwest Passage. Clerke was no Cook, but he was not Furneaux either. Rather than heading home, he resumed the northern quest at the cost of his own life.[51]

On February 16, a small portion of Cook's remains were returned from the shore to *Resolution* by friendly Hawaiian emissaries. The shock of this scene precipitated calls for revenge by the same men who had been grumbling for the better part of two months over Cook's beer and navigational decision making. In Gilbert's telling, the men pleaded with Clerke "that they might go on shore with their Arms" to avenge "the death of their old Commander." Clerke temporarily pacified passions by shifting attention to the importance of "compleating of the Fore Mast." King stated in the postvoyage account that the crew had been "all in a flame" over

Cook's death and related "provocations," such as someone waving Cook's hat on shore. Clerke's dilatory stratagem worked to a limited extent because, as Gilbert reported, "by keeping them thus in suspense for three or four days their Rage began to abate."[52]

Clerke may have averted a full-scale war, but he did not mitigate all hostility. Earlier in the voyage, Cook had thwarted his crew's desire for revenge against the perpetrators of the Grass Cove massacre. Just how skilled Cook was in this regard is clear from King's remorse-filled journal entries decrying the "many reprehensible things" sailors perpetrated in the days after his death. Other crewmembers wrote of beheadings, the bayonetting of noncombatants, and fleeing Hawaiians getting shot in the back. One of the worst perpetrators was Bligh. King allowed that the seamen "were strongly agitat'd at the barbarous manner" by which their captain had been killed. Cook would have been appalled.[53]

More of Cook's remains were returned on February 20. This bundle was more shocking than the first because it contained Cook's severed hand, which had a distinctive scar from an explosion off Newfoundland in 1764. Samwell stated that "those, who looked upon Captn Cook as their father and whose great Qualities they venerated almost to adoration, were doomed to behold his Remains; what their feelings were upon the Occasion is not to be described." On the evening of February 22, the day before the ships departed, Clerke wrote solemnly: "I had the remains of Capt Cook committed to the deep with all the attention and honour we could possibly pay it in this part of the World." Twenty-four hours later, he ordered the anchor "weigh'd and made sail out of the Bay."[54] The expedition left Kealakekua Bay, but in a literal and figurative sense James Cook never did.

Historian John Gascoigne writes that we can discern the high regard in which Cook was held by his men "in the passionate grief that followed his death." King had been on the opposite side of the bay from Cook that fateful day. He knew the cutter had been stolen and that Cook was upset. As he busied himself at the observatory, he heard shots fired from across the water that "roused and agitated our Spirits," making it "impossible to continue on observing." King sent a boat to Clerke to ascertain what had happened, but it did not return as quickly as expected. Instead, King noticed a flurry of boat traffic between the ships at anchor. King's party was "under the most torturing suspence and anxiety that can be conceiv'd." He had "never before felt such agitation as on seeing at last our Cutter coming on shore." Before landing, Bligh called out "strike the Observatorys as quick as possible." King concluded, with gripping effect, that Bligh did not have to announce "the Shocking news that Captn Cook was kill'd, we saw it in his and the Sailors looks."[55]

In *Performances,* Greg Dening writes that in the Kealakekua aftermath, Cook's crew demonstrated "the growing realisation that they had lived with a hero ... They knew they had been present at a moment of some destiny. And they tried in their journals and logs to make sense of it." Their narratives make clear, Glyn Williams notes, that Cook's demise was "an event at once momentous and shocking." According to the master's mate, Henry Roberts, Cook's death occasioned "concern, and sorrow, in every countenance; such an able Navigator, equalld by few and excelled by none, justly stiled father of his people from his great good care and attention." Gilbert, who had once questioned Cook's exploratory vigour, described the scene when the boats returned from shore "informing us of the Captains Death." For half an hour, no one said a word. It was "like a dream that we cou'd not reconcile our selves to for some time. Greif was visible in evry Countenance; Some expressing it by tears; and others by a kind of gloomy dejection; more easy to be conceived then described: for as all our hopes centred in him; our loss became irreparable and the Sense of it was so deeply Impressed upon our minds as not to be forgot."[56]

Samwell wrote of Cook's "never-fading Laurels which his grateful Country ever bestows on those Heros who so eminently distinguish themselves in her Service." After Cook's death, George Trevenen, whose earlier stray criticisms about Cook's stern visage did much to flavour scholarly impressions of the third voyage, confessed he had looked up to him "as our good genius, our safe conductor, and as a kind of superior being." Among his shipmates, Trevenen saw that "universal Gloom, and strong Sentiments of Grief and Melancholy were very observable throughout all ranks on board the Ship on our Quitting this Bay without our great and revered Commander." Similarly, King presciently observed that Kealakekua Bay was destined to become "remarkably famous, for the very unfortunate, and Tragical death of one of the greatest Navigators our Nation or any Nation ever had."[57]

Dening argues that Cook's crew needed to explain, that is historicize, what had just happened to them: "They blamed one another for negligence or incompetence or cowardice. They examined the inconsistencies of their most consistent captain to excuse negligence, incompetence and cowardice on their part, to find a cause of his death in his weariness, his bad health, his crankiness."[58] This provocative hypothesis decodes Beaglehole's diminished-explorer theory, which dominates Cook historiography: it was an exculpatory interpretation created to make sense of the incomprehensible. The best modern analogue to this historical or psychological phenomenon is the propagation of conspiracy theories in the wake of John F. Kennedy's assassination in 1963.

PART FOUR *Sequels*

FOURTEEN

Springtime in Kamchatka

Leaving Kealakekua Bay, *Resolution* and *Discovery* brushed the western coasts of Lanai and Molokai and the north side of Oahu. Whatever might be said about Cook's thoroughness in this voyage, or supposed lack thereof, the absence of anything resembling a survey of the Hawaiian Islands by Charles Clerke is noteworthy. This is especially true of Oahu, which James King described as "by far the finest island of the whole group." Clerke missed Pearl Harbor, the single best anchorage in the archipelago and one of the world's finest. King noted simply: "We saw nothing of the Southern side."[1] *This* is what low-energy, distracted exploring looks like.

A suitable water source eluded Clerke on Oahu, so he hastened to Kauai, where he knew from experience the demand could be satisfied. King supervised the watering party at Waimea Bay under continual harassment. But the detachment weathered the torrent of abuse because this would be their last opportunity to get water before heading north. James Trevenen praised the way King coped with the taunts and missiles the Hawaiians threw at them. His "consummate prudence and knowledge" of Hawaiian culture saved his squad "from a repetition of the Karakakooa scene." Trevenen's passing reference to Kealakekua shows that everyone had been chastened by recent events. Clerke reported that some sailors had been approached "to run away," but the idea of "turning Indian," which had once been prevalent, was now "quite subsided." Indeed, he averred that "you could not inflict a greater punishment" on crewmen than to "oblige them to take that step which 16 or 18 Months ago seemed to be the ultimate wish of their Hearts."[2]

Before leaving Kauai, Clerke "read the Articles of War to the Ships Company," which doubled as an occasion to inform the crew of future steps. David Samwell recorded that the next "rendezvous" was Petropavlovsk, "from whence we are to proceed to the Northward." He was not sanguine about finding a passage and volunteered that this had been "the Opinion of Captn Cook and the officers of both Ships; yet he himself had he lived intended to go to the Northward again this Summer and settle this long disputed point of North East and North West passage beyond the possibility of a doubt." Morale was low. George Gilbert said the expedition left Kauai "with great dissatisfaction, on account of the Death of our unfortunate Commander which still lay heavy upon our minds, as being truly sensible of our loss; this together with the thoughts of the approaching season to the North. The hardships of the last still recent in our memory – and will never be effaced from mine – rendered us quite dispirited." J.C. Beaglehole characterized the forthcoming cruise as "a mission of conscience rather than hope. The gloom of Kealakekua Bay sailed with them."[3]

⚓

Clerke directed the ships west in search of other islands in the chain, which he had learned about from Hawaiian sources. Only one, uninhabited Kaula, was sighted; another outlier, Nihoa, was missed altogether. Clerke mused that he "wish'd much to have seen it as it would have completed the Group." His plan was to sail west on the twentieth parallel. Once Kamchatka's meridian was reached, the ships would head north. Riding the swells driven by the northeast trade winds, the crew kept busy repairing the small boats, restoring sails, spinning yarn (while trading stories), and "picking Oakum," the tedious process of unwinding rope fibres for use as caulk.[4]

When the winds lightened, Clerke deviated from his plan. Near what is now called the Date Line, he veered northwest, taking the hypotenuse toward Kamchatka. There was a geographic side benefit to this revision, because the original course would have taken the ships through previously explored waters. This "new track," Clerke allowed, offered something "in the way of discovery" or, as King put it, "falling in with some new islands on our passage." (This came to naught.) The wind not only carried the ships west but cooled them too. Before the course change, the temperature was a torrid 99 °F. Relief came on April 2 with a "fine Gale for our Northern business," Clerke wrote. Four days later, coats, which had been stored when they left Alaska, were brought out. Clerke's "hasty advance" northward made his crew "complain of the cold confoundedly," but they were as fickle as ever. When the weather warmed over the next few days, the seamen carelessly tossed their fearnought jackets about the deck, and the junior officers had to pick them up.[5]

On April 10, a cask of bread was opened, half of which, Clerke stated, "was so bad as to be wholly unfit for use." To succour the crew, which was coping with this "unfortunate stroke" and the extra work caused by a leak in *Resolution*'s bow, Clerke restarted the daily ration of grog, adding an "extra allowance" on April 15. Soon the ships were in the vicinity of some rumoured (we know now mythical) islands. "Rica de Oro" and "Rica de Plata" appear east of Japan on the "General Chart" published in the Admiralty's account. This is the kind of fantasy Cook revelled in destroying, and it is hard to imagine that these inscriptions would have survived to print had he lived.[6]

The ships hit snow on April 18 while crossing the forty-sixth parallel east of the Kurile Islands. Conditions worsened as the expedition edged toward 50° N. Gripped alternatively in sleet, fog, frost, and snow, the ships and crew experienced temperatures that were, Clerke reported, "exceeding severe and cold," dropping to 29° F. On April 23, a phenomenon that had been routine on Cook's Antarctic voyage occurred on this expedition for the first time. *Resolution* was sheeted with ice, "the Sleet and Snow as it fell having froze to the sails and Rigging." In a little more than three weeks, the temperature had dropped seventy degrees. When the sky cleared and Kamchatka came into view, Clerke saw that its hills were "totally cover'd with Snow." A grey curtain soon fell again, and as the fog worsened, his ship's guns were fired every hour to keep *Discovery* within auditory reach.[7]

On April 24, during a short stretch of fair weather between snow showers, Clerke had the ships approach the coast "to take a view of the Land." It was a "dreary prospect." He had made landfall at 52° 22' N, south of Avacha Bay, which holds Petropavlovsk. It appeared as if "every atom of Earth was cover'd with Snow except the Sides of some of the Hills by the Water side which broke off too abruptly into the Sea for the snow to lay upon their surfaces." A north wind – "a confounded Nipper," in Clerke's colourful parlance – blew *Resolution* and *Discovery* well off land, and during the night, the sleet and snow "froze to the Sails and rigging in such quantities that in the morning we appear'd a fair Mass of Ice." In the icy southern latitudes, Cook had commented on the dangers inherent in such conditions because of eroded manoeuvrability, but this may have been worse. King reported that "the oldest seaman among us, had never met with any thing like the continued showers of sleet, and the extreme cold."[8]

When Clerke came back in sight of land, troubled skies prevented him from establishing latitude. Uncertain of his whereabouts, he wheeled away, declaring the "miserable Weather that is now going would render any attempt to examine it madness." At this juncture, Clerke took the temperature of the crew: "The poor fellows after broiling as they have lately done several Months in the Torrid Zone are now miserably pinch'd with the Cold." Despite a sick list that was getting longer, including several cases of frostbite, he believed the crew had an "abundant allowance of Provision" and appeared to be "in very good spirits." Clerke thought acclimation would "enable us to get the better of this degree of cold very well: its the sudden change that pinches."[9]

On April 28, 1779, the ships pulled into Avacha Bay, 150 miles north of Kamchatka's southern tip on the peninsula's Pacific side. Clerke had a "clearer view of the Country than any we have hitherto had since our making the Coast. We now saw several high Mountains and various ranges of Hills totally immers'd in Snow which indeed at present covers the whole face of this part of the Earth." The highest peak was Avachinsky, an active volcano then and now. (It would soon erupt.) Clerke had "all Hands busied in knocking the Ice off the Rigging which has for some days past been exceeding troublesome to us." He wanted to head north for Petropavlovsk, halfway up the bay's east shore, but ice blocked the way. He sent King and artist John Webber (who could speak German) in a launch better suited to navigating a channel toward the town. They were to pay Clerke's compliments to any governing official and open a line of communication. Though the Englishmen had no other choice because of the ice, this tactic coincided with advice Gerasim Izmailov had provided at Unalaska. In the letter addressed to Magnus von Behm that King was then carrying into town, Izmailov advised "at their arrival in the harbour they should send a boat with the order to present this report, and thereafter they should follow with the ships." If the Englishmen were to sail in "without any notice, they would be shot at with cannon."[10]

The rowers could not get any closer to Petropavlovsk than a peninsula of ice extending a half mile from shore; the rest of the way would have to be covered on foot. King was shocked by the aesthetic prospect. The settlement consisted of a few yurts and seven log houses of "miserable appearance." Looking around the bay, there were "no more huts, no boats, no living Soul. Some few flocks of duck was all that enlightend this very silent solemn, waste prospect." This had not been expected. King's impression, gleaned from Izmailov in the Aleutians, was that Petropavlovsk was "a place of some strength and consideration." (It became so during the Soviet era, when the entire peninsula required military clearance to enter, and it remains the home port for Russia's North Pacific fleet.) King reflected on Cook's decision the previous October at Unalaska on where to spend the winter. Surveying this scene, he recalled "the truth of our late Captns Conjecture, when in a Speech to the Ships Company upon our leaving the Ice, in explaining to them the Necessity there was to put them to short Allowance, said that at Kamschatka he was sure nothing could be got, and that he had so bad an opinion of the Country that he would go and search for some refreshing port to the S[outh]ward." King concluded: "The Very idea of our situation had we wintered here, made us Shudder." As Trevenen phrased it, Cook "saved us from wintering at Kamchatka, where a want of fresh provisions and ... genial climate would probably have proved fatal to many."[11] Cook had been presciently decisive about this in the fall of 1778, over two years into a voyage some have argued he was ill-suited to commence let alone execute effectively. His reward for sparing the crew from overwintering in Petropavlovsk, as envisaged in the Admiralty's instructions, was their impudence when he failed to land in Hawaii on what they deemed an appropriate timetable.

As King's party approached the town, several dogsleds circled them. His original thought was that an official on shore was "sending a carriage," but it was merely a reconnaissance. This suspiciousness surprised King, but it was consistent with what Izmailov had hinted at in Alaska. That was not the worst of it. Trudging landward, the Englishmen sank knee-deep in the slushy ice. They tried running, thinking that would render less weight when their boots contacted the surface, but closer to shore the ice was even more rotten. Before King realized the telltale change in colour, the ice broke, and he fell in the water, nearly drowning. One of the circling Russians threw him a boat hook by which he hauled himself up on firmer ice. Stepping ashore, King was met by a detachment of fifteen soldiers. Both parties drew up: for the British, King and Webber were in the front rank, with their rowers behind them.[12]

King introduced himself by brandishing Izmailov's two letters of introduction, which Cook had presciently procured at Unalaska. King handed one to the head of the Russian unit, "a decent looking personage," and "told him as well as I could that we were English." The soldiers were dismissed, and King's party was "conduct'd to this gentlemans house which we found insufferably hot, otherwise neat and clean." The ensuing conversation was "civil and obliging," and the Englishmen learned their host was a sergeant. (Beaglehole determined his last name was Surgutski.) Izmailov's letter was opened and read but not acted on other than to send the second version "by a special messenger" to Behm, who resided in the regional capital of Bolsheretsk on the other side of the peninsula.[13]

Clerke euphemistically referred to Surgutski as the "governor of this settlement." He fed the visitors roast beef, wild turkey, and fish and lent King some dry clothes. Via Webber, King inquired into the supply and price of flour and beef. Surgutski said nothing could be transacted until

"Major Behm" agreed to it, but his tone was reassuring. A response to that request and, it was implied, the introductory letter that King had carried ashore, was expected in four days. With that, King and the others were transported back via sledge to their launch at the edge of the ice, each man to his own sledge guided by a driver and drawn by five dogs. King was grateful because "it would have been impossible for us to Walk," adding "no boys could be more pleas'd" with their sledge ride.[14]

Clerke had both ships towed by dint of "all hands" through the rotting ice to within a mile of town. A "floating stage" was then laid around *Resolution,* and by suspending water casks over the opposite side, the ship was lifted out of the water so the carpenters could "work upon the Leaks" under the starboard water line. Otherwise, Clerke noted on April 30: "We were now most idly situated." He explained: "We could not possibly have a general connection with the Shore, the passage to and from it being both difficult and dangerous." In addition to food, the ships also needed wood. This task was the more daunting because merely "the weight of a man" caused them to break through the slushy ice. The "wooding business" would have to wait for "the return of a more convenient Season," meaning warmer weather.[15]

⚓

On May 4, a two-man Russian delegation boarded *Resolution* bearing a letter from Behm. Attentive to the language barrier, Behm sent German-speaking Iachan Port as an assistant to the lead emissary, fur trader Vasily Feodositch. Clerke, through Webber, heard a translation of Behm's letter in which the "governor very genteely promis'd an abundant supply to all our Wants" and extended a "very social invitation to myself or any of my officers who would favour him with a visit." Clerke seems to have sanitized his journal, because King noted more sharply that Behm's missive was "merely complimentary" and that one element was very off-putting. According to the surgeon's mate, William Ellis, the letter "was directed to the commander of the English packet-boat." In King's pejorative phrasing, this implied the Englishmen were engaged in a "trading scheme." This seemed to rationalize Behm's decision to send a mere merchant like Feodositch as an emissary, and it explained the latter's surprise at seeing "two ships so much larger than their own Sloops in this part of the World." Ellis correctly intuited that this mistake was Izmailov's, whose introduction referred to the British ships in that manner. But Ellis erred in saying that Izmailov's communication mentioned "that there were no officers on board," which would put the Englishmen in "no better light than a set of sharpers." (In truth, Izmailov listed them all by name.) According to Samwell, Izmailov's epistle conveyed the idea that Cook and his companions were not merely traders but also "Hollanders." This, too, was wrong but could have been seeded during the testing of words from various languages, including Dutch, that had occurred early in the Aleutian island parley. (If this notion gained currency in Petropavlovsk, it provides a more persuasive origin story for the name of Unalaska's Dutch Harbor than the legend that a ship from the Netherlands anchored there in the late eighteenth century.) In any event, Clerke, who was ailing and perhaps thinking it was beneath his station to visit Bolsheretsk, sent King (who could speak French, as could Behm, it was thought) along with Webber to "settle matters."[16]

Peremptorily, John Gore volunteered to join the trek to Bolsheretsk, which would be conducted by Feodositch, Port, two Cossacks, and a handful of Kamchadale porters and oarsmen. The Englishmen were provided with warm clothing and departed on May 7, 1779. This mini-expedition was an admixture of river and land travel across Kamchatkan wilderness, and as such falls as far outside the palm-tree paradigm as any episode in the history of Cook's three voyages. For that reason, it is little known. King's journal is the sole source of information about this exotic adventure, and he gave a vivid account of it.

The brigade's first leg was up the Paratunka River, which was just then breaking up. It was slow-going against the current. King stated that the party camped that night with "the Assistance of [a] good fire and good Punch," which we must infer had alcoholic content. The next segment was expedited by lighter canoes lashed together and a fresh set of Kamchadale rowers enlisted from an inland village. The Englishmen were then transported across a portage by dog sledge, a form of travel "so curious to us that we enjoy'd it prodigiously. We strangers were not permitt'd to drive the dogs ourselves: we had therefore a man setting before us to drive and guide the Sledge, and a very fatigu'ing thing it is, requiring great skill and care to keep the Sledge up, when the road is any way on the side of hills, or the Snow so soft that one side falls in." King's sledge was "upset every mile to the great mirth of the rest."[17]

During the thirty-mile portage, King saw snow "so soft that the dogs at almost every step was up to their bellies. I could not have suppos'd they could have undergone so much fatigue." They then started down the Bolshoya River toward Bolsheretsk. A third night was spent at a village that had the salutary feature of a hot spring dammed to create what King called "a bathing pond." Via translation from Port to Webber, he learned of the medicinal qualities attributed to this pool by the local population, such as soothing "swell'd and contract'd joints." As Cook's protege, King had the presence of mind to bring a thermometer and measured the water temperature at 100 °F (what he called "bloodheat"); the air temperature was 34 °F. That evening, it snowed heavily, and though King's journal is silent on the point it is easy to imagine the Englishmen soaked in the hot water as flakes floated to Earth.[18]

Feodositch told King at the outset of the fourth day that they could expect to reach Bolsheretsk before nightfall if the river's flow was strong. But King reported: "To our mortification we found ourselves terribly empeded by the Shallowness of the water." The team had to portage every half mile. That night, they slept "on the banks," and even by the end of the fifth day it was still hard to "find a place clear of Snow." It was not until May 12 that the expedition reached Bolsheretsk, what King called the "Metropolis of Kamchatka." Approaching town, he saw a crowd gathered to greet them. Two weeks earlier (the Englishmen learned from Feodositch when they first met), Behm had pre-empted a mass exodus out of fear that the visitors in Avacha Bay were French, their great continental enemy. Now, the whole town had turned out. Having travelled for six days and presenting an unbecoming visage of bearded growth, the Englishmen drew up short, sending the Russians ahead with word that they wanted to shave and brush up. They also requested that Behm "not remain waiting for us." Behm "persist'd in his Politeness," so Gore, King, and Webber "were as expeditious as could be, and went and paid our respects to him." No equivalent event of such diplomatic import had transpired since Cape Town. King and his companions felt awkward, "making our first Salutations, bowing and scrapings being marks of good breeding that we had been now for 2 ½ years totally unaccustom'd to." Behm "had forgot his French," but Webber had "the Singular Satisfaction of conversing with him" in German. Dismissing his retinue, Behm conducted them to his house, where they were "introduced to his lady, whose manners and dress were perfectly European, and bespoke good sence and breeding."[19]

Gore, the senior official, acquainted Behm with the object of the British voyage and offered to buy flour and beef. Given what they had seen of the country, the Englishmen were surprised by Behm's confidence in being able to meet their request at all, let alone on their timeline. Gore specified June 5 as Clerke's target for departing Petropavlovsk, a date chosen less because of local conditions than Cook's instructions, which called for reaching the discovery zone during that month. Gore was eager to discuss terms, but Behm wanted to mull over the underlying logistics, so the conversation turned to international affairs, specifically the "American disturbance," as King phrased it. Behm could only tell them that war had broken out. As the session broke up, the Russian soldiers "were turnd out under Arms in Compliment to Captn Gore," and the visitors retired to "a very neat and decent house." They were provided with two sentries, a housekeeper, and a cook to prepare "a Supper according to our Cookery." This "accumilation of Politeness and civility" in such a remote place, King concluded, was "as surprising as it was flattering."[20]

Behm and Gore had a follow-up meeting the next day. The governor agreed to provide what flour and beef that he could, acknowledging that scarcity would reign until the sloops bringing supplies from Okhotsk arrived. Gore explained that Clerke would tender a bill to the Admiralty's "Victualling board," but Behm declined payment, stating it would please Empress Catherine were he to give "every assistance to her friends and Allies the English." King believed Behm was modelling a generous conduct because it had become known locally that he would soon leave Kamchatka permanently, and he wanted to demonstrate to his community of Russians, creoles, and Kamchadales "what manner they ought hereafter to behave to strangers."[21]

By the same token, Clerke modelled behaviour he had seen Cook display with Izmailov at Unalaska. He had supplied Gore with copies of charts from Cook's second voyage as a gift for Behm, presuming the major "was fond of such things." King said Behm "prov'd an Enthusiast in what related to Discoverys, and he was highly pleas'd with the Present." Clerke had also directed that Behm be shown a chart of the current voyage. Behm "was much struck at seeing, in one view," a map of the North Pacific "on the side of Asia as on that of America, of which his countrymen had been so many years employed in acquiring a partial and imperfect knowledge." Behm admired Cook's chart so much that he asked for a copy. King declined but noticed the major's quick uptake "after a Slight hint" that it had been inappropriate to ask for it; he was not offended, King supposed, at the refusal.[22]

At this moment, Behm had an epiphany. The previous year he had been surprised by a friendly overture from the Chukchi, who had heretofore resisted Russian incursions. This news came to him from Timofei Shmalev, his agent in Chukotka, concurrent with his report of two mystery ships in the waters off the northeast extremity of Asia. As King related it, the Chukchi, thinking the congenial Cook was Russian, had subsequently made a "voluntary offer of tribute" that was otherwise inexplicable to Behm. Having been treated with the "greatest kindness" by the British, the Chukchi offered to join "a league of friendship and amity" with Russia. What had been a "perfectly unintelligible" turn of events for Behm now made sense. In the postvoyage account, King grandiosely claimed that the British had schooled the Russians on the proper form of colonial management. The Russian American scholar Evguenia Anichtchenko, whose sleuthing explicated this saga, states: "Although we may doubt King's self-satisfied statement that a single moment of amicable trade with two presumed Russian ships had remedied a century of fighting, Russian sources suggest that Cook's visit to Chukotka did help precipitate a change in the Chukchi's status in the Russian Empire."[23] (As noted earlier, it also provided the British government with a hint about Cook's whereabouts.)

On the way to Bolsheretsk, Feodositch gleaned that the British sailors had a desperate desire for tobacco. Learning this, Behm had four large bags, each weighing over one hundred pounds, delivered to the visitors' house. A separate bundle was prepared for Clerke that contained sugar, tea, butter, honey, and figs plus "some other odd trifles." King said the delegation "tried to oppose his Profusion of Bounty" because it would draw down community reserves. Behm's "constant answer" was that "we must be in distress" from the duration and extent of a journey that "was to them so extraordinary, that it requir'd our Maps, and many corroborating circumstances to make them believe it." At dinner on May 14, Gore disclosed his intention to return to Petropavlovsk. Even though Behm was in the process of closing out his affairs in Bolsheretsk in anticipation of returning to Russia, he told the Englishmen that if they stayed over one more day, he would delay sailing to Okhotsk and return with them to Petropavlovsk. King noted warmly: "There was no resisting, so we agreed." Behm used the extra night to host a ball in honour of his guests. The town's ladies turned out in their finest garments "in the Russian taste." Mrs. Behm unpacked her trunk and appeared "in a rich Elegant European dress." For King, the silks, music, and dancing created "an enchant'd scene in the midst of the Wildest and most dreary country in the world."[24]

For the return to Petropavlovsk, the Englishmen were provided a set of travelling clothes, and their baggage was tended by porters. When the deputation left their lodging, Behm had "soldiers and Cossacks drawn up on one side and the whole of the Male inhabitants of the town dressd out on the other." Preceded by a drum corps, this parade "march'd to the Majors house." Along the way, they were serenaded by a song that Behm told them "was a favourite one of the Russian people in taking leave of their friends." "Madam Behm" and other ladies in their silk cloaks, lined with fur, greeted them at her house – "a most rich appearance," King wrote. After a light meal, everyone proceeded to the waterside. Behm and Gore took salutes, and cannons were fired. King stated that he, Gore, and Webber were so overcome by taking final leave of the town that they hurried to the boat to hide their emotions. With Behm, they pushed around a point and out of sight to the sound of "one cheer more."[25]

On May 16, the day Gore's party left Bolsheretsk, Clerke noted that the ships in Avacha Bay were "perfectly clear of Ice." It was now possible to get them abreast to shore "for the convenience of wooding and Watering." The seine nets were taken out of the hold, and some "fine flat Fish" were hauled in, probably halibut. The Kamchadales also traded cod, trout, and the especially desirable herring. Clerke also procured a steer from Surgutski, which allowed him "to supply the Crews of both Sloops with fresh Beef for their Sunday's Dinner." The temperature edged above 50 °F, so the sides of the ships were varnished.[26]

The Gore party reached Petropavlovsk five days later. Word had been sent ahead that Behm was accompanying Gore, so when the Russian boats came into view, the British ships' launches were sent to greet them. King was relieved to see this gesture, enhanced by the fact that his countrymen were "clean and as well drest as their scarcity of cloathing permitt'd." Behm was welcomed aboard *Resolution* with great enthusiasm. Samwell reported that the officers and scientists conversed with him "very well by means of Mr Webber." Once again, the lead topic was "how affairs stood between England and America." Behm did not have much to offer on that, but according to King, he "was struck at the robust healthy appearance" of the visitors. On this point, Samwell asserted, Behm "supposed that we had just left England, nor was he less astonished to hear that we had lost but such a small Number of Men by sickness." Perhaps no greater compliment was ever paid to the effectiveness of Cook's dietary regimen. Behm reported that the death rate on Russian cruises in the North Pacific was one-third to one-half.[27]

Clerke was not party to this conversation. In worsening health, he had already turned in. Behm chose not to disturb him. (King checked on Clerke in his cabin and was shocked by his deteriorated state.) Behm returned the next day to pay his compliments. As he boarded *Resolution,* he was again "saluted and had the Marines turn'd out for him, and all the distinction shewd him," which, King added, was amply deserved and "all we could pay." The sailors, when told that Behm had gifted them a

A View of the Town and Harbour of St. Peter and St. Paul in Kamtschatka. Engraving by B.T. Pouncy after a painting by John Webber, 1783–84. Cook had planned to stop at Petropavlovsk on his second attempt at northern discovery, and Charles Clerke, after succeeding to command, sailed there dutifully. It was from here that Clerke, through Russian intermediaries, was able to convey word to England that Cook had been killed.

"handsome present of Tobacco," offered in return several days' worth of their grog to the Bolsheretsk garrison since "brandy was scarce in the country."[28]

This was ironic. Cook's delimitation of these spirits the previous December, so that the supply might last another season in the icy latitudes, had created discord. Now the same men were prepared to give it away. Though Clerke admired their generosity, he did not let sentiment govern, and prohibited the gift. Behm was equally uncomfortable about such sacrificial charity. Instead, the officers gave Behm some rum, wine, a spy glass, several quadrants, and, according to Samwell, articles from several South Sea islands with which "he was much pleased and intended to present them to the Empress." King now regretted the casual distribution of plates, glasses, and utensils earlier in the voyage, recognizing how they might have been treasured by these "oblig'ing folks" in need of "comforts of life."[29]

Since Behm was about to leave Kamchatka, Clerke entrusted him with two packets destined for the English ambassador in St. Petersburg. Behm would personally transport a small box containing journals and charts, but because he was not going to take a direct route home (fur business in Siberia required an additional two or three months), they mutually decided that a smaller bundle of more pressing correspondence should be sent by separate "express." Behm projected that it would take seven months for this packet to reach the Russian capital by December 1779, and, indeed, it arrived in London in early January 1780 with the news announcing Cook's death. Given what King cited as Behm's "private and publick Character," Clerke had no hesitation trusting him with these materials, including copies of Cook's journal as well as his own to that point of the voyage. According to King, this decision rested on the risk that "should any misfortune attend us, the Admiralty would have a detaild account of I imagine the principal part of our discoveries."[30]

Taking advantage of Behm's knowledge of the North Pacific, the British officers engaged him in geographic dialogue, much like Cook had with Izmailov. King stated that in Behm's opinion it was "impracticable ever to push farther N than we have done." He knew of no Russian who had ever been north of the sixty-sixth parallel "by sea." Behm volunteered charts for replication but, as King noted, "the outlines of the Coast were so very unlike what we knew it to be, that I did not think it worth the copying." Behm did offer materiel of tangible value, and his subsequent explanation of this gesture to his superiors provides a detailed understanding of British intentions. In September 1779, during a stopover in Irkutsk on his months-long trip home, Behm forwarded intelligence to St. Petersburg about the Englishmen's plan to sail "north in search of a passage, and if they find it, they will sail directly to their own country." He confided further that should the Englishmen fail in the attempt, their proposal was to return to Petropavlovsk, in which case the ships would have to be resupplied again. Because he would have departed by then, Behm informed the capital that at Petropavlovsk he had volunteered whatever cordage or other naval stores that *Resolution* or *Discovery* might require at that time. He had also given Captains Clerke and Gore "a paper." King noted separately that it enjoined "all of that [Russian] Nation we might happen to meet with to give us every assistance." Nevertheless, from Irkutsk four months later, Behm assured officials in St. Petersburg that should anyone have any doubts "about these ships," Petropavlovsk was "armed with the necessary artillery" to fend off "a hostile attack."[31]

Behm's generosity was recognized by his contemporaries and by posterity. In the moment, King called him a "humane man, he gave us every assistance, enquir'd into our distresses, and reliev'd them as far as his means allow'd him." In doing so, "he has done honour to his Royal

Mistress, credit to his Country, and shewn a good example to all people in like stations."[32] This was the highest praise possible in the age of the Enlightenment. Behm might not have been a member of the Republic of Letters, but he embodied its ethos of international cooperation in pursuit of learning. John Webber drew a portrait of the genial governor, and George Vancouver later named a channel after him in Southeast Alaska.

Behm debarked *Resolution* on May 26 to the sight of salutes and the sound of "3 Cheers." King and Webber accompanied their friend ashore and sixteen miles up the Patunka River, where Behm's family was waiting and where the Englishmen were "most affectd at tak'ing our last leave." For King, Behm's "disinterest'd conduct could not fail of making us regret the parting with a man who we had little prospect of ever seeing again." He amplified these sentiments with a reflection that evoked the best of his mentor: "If every country that the stranger may be driven to, had a Behm to preside over their affairs, what honour would redound to their Sovereigns, and to their Country, and what credit it would be to human Nature; we should not then have occasion to make comparisons to the disadvantage of civiliz'd nations, where the exercize of benevolence and humanity, although so well understood, and so much talkd of, is yet so little practiced, whilst the uncultur'd Islander[s] practice these virtues in their full extent."[33] This remarkable testimony came a mere three months after Cook had been killed by "Islanders" and echoed his late commander's ruminations on the autonomous values of Indigenous cultures.

⚓

The expedition's encounter with Magnus von Behm was an uplifting interlude after the joyless sail from Hawaii. Clearly, it restored King's faith in humanity. The mood was also brightened by a celebration of King George III's birthday on June 4. Surgutski and Port were invited aboard as local dignitaries to enjoy "the entertainment of the day." The cattle Behm promised arrived from Bolsheretsk on June 6. Suddenly, the ships' companies were in a land of plenty.[34]

On June 8, Clerke drafted a letter to the Admiralty, which was to go into the express packet for dispatch three days later. In it, Clerke apologizes for its brevity, which was due to his poor health. One-fifth of the way into the letter, he relates the news of Cook's death. Citing his predecessor's view when writing at Unalaska, Clerke references the expedition's dim prospects for finding "a Northern passage into the Atlantic." Fearing that "the impediments are too numerous," he professes that "whatever can be done, shall be done." This was the Cook way. Clerke assigned King the duty of writing the longer navigational history of the voyage, included by attachment, in which King also decries a lack of thoroughness. Had God "preserved the life of our late Commander," King writes, there would have been "no apologies for the imperfect Account I send, which is principally owing to the want of his Assistance."[35]

The ships were now ready to depart Avacha Bay, only a week later than originally planned. On June 13, they were carried out with the ebbing tide and then anchored for the night near the bay's mouth. Shore parties gathered greens while Clerke awaited the optimum conditions of tide and wind for heading into open water. Then something remarkable happened, on a voyage full of such things. On the night of June 14, King noted that the sky "had a very unusual appearance." Before daylight, they were "surprised with a noise resembling distant hollow thunder." At dawn they discovered "the decks and sides coverd with the finest dust," which was likened to emery powder. Avachinsky, one of Kamchatka's contributions to the Pacific's ring of fire had erupted. (It is known for its frequent and destructive eruptions, most recently in 2001.) The "explosions" continued all day, as pea-sized cinders fell on deck. Precipitation filtered through the tephra rained as mud.[36]

This was one of the more remarkable natural phenomena to have occurred during any of Cook's voyages (a volcanic eruption on Tanna during the second expedition had also been witnessed), and many on board commented on it. For some, this was an encounter with the sublime. To King's eye, the atmosphere was cast in "the most dreadful Gloom and dismal appearance I ever saw." The murk, Samwell reported, was "like that of a solar Ecclipse which with the dreadful roaring of the Mountain was altogether very awful and alarming." Nathaniel Portlock observed that "vast Column of Ashes which the mountain had hove forth," coating the decks of the ships with ash. This situation was "so very disagreeable" that walking about one risked "being choak'd by it." Portlock probably spoke for many when he wondered about the fate of their friends back in Petropavlovsk, but he recorded laconically: "We have not had an Opertunity of Learning as we saild the next morning."[37] In the event, Avachinsky did not damage Petropavlovsk, as King would note upon his return the following August.

FIFTEEN

Diminishing Returns

When *Resolution* and *Discovery* sailed from Avacha Bay on June 17, 1779, Charles Clerke resumed his journal, having left off on May 24 because of deteriorating health and the scarcity of reportable activity. Also slowing down was the venerable chronometer that had served for the last three years and during the circumnavigation of the South Pole before that. A repair had been attempted in the still waters of Avacha Bay by a sailor, Benjamin Lyon, who had once worked for a watchmaker. As Clerke and James King reported in a memorandum probably intended for the Board of Longitude, Lyon had "found a piece of dirt between the two teeth of the wheel that carries the second hand, which he thinks to be the principal cause of its stopping." Lyon had cleaned and oiled the mechanisms and made a new spring, but the legendary clock never kept time well again and was retired from use. William Bayly's timepiece aboard *Discovery*, called "K3" by historians, served for the balance of the voyage. On *Resolution*, longitude was thereafter calculated by using astronomical methods.[1]

⚓

From sea, the Kamchatka countryside looked much different than when the Englishmen had arrived. Clerke noted the hills were "mostly clear of Snow with a fine Verdure about them." On June 28, the expedition crossed the sixtieth parallel off the Siberian mainland under a "commanding breeze" that had the ships "under all sail." Clerke was bearing down on the Arctic Circle close to Cook's prescribed schedule – 65° N in June. He was aiming for St. Lawrence Island, which he knew from the previous campaign was aligned with the centre of Bering Strait. As the ships crossed the Gulf of Anadyr, the fogs became more frequent, which made it challenging to keep *Discovery* close by. When the fog burned off, he took in a view of mountains on the Chukchi Peninsula that he deduced "must be very high from our frequently seeing them at so great a distance."[2]

St. Lawrence appeared out of the haze on July 4. King and Bayly took separate readings for latitude and longitude, which, Clerke was pleased to report, were "just the same now as when we settled it last Year." From St. Lawrence, at 63° 30' N, the ships stood toward the strait, running equidistant from the extremities of both continents. King described this exotic moment: "At ten at night, the weather becoming clear, we had an opportunity of seeing, at the same moment, the remarkable peaked hill, near Cape Prince of Wales, on the coast of America, and the East Cape of Asia, with the two connecting islands of Saint Diomede between them." There was no puzzling over geography; both Staehlin and Cook were ghosts now. Clerke felt "perfectly acquainted with our situation," but an intimation of future events came when he noted "there is a great deal of Ice adhering to the shores of this part of Asia." King stated this greatly discouraged "hopes of advancing much farther Northward this year, than we had done the preceeding." David Samwell also noted in his journal that the ice was more extensive off the Siberian coast, but he was more optimistic because there was "none in Sight" to the east. Clerke was silent on this point, much in the mode of Cook. Approaching the polar circle, there were "small pieces of Ice" floating about, some with smudges of dirt, which he concluded had "lately broke from the shores" of one continent or the other, a potentially positive sign.[3]

Just shy of the sixty-seventh parallel, Clerke bore east, notionally toward Baffin Bay. This would have been his natural inclination, as it had been for Cook, but ice crowding the Siberian coast made it mandatory. On July 7, he encountered "an extensive field of loose drift Ice, which appear'd very rotten and decay'd," another encouraging sign that the ice was breaking up in the long daylight of midsummer. Some pieces were big enough to bear the weight of their old Arctic friends, walruses. As the day wore on, larger pieces drifted past, but it soon became apparent that the ships were at the outer edge of a "firm body" of ice, which to Clerke appeared "connected with the Continent." The ice was relatively low, so King was able "to see over a great extent of it. The whole presented a solid and compact surface, not in the smallest degree thawed." The ice here, as on the Siberian coast, appeared "to adhere to the land." Clerke concluded "it was clear we could not possibly make any way to the N[orth]ward upon the American coast." At 67° 15' N, he was still three degrees south of Cook's northernmost station.[4]

In response, Clerke bore west along the edge of "this immense body of Ice to which we could see no bounds from our Masts Head." The strategy, as it had been the previous summer, was to "find some opening to the N[orth]ward before we reach'd the opposite shore," that is, Asia. Keeping the ice, as King phrased it, "close on board," the ships momentarily rounded the pack's western extremity and found it "trending nearly North." Now farther from the smooth, low, landfast ice near the Alaskan shore, the pack here was rough and uneven, rising in some places twenty feet above the water. This was the main body of the Arctic's multiyear ice, stressed and recompressed over time to create ridges and hummocks. Notwithstanding cautionary measures, floes broken off the pack were driven into the ships.[5]

Clerke mused about having "all the Season before us," but July 8 broke contrarily with what Samwell called a "heavy fall of Snow." According to King, after the weather cleared, "we found ourselves close to an expanse of what appeared from the deck solid ice," but from the masthead "several pieces were seen floating in vacant spaces of water." Pushing northwest, the ships gradually reached 69° 25' N, two full degrees higher than the initial contact with the ice. This was a first for the expedition, because Cook had never reached a station higher than he did the day he saw the blink.[6]

By July 9, the ships had sailed 120 miles to the northwest from where they had left the North American coast. In the postvoyage account, King said they went that entire distance "without seeing any opening, or a clear sea to the Northward beyond it." In his journal, Clerke reported that the pack was not only "contiguous" but also "as compact as a Wall." The ships kept dodging "immense quantities of drift Ice" well into the night, which was fully lit at that time of year, and they took care to keep the "main body well in sight" at all times. Clerke carried farther west, reaching the same meridian as the eastern tip of Siberia, but considered his prospect for getting farther north "a very poor one, for could we have clear'd all this drift Ice that solid body was a most effectual barrier." Endowed with more time than Cook had had, because of an earlier arrival in the discovery zone, Clerke headed southeast, the only quarter clear of ice, to minimize damage to the ships and to "pass a few days." Clerke was waiting to see if seasonal warming would have any effect on "these confounded impediments to our business."[7]

Clerke knew that standing off, hoping an opening would emerge, was a long shot, but like Cook he wanted to exhaust all recourses. King's account reflects his understanding that Clerke intended to head back through Bering Strait to St. Lawrence Bay, home to "our Tschutkski friends," a diversion that would necessitate an extended withdrawal from the ice edge. Clerke's failing health probably informed this strategy. Now the expedition truly *was* being led by an explorer with diminished capabilities. But instead of retreating all the way to Chukotka, Clerke dropped only two parallels. Nevertheless, this manoeuvre failed to achieve the desired result because he could not separate from the ice. It chased him, just like the pack had almost marooned Cook at Icy Cape the year before.[8]

Clerke made the best of this stalemate by hoisting out the small boats to hunt walrus. In a compelling but virtually unknown piece of natural history (because it has been eclipsed by the palm-tree paradigm), King wrote:

> The gentlemen who went on this party were witnesses of several remarkable instances of parental affection in those animals. On the approach of our boats toward the ice, they all took their cubs under their fins, and endeavoured to escape with them into the sea. Several, whose young were killed or wounded and left floating on the surface, rose again, and carried them down, sometimes just as our people were going to take them up into the boat; and might be traced bearing them a great distance through the water, which was coloured with their blood: we afterward observed them bringing them, at times, above the surface, as if for air, and again diving under it with a dreadful bellowing. The female, in particular, whose young had been destroyed, and taken into the boat, became so enraged, that she attacked the cutter, and struck her two tusks through the bottom of it.[9]

On July 11, Clerke got the ships out of a maze of ice floes and into water "perfectly clear of all incumbrances." It had been as difficult to make way south as it had north. Clerke did not fear getting trapped because it was still early summer, but nor was he sanguine about finding the Northwest Passage. The ice was simply too "abundant and compact" to think otherwise. Nevertheless, he redirected the ships north because, given the vagaries of the wind, the sea was now more clear of floating ice in that direction than to the south. Nature seemed to be toying with him. Clerke aimed to "push to the N[orth]ward before the Season is too far advanc'd upon us," but by July 13 the ships were again, as King phrased it, "close in with a solid field of ice." Informed by many sorties around polar ice packs with Cook combined with his own experience, Clerke feared "we shall find it an obstinate and unconquerable barrier to all our hopes and Endeavours."[10]

Before the expedition's final assault on the ice, Clerke lowered a boat so the "time keepers" (King and Bayly) could compare notes aboard *Discovery*. The winds were light and the weather fair, which allowed a distant view of the horizon from the masthead. The dreams of three centuries past hung in the balance. Looking north, King recorded: "We could see no limits" to the ice. "This, at once dashed all our hopes of penetrating farther; which had been considerably raised, by having now advanced ... through a space, which, on the 9th, we had found occupied by impenetrable ice." Since getting northward had been precluded, King stated that "Clerke resolved to make one more, and final attempt on the American coast, for Baffin's Bay, since we had been able to advance the farthest on this side last year."[11]

The ships ran east on the sixty-ninth parallel, twenty miles north of their westerly retreat the week before. Visibility was so limited by fog and showers of sleet and snow that *Resolution*'s swivel-gun was fired every half hour to keep *Discovery* within hailing distance. As both vessels inched ahead Clerke tried to observe any difference in the state of the ice but "could see none." There were pools of standing water here and there but generally the ice pack presented "a very firm solid Mass." Clerke was finally able to break 70° N, but soon was caught "in a large Bay form'd by the solid Ice." King found "one compact body surrounding us on all sides, except on the South quarter." Clerke reefed the topsails and backed away from the edge "as fast as I could."[12]

After waiting out a squall of snow and sleet, the crew reposted the yards, and the ships again made sail northeast. On July 18, Clerke recrossed both the seventieth parallel and Cook's track leading to the blink. When the fog lifted, he saw "ice ahead." The pack seemed to be "still in its old state." Clerke feared "it will continue long enough to frustrate all the attempts to proceed to the N[orth]ward." Nevertheless, he observed: "We must examine its boundaries and try what is to be done." This was the Cook way, and on this next two-day operation Clerke reached his Farthest North, 70° 33' N, eleven minutes shy of Cook's record though one degree west of that station.[13]

With the ships so close to the ice, King thought it odd not seeing walruses on the main pack, "yet they were in herds, and in greater numbers on the detached fragments than we had ever observed before." He failed to make the connection, but in his next sentence he recorded that "a white bear was seen swimming close by the Discovery; it afterward made to the ice, on which were also two others." Samwell also chronicled "a White Bear swimming a head of the Ship so near that we might easily have killed it, however we suffered it to escape on the Ice."[14] (The polar bear was a relatively recent arrival in the region, dating to the Pleistocene Age and its episodic advances and retreats of glacial ice. It evolved from a brown bear cluster from Siberia that became isolated during the Ice Age. One of the largest aggregations of polar bears can still be found in the area where the sailors aboard *Resolution* and *Discovery* sighted them – at

A WHITE BEAR.

A White Bear. Engraving by Mazell, after a painting by John Webber, 1783–84. Cook never spotted this icon of the North, but his crew did during the expedition's second season in the Arctic. These magnificent bears were seen swimming among the ice floes, and Webber sketched one.

the convergence of the Chukchi and Beaufort Seas. They are now tragic emblems of global warming, and their prospects for survival are alarming.)

⚓

What Beaglehole would term the "moment of final defeat" for the third voyage's quest came in the early light of July 19, 1779. In his postvoyage narrative, King recorded that "at one in the morning, the weather clearing up, we again steered to the North East, till two, when we were a second time so completely embayed, that there was no opening left, but to the South; to which quarter we accordingly directed our course, returning through a remarkably smooth water ... by the same way we had come in." *Resolution* and *Discovery* had come within fifteen miles "of the point to which we advanced last season."[15]

Retreating south, the prospect of securing bear meat could no longer be resisted. Samwell wrote: "We saw two white Bears swimming a head of the Resolution; a Boat was sent after them." Both were killed and brought on board to be weighed, measured, and butchered. "These animals," King wrote, "afforded us a few excellent meals of fresh meat. The flesh had indeed a strong fishy taste, but was, in every respect, infinitely superior to [walrus]; which nevertheless our people were again persuaded, without much difficulty, to prefer to their salted provisions." Samwell noted that these "were exactly the same as the Bears caught in the Greenland Seas."[16] Cook's ships were not going to reach those Atlantic waters, but his final voyage pieced together another commonality of the Arctic Basin.

Two days later, Clerke offered summary reflections about trying to get north of America. His July 21 entry reads:

> It is now clearly impossible to proceed in the least farther to the N[orth]ward upon this Coast and it is equally as improbable that this amazing mass of Ice should be dissolv'd by the few remaining Summer weeks which will terminate this Season but it will doubtlessly remain as it now is a most unsurmountable barrier to every attempt we can possibly make. I therefore think it is the best step I can take for the good of the service to trace the Ice over to the Asiatic Coast, try if I can find a Hole that will admit me any farther North, if not see whats to be done upon that Coast where I hope but cannot much flatter myself with meeting with better success, for this Sea is now so Choak'd with Ice that a passage I fear is totally out of the question.[17]

This strong and evocative paragraph with memorable phrasing is noteworthy for mentioning Cook's strategy, and Clerke's emulation of it, from the previous year. It is also the end of Clerke's journal.

Consumed by tuberculosis, Clerke would be dead in a month's time. Entirely to his credit, he knew that with his condition the Arctic's cold weather would kill him, but he chose to proceed for "the good of the service." His pessimism was shared by others. As King defined the moment, compact ice rendered "every effort we could make to a nearer approach to the land fruitless." For him, this was the final farewell to the idea of a "passage to Old England." Samwell fatalistically observed that having determined the ice was fixed to North America, "our next business is to follow it to the Westward till it joins to that of Asia."[18]

The ships bore across the Chukchi Sea in one last Cook-like attempt at the Siberian coast. Dodging the ice on July 21, they were temporarily embayed just above the sixty-ninth parallel on the meridian of its eastern-most cape. The Englishmen worked out of this predicament, as before, but this occasion was dispositive. Summarizing the expedition's recent experience, King wrote: "Since the 8th of this month, we had twice traversed this sea, in lines nearly parallel with the run we had just now made; that in the first of those traverses, we were not able to penetrate so far North ... as in the second; and that in the last we had again found an united body of ice, generally about five leagues [from] its position in the preceding run." This proved "that the large, compact fields of ice ... were moveable," a finding that replicated what Cook had discovered near Antarctica. King concluded there was no "well-founded expectation of advancing much farther in the most favourable seasons."[19]

Criss-crossing the Arctic took a toll on the ships. Samwell's *Discovery* "received many heavy blows on her Bows" from the floes, doing her "much damage." *Resolution* sailed south into clear water, but *Discovery*'s exit was blocked. Being hemmed in was bad enough, but *Discovery* happened to be trapped close enough to the ice edge, which, during a heavy swell, risked "heaving her violently against the ice" and damaging the ship's planks. In a calm Cook-like manner, John Gore ordered ice hooks sunk into a large piece, and by that means the crew was able to warp the ship into an oasis of tranquil water surrounded by a wall of ice that offered protection from the troublesome swell.[20]

The officers on *Resolution*'s deck viewed all this with "gloomy apprehensions," King wrote, but the situation worsened when the weather became "thick and hazy" and they lost sight of *Discovery*. In the fog, *Resolution* reversed course and bore back toward the icy bay, firing a cannon every thirty minutes in the hopes of staying in aural range until *Discovery* escaped. Eventually, a northern breeze sprang up, and Gore's ship broke free by hoisting sails to catch the wind and pressing herself through the ice. Just as she emerged, Samwell and his mates "heard a Gun from the Resolution at a great distance off, which we immediately answered." King stated that after more than four hours of separation, "we heard her guns in answer to ours."[21]

After the reunion, Gore was surprised to see that *Discovery* was not in worse shape. Nonetheless, she had become, as King phrased it, "very leaky." This was a chastening experience, so to avoid a replay, the ships dropped below the sixty-eighth parallel to get clear of the floes. This was making a virtue of necessity. On the meridian of Bering Strait's centreline, Samwell reported: "It appears impracticable to proceed to the Westward for the Coast of Asia except in a lower latitude." The ships had barely made two-thirds the distance from Icy Cape to Cape Shmidta that Cook had conducted on the sixty-ninth parallel the previous year. After descending to 68° N, the ships tried venturing west, but even here, King stated, "the ice again shewed itself." After "standing backward and forward, in order to clear ourselves of different bodies of ice," he could discern the Asian mainland to the southwest, but it was "surrounded by loose masses of ice, with the firm body of it in sight ... as far as the eye could reach." Cook had reached six degrees of longitude farther west along the Siberian coast before he turned for Bering Strait. To Samwell, this was the voyage's defining moment, one that "put an end to our Navigation in these Seas in Search of a North East and NWest Passage and here we may say that the Business of our Voyage is concluded."[22]

Samwell's journal and King's published account suggest that it was July 27, 1779, not July 19, as Beaglehole has it, that the futility of the northern quest was realized with finality. Describing both northern seasons and the "absolute impracticability" of the voyage's mission, Samwell deemed that what "Captn Cook had sufficiently proved last Year, we have now confirmed by joining the Ice from one Continent to the other between the Latitude of 67 and 70. This Ice, which probably fills up the Space between Asia and America in the same Latitudes every summer that we found it in, is a perfect bar to the Passage supposing that the two Continents do not join, which however is far from being clear."[23]

Samwell's last point revisited a question Cook had once pondered. Cook never recorded a final conclusion on whether the continents were connected, but Henry Roberts's "General Chart," published with the Admiralty's account in 1784, hints at the great navigator's thinking. Roberts's North American coastline becomes indistinct above 75° N, but the trend suggests the easterly trajectory it carries in reality. Similarly, the Siberian coast is delineated for sixty degrees of longitude west of the strait with no intimation of a connection to North America above 75° N, the uppermost parallel projected. In the late eighteenth century, the question of an intercontinental bridge was moot, for as Samwell observed, the ice pack filling "the Space between them the whole year, from our having found it there the latter end of last Summer and the beginning and middle of this, forms an Obstruction that will ever render a navigable Passage this Way totally impracticable."[24] "Forever" in this case lasted two hundred years.

In James Burney's estimation, the "approaches to either Continent obstructed by the Sea being so full of Ice, it was judged fruitless to make any more attempts." Contributing to this view was the condition of the ships. *Resolution* had been in a somewhat ragged condition the entire voyage, but now *Discovery* was stressed. King, describing its state of repair to the bedridden Clerke, said she was now increasingly at risk from "rude strokes" against the ice. A leak was producing a foot of water every four hours. Gore, unqualifiedly, had had enough. In a letter to the Admiralty inscribed at Petropavlovsk after succeeding to overall command, he stated that despite a "zig-zag" course aimed at avoiding the ice, "both Ships received so much damage (in particular the Discovery), that it became absolutely necessary to proceed to some Port to repair."[25]

King was also ready to retire to the south. Indeed, his journal records the actual tipping point. Citing "the very indifferent prospect of getting so high as the last year, or at all events till a period to[o] late to have the smallest chance of accomplishing our object," he presumed Clerke would "bid adieu the Ice." This sentence reflects a key strategic realization derived from two years' exploration at slightly varying times in the boreal summer. If they arrived too early, the ice was impenetrable, yet reaching the same station in September did not allow enough time to reach Baffin Bay. King followed up with the psychological consequence: "If a stranger was to see the countenances of the Inferior officers and men he would in compassion to them be for leaving this fruitless search." He averred that anyone considering "the situation we have been in for 3 weeks, and at the end left without a ray of hope, will then not be surpris'd at our desire of leaving this part of the world; nor after a 3 years absence of feeling some inclination of seeing our native country." This experience had an enduring effect on the Englishmen, as was recorded by Timofei Shmalev, Behm's successor, who encountered the expedition on its return to Kamchatka. He reported to St. Petersburg that "during the whole month of July they lived in danger and despair; and finally in great need."[26]

For the postvoyage account, King would compose a compelling apologia, appended by calendar to July 27, 1779, that marked the true end of the voyage's discovery phase. He recapitulated information from his journal regarding the inability to approach either continent, the toll ice took on the ships, plus his and Gore's "representations" to Clerke. In light of their recommendation, King wrote, Clerke determined "not to lose more time in what he concluded to be an unattainable object, but to sail for [Avacha] Bay, to repair our damages there; and, before the winter should set in, and render all other efforts toward discovery impracticable, to explore the coast of Japan."[27] This was an attempt to derive some, perhaps a token, navigational reward from Cook's decision to extend the voyage for another year. It is probably what Cook would have done had he lived.

As for the emotional response to Clerke's decision, King wrote:

I will not endeavor to conceal the joy that brightened the countenance of every individual, as soon as Captain Clerke's resolutions were made known. We were all heartily sick of a navigation full of danger, and in which the utmost perseverance had not been repaid with the smallest probability of success. We therefore turned our faces toward home, after an absence of three years, with a delight and satisfaction, which, notwithstanding the tedious voyage we had still to make, and the immense distance we had to run, were as freely entertained, and perhaps as fully enjoyed, as if we had been already in sight of the Lands-end.

King outlined the bracing geographic realization about reaching home: "We have to run more than half the Globe in Longitude, and in Latitude farther than the distance of the two Poles."[28]

⚓

The ships passed Bering Strait on July 30, steering for the Asian coast. At this juncture in the Admiralty's account, King appends an extensive reflection on the season of exploration just concluded in comparison to the previous year's effort, the nature and extent of the Arctic ice pack, and what these conclusions portended for the Northwest Passage and Cook's legacy. He was emulating Cook, who, at the end of the Antarctic voyage's discovery phase, composed an authoritative summary of the expedition's object and findings. King starts with the question of Siberian geography, an issue that involved problematic cartographic imagery emanating from Russian sources. These maps, King explains, had often proved unreliable because they seemed to be drawn "more according to the fancy of the compiler, than on any grounds of more accurate information," although he had found their projections of continental Asia generally accurate. Still, some questions remained. King writes: "Captain Cook was always strongly of opinion, that the Northern coast of Asia ... has hitherto been generally laid down more than two degrees to the Northward of its true position." If Cook was right, then it was probable "that the Asiatic coast does not any where exceed 70° before it trends to the Westward; and consequently, that we were within 1° of its North Eastern extremity." Cook was right. The shore of the East Siberian Sea is within the range Cook specified for forty degrees of longitude west of where he last saw it at Cape Shmidta.[29]

King then declares that had Cook "lived to this period of our voyage, and experienced, in a second attempt, the impracticability of a North East or North West passage from the Pacific to the Atlantic Ocean, he would doubtless have laid before the Public, in one connected view, an account of the obstacles which defeated this, the primary object of our expedition, together with his observations on a subject of such magnitude, and which had engaged the attention, and divided the opinions of philosophers and navigators, for upward of two hundred years." After declaring himself unequal to a task he believes should have been reserved for his mentor, King offers a few summary observations. First, he affirms the wisdom of the Admiralty's supposition, informed by Hearne's overland expedition, that any Northwest Passage had to run above 65° N and, if it existed, could only emerge in Baffin Bay or above Greenland. The voyage had also proved that Bering Strait was the Pacific basin's only point of access to the Arctic's icy latitudes. This now normative geographic understanding is a major outcome of Cook's final voyage.[30]

As for ice season's duration, King writes that he noticed that the Arctic Ocean "is clearer of ice in August than in July, and perhaps in a part of September it maybe still more free." We may infer that this surprised King (and Cook) because the Admiralty had instructed a late-June arrival in the discovery zone. This was another important finding. The pattern King describes guides the movement of ocean-going craft in the Arctic to this day. Anticipating or, perhaps more accurately, responding to criticism already circulating in London (he was writing in 1781–82) that Cook and Clerke should have gotten to 65° N earlier than they did (to provide transit time to the Atlantic), King cites Avacha Bay. That body of water, in much lower latitudes, was still frozen in May. He suspected the same for Bering Strait because in Petropavlovsk he had heard that one could cross "from one continent to the other" in winter. King points out that most of summer had to pass before the Arctic's ice would recede, and it did just as the fall equinox loomed. (The annual ice minimum is indeed approximate to September 21.) Accordingly, it would have been "madness to run from the Icy Cape to the known parts of Baffin's Bay ... in so short a time as that passage can be supposed to continue open." Unstated, but implied retrospectively, is Cook's fortune in seeing the blink of compact ice as early and as far to the west as he did. To have ventured farther east would have risked being marooned (as happened to several Arctic explorers in the nineteenth century). King believed the Northeast Passage was even less probable, an opinion based on infrequent Russian exploration in that zone and the Taymyr Peninsula's protrusion at mid-continent to 78° N. That parallel, he explains, was close to the perceived limit of northern navigation, evidenced by Constantine Phipps's ice strike at 80° N in the Svalbard Archipelago.[31] But King was wrong. The Northern Sea Route is a far more practicable interoceanic passage than its Canadian counterpart.

King then turned to a discussion of sea ice and its formation. The hypothesis that saltwater did not freeze was "suspect," he argues. King claims that Cook's view of the matter "formerly coincided with that of the theorists," but then he "found abundant reason, in the present voyage, for changing his sentiments." This is an overstatement, probably borne of King's postvoyage exposure to Johann Forster's 1778 treatise. Cook's thinking went through its most significant transformation during the Antarctic voyage, not the third. At the start of the former, he fully agreed with the "theorists" that saltwater did not freeze, but by its end he had grave doubts. In his own writings, over two voyages to the icy latitudes, Cook never came to the unqualified conclusion that Forster did – saltwater froze – but he was approaching that position. He certainly rejected the notion that the great volumes of ice floating in the world's oceans were riverine outflows. Reaffirming that thinking, King cites what Hearne said of the Coppermine, adds his study of Siberia's rivers, and concludes that "the ice we have seen, rises above the level of the sea to a height equal to the depth of those rivers," not even accounting for the extent of ice below water. King also disagrees with those who "imagine land to be necessary for the formation of ice," citing the ice-choked waters north of Bering Strait far distant from either continent. He refers the inquisitive reader to Forster's *Observations*, "where he will find the question of the formation of ice, fully and satisfactorily discussed, and the probability of open polar seas disproved by a variety of the powerful arguments."[32]

Finally, King compares the second season of Arctic exploration with Cook's first. In his estimation, Clerke did "little more than confirm the observations we had made in the first." The ships met with compact ice three degrees of latitude south of where Cook did. The Asian coast during the second season, he continues, was far more daunting than it had been the year before, for Clerke met the ice sooner (as measured by latitude) "and in greater quantities." The most significant lesson drawn from Clerke's follow-up, he argues, was derived by retracing the ice line from side to side, during which it was discerned that the Arctic's compact ice drifted across significant distances.[33]

⚓

When the ships exited the strait, Clerke was near death. Burney, reminiscing years later, stated that by this point he had been reduced to "an absolute skeleton." Sensing the end, Clerke devoted his remaining strength to dictating a letter addressed to Joseph Banks, dated August 10, 1779. "The disorder I was attacked with in the King's bench prison has proved consumptive," he begins. Clerke then tells Banks that the disease had progressed such "that I am not able to turn myself in my bed, so that my stay in this world must be of very short duration." He introduces "Mr King ... who is so kind to be my amanuensis on this occasion," calling him "my very dear and particular friend and I will make no apology in recommending him to a share of your friendship." Clerke then bids his "honoured friend ... a final adieu; may you enjoy many happy returns years in this world, and in the end attain that fame your indefatigable industry so richly deserves." He finishes the letter by signing it in his own hand. Beaglehole later deemed Clerke's missive to Banks "perhaps the most moving document in the whole history of Cook's voyages."[34]

King, who had become *Resolution*'s de facto commander, was touched by Clerke's "fortitude and good spirits" and expressed astonishment that he had hung on "till now." "Never was a decay, so melancholy and gradual," he wrote. Clerke died on August 22 off the coast of Kamchatka as the ships approached Avacha Bay. King offered a sombre testimonial, "lamenting the death of a man in the prime of his Age, whose life had been mostly spent at Sea, with few intervals of quiet or the enjoyment of satisfactions only to be met on land, amongst ones Relations and friends." For Clerke's resumé, King cited three circumnavigations of the world and an attempt at "a fourth, In the pursuit of which for the last half year, he commanded the expedition, nor did he swerve in any instance from persevering on account of his health, preferring his duty to his Country, to even his own life."[35] Though little remarked on in the literature of the expedition, Clerke spent as many days north of Bering Strait (twenty-six) as Cook did.

Samwell's reflections on Clerke's career were a compound of admiration and reservation. He noted Clerke's service during the Seven Years' War and his round-the-world voyages. Like King, Samwell allowed that Clerke's "perseverance in pursuing the Voyage after the death of Captn Cook, notwithstanding his own bad state of Health, will ever reflect Honour upon his Memory." Ungenerously, Samwell averred that "Captn Clerke was a sensible Man and a good Sailor, but did not possess that degree of Firmness and Resolution necessary to constitute the Character of a great Commander." This was an indirect compliment to Cook and points to a possible source for the many disciplinary problems aboard *Discovery* in Polynesia.[36]

With Clerke's death, Lieutenant John Gore assumed command. Two days later, the ships arrived in Avacha Bay, their ensigns at half-mast. At Petropavlovsk, Gore moved into *Resolution*'s great cabin while King relocated to *Discovery* as her captain and second in overall authority. Becoming reacquainted with the town, the Englishmen learned that Sergeant Surgutski was still commandant, and from him they learned that Magnus von Behm had left for St. Petersburg from Bolsheretsk the previous June, as planned. Always curious as to the state of affairs with the American colonies, they discovered that France had allied with the rebels. Also edifying was the transformation of the landscape. To Burney: "The Country has now a most beautiful appearance; Kamtschatka in Summer and Kamtschatka in winter one would Scarcely call the same place." Samwell was more expansive. The scene at Avacha Bay formed "a perfect Contrast to what it was on our first Arrival here in May, being then entirely covered with Snow; whereas now Hill and Dale is adorned with the most delightful and cherishing Verdure." One particular scene captured Samwell's attention. The summit "of a very high Mountain far inland" was "covered with Snow, which rising behind some Hills of a moderate height covered with Verdure presents through a Vally a most delightful and grand Picture, exhibiting the Image of Summer and Winter at one View."[37]

As much as the Englishmen were absorbed by the new look of the landscape, they were shocked by the appearance of Petropavlovsk's residents. To King, "the country was much improved in its appearance since we were last here," but "the Russians looked, if possible, worse now than they did then." The Russians said as much about the Englishmen. Since "neither party seemed to like to be told of their bad looks," King stated, "we found mutual consolation in throwing the blame upon the country, whose green and lively complexion, we agreed, cast a deadness and sallowness upon our own."[38]

Repairs and provisioning ensued, aided by Captain Shmalev who, like his predecessor Behm, came over from Bolsheretsk. Many of *Discovery*'s hull planks needed replacement, and repeated encounters with the ice had skewed some of her timbers. While the carpenters worked on the ship, all other hands were put to work fishing, gathering greens, drying gunpowder, securing new ballast, and repairing pumps, sails, and rigging. Pitch, tar, cordage, flour, and thirty-six head of beef were procured, everything the Englishmen wanted except canvas.

One exceptional duty stood out. A work gang from *Discovery* cleared a place for Clerke's interment and dug his grave. The crew's clothes were washed and mended for the funeral. On the afternoon of August 29, Clerke's corpse was ferried to shore while both ships tolled their bells. On landing, the ships' cannons roared with a twenty-gun salute. "The officers and men of both ships walked in procession to the grave," King wrote. An Orthodox priest accompanied them, as did the entire Russian garrison, who "attended with great respect and solemnity." The marines fired three volleys when Clerke's corpse was lowered into the ground.[39] La Pérouse visited the grave in 1787 and replaced its wooden marker with one made of copper.

While repairs proceeded, King went on two bear-hunting expeditions. His engaging description of these adventures in the published account cements his identity as a sensitive naturalist. Touting the "natural affection" bears show one another, he learned that the Kamchadales never fired on a young bear with its mother near "for, if the cub drops she becomes enraged to a degree little short of madness; and if she gets sight of the enemy, will only quit her revenge with her life." For that reason, sows were shot first, then "the cubs will not leave her side, even after she has been dead a long time; but continue about her, shewing, by a variety of affecting actions and gestures, marks of the deepest affliction, and thus become an easy prey to the hunters." The sagacity of the bears was not "less worthy to be remarked." King saw them toppling boulders onto smaller animals on hillsides below them.[40]

King also provided a colourful study of Kamchatka's canines. To him, they resembled Pomeranians "in shape and mien" though they were larger and mostly a dirty cream colour. Turned loose to fend for themselves in the summer, a large number roamed Petropavlovsk. Only neutered males were used to haul sledges. They were trained as pups by tying them to stakes with leather straps that could stretch. With "their victuals placed at a proper distance out of their reach [and] by constantly pulling and labouring, in order to come at their food, they acquire both the strength of limbs, and the habit of drawing, that are necessary for their future destination."[41]

While King pursued his muses, Gore was proving to be an unpopular leader. Within weeks of Clerke's death, there was grousing about his management. James Trevenen considered Gore "that old conceited American." Samwell was astonished to see Gore "degrade himself" by allowing Surgutski, a mere sergeant, to sit at "his table." Clerke, after all, had sent emissaries to meet with the far more distinguished Magnus von

Behm. Samwell also felt Gore was lingering at Avacha Bay two or three weeks longer than necessary. Yet another example of diminished leadership occurred in late September as the date of departure neared. Gore consulted, in a manner of speaking, with the officers from both ships. In a letter, he sought their insights on how the expedition might best proceed; more pointedly, he asked for signed, written responses. Though Cook occasionally secured advice from fellow officers, one strains to imagine him doing something like this. The great navigator never conducted his voyages by consensus. For good reason. Gore might have thought his process would provide independent thought, but the officers simply sought out one another's opinion and conformed their responses – the very definition of groupthink.[42]

William Harvey's letter is representative. Given the poor condition of the ships, he said the only navigational objective worth pursuing was settling "the Latitudes and Longitudes of the Principal places of the Japan Isles," though not a "compleat Survey of them." He recommended proceeding to China for resupply in advance of the push across the Indian Ocean to Cape Town. Harvey made a point of bringing up one of the major lessons from Cook's *Endeavour* voyage: avoiding "that unhealthy place Batavia." Samwell recorded that "all agreed, that going to the Eastward of Japan and touching for Refreshments at Macao in China would be the most eligible Plan for us to pursue in making our Passage home."[43]

This was the plan adopted and attempted but probably not the one a vigorous and ambitious Cook would have conducted. Years later, Trevenen told the Russian ambassador to Great Britain that before he died Cook spoke of his intention to survey the coast of China and Japan if he could not find a northern passage to the Atlantic. This would have been pursuant to his instructions to get "back to England by such Route as you may think best for the improvement of Geography and Navigation."[44] Had Cook lived, surely the composite map of his voyages would have resulted in a more accurate depiction of these East Asian coastlines than that produced after his death. This loss to geographic knowledge was eventually recouped by La Pérouse, who exploited one of the few opportunities that Cook left remaining in the Pacific basin, and only then because he died prematurely. In any event, the expedition's visit to China would loom large in the history of the Pacific Northwest and the fur trade and, consequently, the evolution of the Northwest Passage as a cartographic image.

Prior to departing Petropavlovsk, the crew brought the extra cannons stored in the hold to the quarter deck. As King explained, the ships were "now about to visit nations, our reception amongst whom might a good deal depend on the respectability of our appearance." These preparations precipitated a desertion, something Cook always had to contend with in the South Pacific but was now an issue in Kamchatka, of all places. The culprit was the marine drummer, Jeremiah Holloway, a man King described as "long useless to us, from a swelling in his knee, which rendered him lame, yet this made me the more unwilling he should be left behind, to become a miserable burthen, both to the Russians and himself." Holloway's Kamchadale consort "had often been observed persuading him to stay behind." Nonetheless, the last few days in port were dominated by festivities. The Englishmen hosted a party on shore to mark the anniversary of Catherine the Great's coronation, which was capped off by a twenty-one-gun salute from the ships at anchor. What Gilbert called "a tedious stay of seven weeks" concluded on October 10. For the second time, the expedition left the remains of its commander behind. As *Resolution* and *Discovery* sailed out of Avacha Bay, a keg of beer King had brewed using a local pine species was tapped and served to the crew to celebrate the commencement of the voyage home.[45]

SIXTEEN

Seeding the Fur Trade on the Voyage Home

Under Cook's command, the crew had procured sea otter pelts at Nootka Sound, not as a prospective trade material, but solely for protection against the cold climate in the exploration season ahead. Cook was cognizant of the Russian fur trade in the North Pacific, but he had concluded that it was impractical for his countrymen to compete in Alaskan waters because the hunting zone was so distant from British ports. The expedition's two stops at Petropavlovsk had accentuated the crew's awareness of the value of these furs, but the true epiphany came during *Resolution* and *Discovery*'s stay in China during the winter of 1779–80, after Cook was long dead. This is why it is a mistake to credit him as the originator of the Northwest Coast fur trade.[1]

The value Chinese merchants placed on sea otter fur was revelatory to Cook's men, and intelligence about where and how they had been procured was avidly sought by the British merchantmen they met in China. English traders soon reversed the track John Gore and James King had taken from the North Pacific in search of that wealth. For the final section of the account published in 1784, King penned a detailed prospectus on how the fur trade might be conducted, inaugurating a rush to the Northwest Coast from British and American ports that complemented those emanating from China and India. Thus, it was King, not Cook, who outlined how Euro-American traders might join in the bounty.

King opens his discourse by noting that sea otters were already much diminished where the Russians had reached but could still be found "in great plenty" on the Northwest Coast. At Petropavlovsk, where King had been introduced to the trade's commercial dynamics, he had been startled by "the quantity of specie in circulation in so poor a country." The Englishmen "were not less astonished than delighted" to sell furs acquired on the American coast for a material pittance in exchange for being paid in silver – thirty rubles per skin. This proved an embarrassment of riches because the outpost had "neither gin shops to resort to, nor tobacco, or any thing else" the seamen wanted. The coins "soon became troublesome companions," strewn about on deck.[2]

King came to learn in Kamchatka that the grand Russian fur-trade mart was at Kiahkta, on the modern border of Mongolia and Russia, four hundred miles south of Irkutsk. Because Russia, unlike Britain, Holland, Spain and Portugal, was not a maritime power, Kiahkta was its sole point of entry and exchange into and with China. Importers there sold furs "at Pekin at a great advance," and they were then redistributed, including to the Japanese market. This presented a mercantile opportunity of the first magnitude. In his prospectus, King asks: "If, therefore, a skin is worth thirty roubles in Kamtschatka, to be transported first to Okhotsk, thence to be conveyed by land to Kiachta, a distance of one thousand three hundred and sixty-four miles, thence on to Pekin, seven hundred and sixty miles more, and after this to be transported to Japan, what a prodigiously advantageous trade might be carried on between [Petropavlovsk] and Japan, which is but a fortnight's, at most, three weeks sail from it?"[3] This question was the first (published) inkling of British interest in the North Pacific fur trade.

King then explains how Russian packet boats would wait while "frozen up" in Avacha Bay and, "as soon as the season would permit," sail to Unalaska. He points out that the most "valuable part of the fur-trade is carried on with the islands that lie between Kamtschatka and America," discovered by Bering – that is, the Aleutian and Shumagin chains. Drawing on insights derived from Gerasim Izmailov at English Bay and in Kamchatka, King explains how the cartography that had bedevilled Cook could be so wildly off the mark. On Russian charts, "the whole sea between Kamtschatka and America is covered with islands." This distorted profusion was created by fur-trading "adventurers" striking "land, which they imagined did not agree with the situation ... laid down by preceding voyages." Then every trading venture "immediately concluded it must be a new discovery, and reported it as such on their return; and since the vessels employed in these expeditions were usually out three or four years, and oftentimes longer, these mistakes were not in the way of being soon rectified." These same "adventurers," (a put-down that contrasts their status with the importance of scientific explorers such as himself) were sure to "make a proper use of the advantages we have opened to them," a reference to Cook Inlet and Prince William Sound. This would have happened anyway over time as the otters were sequentially trapped out, but Cook's third voyage unquestionably expedited Russian expansion to the east.[4]

⚓

When the expedition left Avacha Bay in October 1779, the condition of the ships generally, and their sails and cordage specifically, made it too dangerous to sail between China and Japan, where winter storms might drive them onto a lee shore. All the officers who expressed an opinion to Gore were in agreement on this point, but it would come at a cost to geography. King believed a cruise west of Japan would afford "the largest field for discovery," but *Resolution* and *Discovery* sailed instead on the Pacific side. As a substitute, Gore intended to explore the Kurile Islands, whose northern extent had only been hazily charted. It was also uncertain whether the islands were controlled by Japan or Russia. The practical goal, reflecting the emerging mercantile tenor of the voyage, was determining if the Kuriles had serviceable harbours. As King phrased it, they might serve "either as places of shelter for any future navigators, who may be employed in exploring these seas, or as the means of opening a commercial intercourse among the neighbouring dominions of the two empires." The navigational impulse came right out of the Cook playbook, but the commercial inclination was the first for a Cook voyage. Assuredly, Cook had engaged in a form of international trade at his Cape Town supply station and with Indigenous peoples in the Pacific basin, but he never outlined or sought a commercial prospect.[5]

Gore's idea about surveying the Kurile Archipelago was literally blown away by gale force winds that took the ships far to the southeast of Hokkaido. A token effort was made to regain the Kuriles, but it was more to pursue an ethnological oddity than to achieve the previously stated navigational goal. The Russians in Petropavlovsk had regaled the Englishmen with stories of a race of "remarkably hairy" people who lived in these islands. This piece of exotica was a Cook-voyage bookend to the supposed giants of Patagonia that had been discussed aboard *Endeavour* on the way to Tahiti. In any event, the opportunity Gore missed would be seized by La Pérouse in 1786. As historian Robin Inglis points out, the French explorer's definition of the East Asia coastline "was his single most important accomplishment." That was only possible, La Pérouse recorded, because the "indefatigable activity of Captain Cook" was brought to a premature close by "the sad event that ended his life."[6]

Contrary winds forced Gore to forsake the Kuriles, so he settled on a course toward Japan. Within a few days, the ships secured a sublime view. In King's words: "We now discovered to the Westward a remarkably high mountain, with a round top, rising far inland." This was Mount Fuji, which

David Samwell found in "the Shape of a Sugar Loaf, the top of it is flat like the Crater of a Volcano." This enchanting scene was dissipated by the threat of stormy weather. The ships stood offshore, King reported, to "prevent our being entangled with the land." The gale carried them 150 miles to the east. The combination of strong currents and the North Pacific's unsettled weather at that time of year prompted the expedition to "leave Japan altogether, and prosecute our voyage to China."[7]

The cartographic consequences of Gore's initial timidity leaving Kamchatka and these storms are evident on the master chart of Cook's three-voyage track. Japan's depiction is the most recognizable deviation from our normative understanding of Pacific basin geography. This loss was realized in the moment. Gore was reluctant to abandon a track that Europeans had never explored. The only thing British navigators learned was that (contrary to maps they were familiar with) Japan was not one large island but instead a "cluster" about the size of Great Britain.[8]

Redirecting toward the East China Sea in mid-November 1779, *Resolution* and *Discovery* passed a group of small, barren islands, uninhabited at that time but known in European trading circles as the Volcano Islands, today's Bonin group. King saw "prodigious quantities" of pumice floating on the surface, which led him to conclude that "some great volcanic convulsion" had occurred. One island had a "high conical hill" at its southern tip that was "flattish at the top." This was Iwo Jima's Mount Suribachi. An odoriferous plume led Gore to call it "Sulphur Island." The ships proceeded between Formosa and the Philippines aiming for Macao (Macau), the Portuguese harbour across the Pearl River Delta from Hong Kong. They had trouble fetching that port, raising the fear that if it was missed the expedition might have to bear away for Batavia. This, King recorded, was "a place we all dreaded exceedingly, from the sad havoc the unhealthiness of the climate had made in crews of the former ships that had been out on discovery, and had touched there." Foremost among these was *Endeavour.*[9]

The expedition encountered several Chinese fishing boats on November 29, which seemed to offer a way-finding solution but, King noted, their sailors "eyed us with great indifference." However, the next day another craft approached *Discovery* and offered to send a pilot aboard. King declined because protocol dictated that the consort should follow Gore's lead. To his relief, *Resolution* soon took on a pilot, which meant they were guaranteed to make Macao and could avoid "being forced to Batavia." The "strong and eager desires of hearing news from Europe" added to his joy.[10]

On the final approach to Macao, George Gilbert noted, the commanders secured all the "journals, charts, drawings and remarks of all kinds relative to the Voyage" from the "Gentlemen" (junior officers, midshipmen, and scientists), and "a diligent search was likewise made amongst the sailors." This was standard Admiralty practice, and Cook had executed it at Cape Town near the end of his first two voyages. Samwell's journal ends at this point with a note that his chronicle was among those "given up" to preserve "secrecy with respect to our Discoveries." King recollected that this order required "some delicacy" because most of the officers and several seamen "had amused themselves with writing accounts of our proceedings for their own private satisfaction, or that of their friends." King feared some of these texts, "from carelessness or design, might fall into the hands of printers, and give rise to spurious and imperfect accounts of the voyage."[11]

By 1784, when those words appeared in print, several spurious and imperfect accounts had already been published. Cook had been much more successful in this regard. Unauthorized narratives from his first two expeditions were rare. His death, and the loss of command and control it represented, combined with the antiauthoritarian zeitgeist and the increasing popularity of travel literature, resulted in a plethora of illicit books on the third voyage. King insisted he had done his best, having assembled *Discovery*'s crew on deck, where he "acquainted them with the orders we had received, and the reasons which, I thought, ought to induce them to yield a ready obedience." He offered to keep such records in his own custody "till ... the publication of the History of the Voyage ... after which, they should faithfully be restored back to them." King was confident he had met with the "cheerful compliance" of his crew and had persuaded himself "that every scrap of paper, containing any transactions relating to the voyage, were given up." Events proved otherwise. The incipient authors anticipated the embargo and offered up duplicate copies of the narratives they intended to share with their "friends."[12]

In early December 1779, *Resolution* and *Discovery* anchored at Macao to the crew's "inexpressible joy, and satisfaction," wrote Gilbert. Repair and provisioning commenced, as did much imbibing of intoxicating beverages by seamen on shore leave. But the most eagerly sought item was news from home. Other than what little Gerasim Izmailov and Magnus von Behm had been able to communicate, the crews had been incommunicado since leaving Cape Town three years earlier. A Portuguese official directed King to "English gentlemen" living in port. King's mind was filled with "anxious hopes and fears, the conflict between curiosity and apprehension" as he and several officers walked in a "state of agitation" toward the domiciles of their countrymen. The British merchants were first queried about "objects of private concern," which proved fruitless, "as was indeed to be expected." On the other hand, "events of a public nature, which had happened since our departure, and now, for some time, burst all at once upon us, overwhelmed every other feeling, and left us, for the first time, almost without the power of reflection." King declared that after returning to their ships the officers "continued questioning each other about the truth of what we heard."[13]

Gilbert's journal contextualizes King's remark. He comments that the war with France "was a very unexpected event to us; as in general we were of opinion that the Rebellion in America wou'd have been quell'd long before that time." (George III believed the same.) King stated that he and his peers were struck "by the most poignant regret at finding ourselves cut off, at such a distance, from the scene, where, we imagined, the fate of fleets and armies was every moment deciding." The news "made us the more exceedingly anxious to hasten our departure as much as possible." In practical terms, this meant conducting a passage to the nearby port of Canton (Guangzhou), where the naval stores of the East India Company could be secured on credit, as opposed to the upfront payment that Portuguese merchants required.[14]

Gore dispatched King and a few men in a sampan to Canton, seven miles distant. King was surprised by the limited supply of provisions available there. Gilbert sarcastically noted that King "did not get so good a supply, as we did at Kamtchatka."[15] In the big picture, this mattered little, certainly in contrast to another commercial transaction King conducted on the side while in Canton, one he made relatively brief mention of in his journal. But King's published account of his sojourn there contains a series of paragraphs that were among the most consequential of any found in the literature on Cook's three voyages.

"As Canton was likely to be the most advantageous market for furs," King begins, "I was desired to by Captain Gore to carry with me about twenty sea-otter skins, chiefly the property of our deceased Commanders, and to dispose of them at the best price I could procure." From his experience at Kamchatka, King was confident he would meet eager buyers willing to pay a premium price for what had been procured for either a hatchet, a saw, or a metal item of even less substance. The English merchants who counselled King directed him to a particular Chinese retailer, who, by reputation, "would at once offer me a fair and reasonable price." Laying out his "goods," King was told they were worth Spn$300, then the dominant international currency based on the productivity of Spain's Central American mines. King was stunned by the low bid, and insisted, on the basis of the price pelts sold for in Kamchatka, "that he had not offered me one half their value." A series of counteroffers ensued. King started at Spn$1,000, but when he began to feel ill and tired of the contest, he agreed on Spn$800, roughly Spn$40 apiece and below what he suspected (and would soon confirm) was their actual market value. Departing Canton, he found solace in "a large collection of English periodical publications" secured from his countrymen living there. These magazines were a valuable acquisition because they "enabled us to return not total strangers to what had been transacting in our native country" and helped pass time "during our tedious voyage home."[16]

Returning to Macao, King discovered that "a brisk trade had been carrying on with the Chinese, for the sea-otter skins, which had, every day, been rising in their value." One seaman had sold his small bundle for Spn$800, equal to what King had wrangled in Canton. A clean, well-preserved pelt fetched Spn$120. King estimated that the Englishmen sold £2,000 worth, much less than what could have been possible had the crews kept "the quantity we had originally got from the Americans," meaning the Indigenous peoples of Nootka and Prince William Sound. Two-thirds had been sold, traded, or simply given away at Kamchatka.[17] The modern value of the cumulative sales King described would approach US$450,000.

The after-effect of these transactions is impossible to overestimate. They reshaped the history of North America by setting in motion a series of events that proved profoundly tragic for the Indigenous peoples of Alaska and the Northwest Coast and led to Anglo-American hegemony over that quadrant of the continent. All this was mediated through a few lines of text from King's account:

> When ... it is remembered, that the furs were, at first, collected without our having any idea of their real value; that the greatest part had been worn by the Indians, from whom we purchased them; that they were afterward preserved with little care, and frequently used for bed-clothes, and other purposes, during our cruize to the North; and that, probably, we had never got the full value for them in China; the advantages that might be derived from a voyage to that part of the American coast, undertaken with commercial views, appear to me of a degree of importance sufficient to call for the attention of the Public.[18]

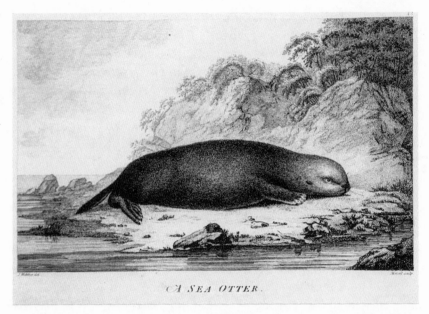

C A SEA OTTER.

A Sea Otter. Engraving by Mazell, after a painting by John Webber, 1783–84. In April 1778, Cook's crews traded with the Mowachaht of Nootka Sound for furs as part of their preparations for sailing into the cold and damp of the Far North. After Cook's death, on the expedition's way home, the same furs were found to fetch a high price in China. The prospect of a lucrative fur trade between the Northwest Coast and Asia was outlined by Lieutenant James King in his account of the voyage's proceedings subsequent to Cook's death.

This long sentence was a global seismic event, culturally and economically, and arguably of greater impact than anything James Cook had written. It precipitated the rush of British and American fur traders to the Northwest Coast in search of soft gold.

Even in the moment, this prospect created tumult within Gore's ranks. He and King dealt with dissension that the latter alleged "was not far short of mutiny." The sailors were in a "rage ... to return to Cook's River, and, by another cargo of skins, to make their fortunes." King confessed he could not help indulging similar thoughts, but his reverie of adventure and wealth closed when he reflected on "the situation of affairs at home," an allusion to the American rebellion and the allied war with France.[19]

King scripted a plan for others to pursue. He envisaged two ships from the East India Company's inventory in Canton "fitted out for sea, with a year's pay and provision." "The expence of the necessary articles for barter" with the Indigenous peoples of North America, he added cavalierly, "is scarcely worth mentioning": five tonnes of wrought iron plus a smithy and an apprentice, who could bend and cut the metal to meet local tastes. Although trifles such as beads and mirrors might work in trade, he noted that "iron is the only sure commodity for their market." After throwing a few gross of knives into the hold plus "some bales of coarse woollen cloth," the incipient Northwest Coast maritime fur trader would be equipped for action. The second ship would be necessary for security and protection against "an untoward accident."[20]

King even mapped out the prospective trading route. He advised sailing from Canton with the spring monsoon and tracking through the East China Sea to the Sea of Japan so as to fall in with the Kurile Islands by the end of June. Traders could procure wood and water in Kamchatka before shaping "their course for the Shumagins, and from thence to Cook's River, purchasing, as they proceed, as many skins as they are able." The path he outlined consciously avoided the Russian trading zone in the Aleutians. King advised against tarrying in the Gulf of Alaska, suggesting that the North American coast between 56° N and 50° N would be doubly fruitful. Sea otters could be found there in plenty and, since Cook's ships had been blown out to sea in these latitudes, productive additions to geography might also result. (This last recommendation proved unintentionally detrimental to Cook's reputation because incipient maritime fur traders who had read King later discovered several inlets in those latitudes that called the great navigator's thoroughness into question. Cook's supposed nonfeasance along the Pacific Slope resurrected the Northwest Passage as a cartographic image until the publication of George Vancouver's master chart in 1798 settled the matter.) King believed the ideal time for a trading tour of the Northwest Coast would be June through September with a return to China before October was out. Though his text carefully couches the proposed follow-up to the North American waters first visited by Cook as a geographic venture, this part of King's narrative is actually a business plan.[21]

⚓

Resolution and *Discovery* were readied for departure from Macao in early January 1780. Since the ships were about to sail into the imperially contested waters of the Indian and Atlantic oceans, cannon were brought to the top deck, and the upper works were strengthened "to give our small force as respectable an appearance as possible," King related. These steps were taken even though the periodical literature King had procured, complemented by verbal reports from English factors working in the Chinese ports, stated that Cook's ships had been exempted from any "molestation" by the French and American governments. King made note of Gore's appreciation of the "liberal exceptions" to wartime footing but also that Gore wished to preclude even inadvertent opportunities for capture.[22]

In this period of preparation, two men successfully mutinied. King assumed "these people had been seduced by the prevailing notion of making a fortune, by returning to the fur islands." The ships sailed for Cape Town on January 13, two days after the news of Cook's death had been published in London. After a short stop for water and buffalo meat at Con Son Island off the Mekong River delta, the ships recrossed the equator on February 2. Entering Sunda Strait, which separates Java from Sumatra, two European ships were spotted. "Not knowing to what nation they might belong," King testified, "we cleared our ships for action." Finding them stationary, they were hailed, and they proved to be two Dutch East Indiamen. One was bound for Europe, the other was a packet ship that trafficked in the vicinity of Batavia and was stationed offshore because of that town's "extreme unwholesomeness."[23]

The Dutch traders apprised the Englishmen that safe water could be procured from the island of Krakatoa at the western entrance to the narrows. This place, King penned, was "esteemed very healthy" compared to neighbouring districts, which did not prove to be entirely true. The only hint of volcanism, for which the island became famous, was "a very hot spring" on a neighbouring island that was used for bathing. Sensibly, Gore did not dwell in these waters for long. The memory of *Endeavour* still seared. Gilbert termed this zone the hottest "and most unhealthy place in the world," though the antics of Krakatoa's innumerable "monkies" amused some, though not all. To the commanders' great "annoyance," wrote King, most of the sailors provided themselves "with one, if not two of these troublesome animals" as pets. (The tradition of monkeys as maritime mascots is generally considered to have originated in the Napoleonic or Victorian eras, but here is proof to the contrary.) In any event, by February 16, the watering was complete, and with both ships "ready for sea," they sailed into the Indian Ocean.[24]

At this juncture during the *Endeavour* voyage, Cook began to litter his journal with obituaries for those who died from malaria and dysentery. King reflected on this, noting that when the ships had entered the Java Sea his crew "began to experience the powerful effects of this pestilential climate." Two of King's men on *Discovery* "fell dangerously ill of malignant putrid fevers; which however we prevented from spreading, by putting the patients apart from the rest" in the driest berths. Others complained of "violent pains in the head; and even the healthiest among us felt a sensation of suffocating heat, attended by an insufferable languor, and a total loss of appetite." Gore and King were uneasy, but having avoided Batavia, they eluded the worst of what King termed "these fatal seas" without loss of life. King credited this "to the vigorous health of the crews, when we first arrived here, as well as to the strict attention, now become habitual in our men, to the salutary regulations introduced amongst us by Captain Cook."[25]

Nevertheless, according to King, "the effects of the noxious climate we had left" lingered in the form of "obstinate coughs and fevers" and "fluxes" that afflicted some crewmen "till our arrival at the Cape." This latter condition, probably dysentery, was avoided on *Discovery*. King deduced that Gore's sailors on *Resolution* had "caught this disorder" from the Dutch vessels at Krakatoa. King, more cautious than Gore, had instructed his men that if any Dutch ships were encountered "not to suffer any of our people, on any account whatever, to go on board." Having seen the sister ship deviate from this protocol, he employed a prophylactic: "Whenever we had afterward occasion to have any communication with the Resolution, the same caution was constantly observed."[26] One imagines Cook employing this form of social distancing. The Hawaiians, who had once inferred that King was Cook's son, were not far off.

The voyage to the Cape of Good Hope was uneventful and conducted in mostly favourable weather. Gore toyed with the idea of bypassing the cape and sailing directly to St. Helena in the South Atlantic, but *Resolution*'s steering became problematic, so he headed for what King termed "the most eligible place, both for the recovery of his sick, and for procuring a new main-piece to the rudder." As the ships approached the African coast, sails were seen to the southwest. Fearing an encounter with French warships on their way to Mauritius, Gore had his vessels cleared for action while he maintained a safe distance. The threat disappeared in the haze, but based on intelligence gathered later in Cape Town, the mystery ship was determined to be a British East Indiaman. On April 13, 1780, *Resolution* and *Discovery* pulled into the shadow of Table Mountain.[27]

Christoffel Brand, a Dutch merchant whom King referred to as "the Governor of this place," rushed aboard. Brand expressed to King his "great affection for Captain Cook, who had been his constant guest, the many times he had visited the Cape; and though he had received the news of his melancholy fate some time before, he was exceedingly affected at the sight of our ships returning without their old Commander." Indeed, Brand was surprised to see the ships at all. A Dutch East Indiaman had seen them at Macao and, beating the British ships to Cape Town, carried not only the news of Cook's death (proof of how quickly this news spread) but also the report that his former crews "were in a most wretched state, having only fourteen hands left on board the Resolution" and seven on her consort. Brand was stunned to see the British sailors in both number and "so stout and healthy a condition." Cook would have been proud. King was at a loss to explain why the Dutch traders had propagated "so wanton and malicious a falsehood."[28]

In Cape Town, the Englishmen heard reaffirmations that "all the powers" were at war with Great Britain (technically, the Dutch were allies, but they acted as neutrals). Since China, they had been under the impression that Britain's rivals had all issued orders to their captains at sea to let Cook's ships pass unmolested, but this was not true. When officials in Madrid had learned of Cook's voyage to the North Pacific, alarm bells had rung because Spain had more to lose in this equation than any other nation. The Spanish viceroy of New Spain, Antonio de Bucareli, was directed by his superior, José de Gálvez, to impede Cook "in every way possible without using force but by taking measures to restrict assistance and supplies." (Recall that Cook had been advised to avoid contact with Spanish outposts on the Pacific Rim in anticipation of this.) Bucareli demurred, citing the dubiety of the Northwest Passage that Cook had been sent to find. Gálvez insisted that Cook be "detained, imprisoned and tried in accordance with the Laws of the Indies."[29]

Of the French commitment there was little doubt; King had first learned of it in Canton. Brand showed the Englishmen a copy of a letter from Philip Stephens, secretary of the Admiralty's board, which directly cited French policy. "With respect to the Americans, the matters still rested on report," King stated. This was properly cautious because, when the Continental Congress learned of Benjamin Franklin's pre-emptory act emulating the French, they disavowed it. (Joseph Banks later arranged to have the Royal Society recognize Franklin's gesture by awarding him with the organization's prestigious gold medal.) Gore intended to reciprocate these gestures, real and imagined, by acting as if he commanded a neutral party. Thus, he declined to join a flotilla of British East Indiamen returning to England, fearing that if this convoy fell in with enemy ships it could result in what King called "a very difficult and embarrassing situation."[30]

⚓

Resolution and *Discovery* sailed for England on May 9. They passed the equator on June 12 in a perfunctory manner, putting them on a timeline to arrive home in four weeks. Tragicomically, strong winds and currents kept Gore from sailing into the English Channel. So he stood to the west of Ireland on what turned out to be a lengthy clockwise circumnavigation of the British Isles. If Cook had been unpopular for what was perceived as his delay in landing at Hawaii, Gore received a larger measure of scorn for not getting to London on a timely basis. The trip from Cape Town to the Orkney Islands north of Scotland was the longest stretch without seeing land during the entire voyage. Gilbert called it "a very tedious passage of three months two weeks and three days" (not that anyone was counting). King was fortunate to disembark at Stromness in the Orkneys on August 22 with orders to make his way to London with the expedition's journals and charts. James Trevenen wanted to accompany King, but Gore would not let him. In a letter to his mother, Trevenen sarcastically asserted that Gore was "the only person in the fleet that does not eagerly wish to get home." Indeed, two men died during Gore's temporizing. What made matters worse, in Trevenen's estimation, was that the war with the colonies and France was raging, but he and fellow officers were lying about in "now useless hulks of ships, hitherto ... so active and vigorous in their operations" – that is, when they had been "actuated by the sublime and soaring genius of a Cook."[31]

Gore waited another month at Stromness before favourable winds took him to home port. The ships sailed up the Thames and were lashed to the docks on October 7, 1780, four years and three months after Cook had departed Plymouth. The last words in Gilbert's log recount the death due to sickness of a mere seven persons aboard *Resolution* and three, by accident, on *Discovery;* he pointedly excluded "those that were killd with our great and unfortunate Commander." Near the end of the Admiralty's account, King also emphasizes the low mortality rate from sickness. He cites the expedition's "unremitting attention" to dietary regulations "established by Captain Cook, with which the world is already acquainted." The "baneful effects of salt provisions" had been mitigated "by availing ourselves of every substitute, our situation at various times afforded." By imputation, one such provision was walrus meat, one of several "articles, which our people had not been used to consider as food." Objections to it were only overcome, he notes, by "the joint aid of persuasion, authority, and example," code for Cook's leadership attributes.[32]

King also deemed it worth mentioning that the only times *Resolution* and *Discovery* had lost sight of each other for more than part of a day had been during a storm off the coast of Hawaii in January 1779 and in the fog of Kamchatka's Avacha Bay. "A stronger proof cannot be given of the skill and vigilance of our subaltern officers, to whom this share of merit almost intirely belongs," he added, generously ending the narrative of Cook's final voyage.[33]

Conclusion

After the final Cook expedition reached port in October 1780, David Samwell wrote his friend Matthew Gregson. In the letter, he apologizes for the "very few natural Curiosities" brought home, a paucity he attributes to the voyage's length, during which most of what was collected was "destroyed one way or other." As for "artificial curiosities," meaning Indigenous material culture, he writes that the expedition was "not so badly off." Then comes the clincher: "Geography has been the grand Object of our voyage, there we shine." The journey had been anticlimactic after Cook's death, but "from the various Nations we have seen and the Discoveries we have made, I may venture to say that this Voyage is the most curious and important by far of any of the late Expeditions to ye South Sea."[1]

Samwell informed Gregson that the expedition had "made so many Discoveries that all ye old Charts or Maps of ye world are now of no use. You must buy none till our Voyage is published."[2] The master chart of Cook's three voyages, prepared by Henry Roberts, would be worth waiting for, but it was also long in the making. Before it was published, Roberts disclosed that "after our departure from England, I was instrusted by Captain Cook to complete a map of the world as a general chart." Cook provided him "the best materials he was in possession of for that purpose; and before his death this business was in a great measure accomplished." This was the first modern map of the world. Anyone looking at it today will recognize it as a conventional projection of the continents, especially the oceans separating them. The path-breaking image of the North Pacific, derived solely from Cook's third voyage, was especially noteworthy in this regard. As Lieutenant James Burney phrased it: "The ability and diligence exercised" during the voyage "will best appear by comparing the Map of the World drawn immediately after, and by keeping in mind, that the addition of so large an extent of intricate coast, before unknown, was effected by the labour of a single expedition in little more than half a year."[3] That is, by combining Cook's survey from California to Alaska's

Arctic coast with his preceding delineation of the South Pacific's terrestrial holdings, Roberts had accurately framed the world's largest basin.

Before Cook, maps were populated with speculative continents, islands, and seas. Many were festooned, as John Gascoigne phrases it, "with winsome decorations of mermaids and sea-beasts" and other fanciful images drawn from maritime legend. Roberts's map shows Cook's three tracks across a global projection that is pristine in its strict Enlightenment adherence to the grid of latitude and longitude. Otherwise, it is unadorned. The map is also significant for its placement of the Pacific basin in the centre of the image. The northeast quadrant accurately depicts the Pacific trend line of North America and the subcontinental dimensions of Alaska for the first time. Great Britain is demurely situated in the peripheral upper left of the image.[4]

The account of Cook's third voyage was published in June 1784, nearly four years after the expedition ended. The original edition, in three quarto volumes (the first two under the nominal authorship of James Cook, the third by James King), was edited by Canon John Douglas. This publication process was more complicated for Douglas than the account of the second voyage because key figures – Cook, Charles Clerke, and William Anderson – were dead. Compounding matters, King shipped out to the West Indies in 1781. After returning the following year, he prepared the narrative of events following Cook's demise, having been recommended to Douglas by Joseph Banks because of his literary ability.

Publication was also held up by the large number of plates that needed to be engraved and an extended search for special paper to print them on before they could be bound into the text. At the request of Lord Sandwich, Alexander Dalrymple, by this time the chief hydrographer of the East India Company, supervised the production of eighty-seven landscapes, portraits, and coastal profiles and all but one chart. The exception was Roberts's masterwork, which, because of its cost and importance, was prepared under the Admiralty's direction. These delays provided an opportunity for Lieutenant John Rickman (1781), Heinrich Zimmerman (1781, in German), William Ellis (1782), and John Ledyard (1783) to publish accounts before the official narrative appeared. It has been argued that interest in the Admiralty's account was dulled by these predecessors, but all two thousand copies sold out in three days. Demand was so high that in the secondary market eager buyers drove the value to double the original price, which was just shy of £5. Five additional English editions were printed that same year, and the first French copy appeared

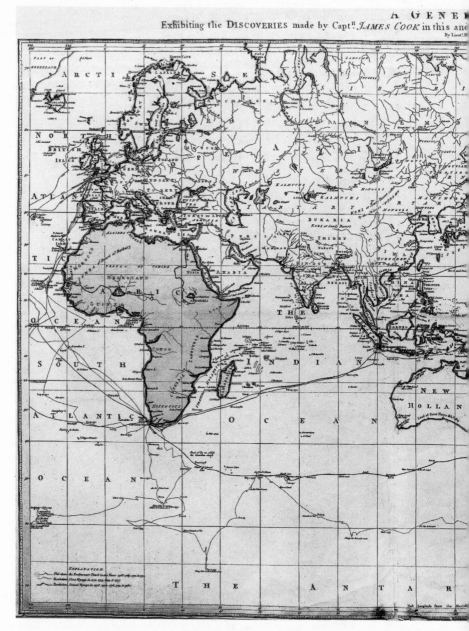

Henry Roberts, *A General Chart Exhibiting the Discoveries Made by Captn James Cook in This and His Two Preceding Voyages; with the Tracks of the Ships under his Command, by Lieut. [Henry] Roberts of His Majesty's Royal Navy.* Engraved on two sheets by John Lodge, 1784. During the third voyage, Cook requested that Henry Roberts, who had proved to be a master draughtsman, draw a map of the world as refined by Cook's own work over the course of his several expeditions. The work was largely complete

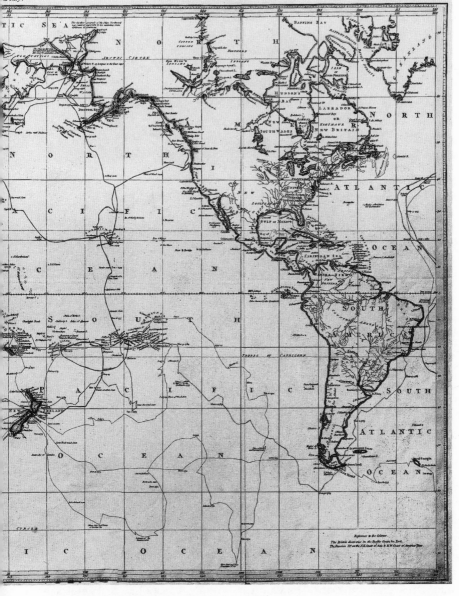

before his death. After the voyage, the Admiralty assigned Roberts with the task of completing a general chart, to be published with the official account of the voyage. The principal addition was the introduction of Cook's track for all three voyages. Arguably, the Cook-Roberts "General Chart" is the first modern map of the world. Its effectiveness as a cartographic expression is based on strict Enlightenment-era reliance on the grid of latitude and longitude.

in 1785. Another fourteen editions, in various languages, were published by 1800.[5]

The full title is probative: *A Voyage to the Pacific Ocean. Undertaken, by the Command of His Majesty, for Making Discoveries in the Northern Hemisphere. To Determine the Position and Extent of the West Side of North America; Its Distance from Asia; and the Practicability of a Northern Passage to Europe.* The serial subtitles, by their ordering, defined the actual zone of discovery to be *north* of the equator but diminished the search for the Northwest Passage as an objective. By positioning the book in this way, the Admiralty was assuredly engaging in a public-relations gambit. In Cook's instructions, finding a route to the Atlantic from the Pacific was clearly the prime directive. Even the volume's unattributed inscription to Cook's memory adds to this deception by failing to refer to the Northwest Passage explicitly as the point of his last mission.

Nevertheless, it was the exploration story, not the policy aims that had originated the venture, that drove public interest. According to Lynne Withey, the typical reader would not have been drawn to maps, disproving cartographic myths, or botany – those were "the sort of accomplishments that Cook himself valued." She asserts that Cook's appeal outside the meeting rooms of the Royal Society and the Admiralty lay in the fact that he had "discovered new lands without the violence that had marked earlier periods of exploration ... opening vast regions of the globe to the spread of European civilization by peaceful means."[6] Translated to a modern marketing idiom, Cook's brand was the application of Enlightenment idealism to the work of discovery.

Douglas confessed to taking "more liberties" than he had with the account of the second voyage but asserted that he still "faithfully represented the facts." His goal was to clothe Cook's journal "with better Stile than fell to the usual Share of the Capt." Nevertheless, as Ian MacLaren shows in his incisive case study of the expedition's North American cruise, Douglas's additions were often intrusive, with serious historiographic consequences. As Noel Currie points out, postcolonial scholars have been quick to criticize the erasure of Indigenous voices from the encounter story, but Cook's responsibility for this elision has been exaggerated. She maintains that this critical reaction is actually a response to Douglas's published text and his heroizing of the captain, not to what Cook himself actually wrote. For example, she cites comments attributed to "Cook" in the Admiralty's edition regarding the inferiority of Mowachaht and Muchalaht technology at Nootka Sound that are not found in the original manuscript annotated by J.C. Beaglehole. Extending her thesis that "the historical figure known as 'Captain Cook' has been largely a textual construction," she asserts that Douglas's "national hero" often differs from the person we find in Beaglehole's version "so significantly ... as to raise the question of who speaks in the 1784 edition." This question looms large in regard to the issue of cannibalism at Nootka, a Douglas invention. According to Currie, Douglas produced a "composite" text, and how it differed from Cook's account has not always been fully grasped.[7]

Douglas was on more solid ground interpreting Cook as a geographer. He envisaged his introduction to the first volume as an "epilogue to our Voyages of discovery." Making the best case for an expedition that did not meet its stated objective, and whose commander was killed during it, he argues that Cook "discovered a much larger proportion of the North West Coast of America than the Spaniards, though settled in the neighbourhood, had, in all their attempts, for above two hundred years, been able to do." Furthermore, Cook had ascertained "the true position of the Western coasts of America" up to 70° N and "the position of the North Eastern extremity of Asia." This established "the relative situation of Asia and America, and discovering the narrow bounds of the strait that divides them" threw "a blaze of light upon this important part of the geography of the globe, and solved the puzzling problem about the peopling of America."[8]

Expanding on a comment in King's journal, Douglas maintained that Cook's "great discovery" of Asia's proximity to America precluded Enlightenment thinkers from being "any more ridiculed, for believing that the former could easily furnish its inhabitants to the latter." To Douglas, Cook's last voyage had undermined a tenet of fundamental Judeo-Christian thought, or what the churchman termed "the credibility of the Mosaic account of *the peopling of the earth*."[9] This was among the first published postulations of the so-called Siberian land-bridge theory.

Notwithstanding his approbation of Cook's contributions to North Pacific geography, Douglas still had a major interpretive problem to tackle. He addressed it head on using the "f" word: "Though the principal object of the voyage failed, the world will be greatly benefited even by the failure, as it has brought us to the knowledge of the existence of the impediments, which future navigators may expect to meet with in attempting to go to the East Indies through Beering's strait."[10] This proved to be both a vast underestimation of Arctic realities and an overstatement of the potential of future voyages. Cook's experience merely hinted at the tribulations that would be visited on Arctic explorers, from Sir John Franklin's disappearance in the 1840s to the tragic fate of the USS *Jeannette* in 1881.

No explorer even attempted to follow Cook into Bering Strait until Otto Kotzebue, sailing for Russia, did so in 1816, and no Englishman followed him until F.W. Beechey did so in 1826. Indeed, no one made it through the Northwest Passage until Roald Amundsen made his ice-embedded voyage between 1903 and 1906. It might be asked, even now when seasonal transits of the passage are common, whether anyone has yet sailed specifically to "the East Indies" from Europe through Bering Strait.

In a provocative footnote to his introductory essay, Douglas noted: "The fictions of speculative geographers in the Southern hemisphere, have been continents; in the Northern hemisphere, they have been seas. It may be observed, therefore, that if Captain Cook in his first voyages annihilated imaginary Southern lands, he has made amends for the havoc, in his third voyage, by annihilating imaginary Northern seas, and filling up the vast space, which had been allotted to them, with the solid contents of his new discoveries of American land farther West and North than had hitherto been traced." This was a forceful metaphor for understanding Cook's contribution to geography and an indictment of several preceding generations of inventive cartographers. Douglas referred to Cook's "fruitless traverses through every corner of the Southern hemisphere," where instead of the "promised fairy land" of Terra Australis Incognita, he "found nothing but barren rocks, scarcely affording shelter to penguins and seals; and dreary seas, and mountains of ice, occupying the immense space allotted to imaginary paradises." If the second voyage, celebrated in Cook historiography for its effectiveness, was at the time deemed "fruitless," how then to calculate the value of the third? Douglas asked, rhetorically: "What numbers of new bays, and harbours, and anchoring-places, are now, for the first time, brought forward, where ships may be sheltered, and their crews find tolerable refreshments?" Within a year of those words appearing in print, British fur traders would start navigating their way to the North American harbours Cook had delineated. Douglas anticipated as much, having secured insights about the promise of what he called "fresh branches of commerce" from what King had recorded.[11]

King, who died in October 1783 prior to publication of *A Voyage to the Pacific,* also addressed Cook's legacy in the third volume. First, he outlined the accomplishments of Cook's first two expeditions: a full scoping of Tahiti and neighbouring islands, discernment of the dual insularity of New Zealand, the survey of the east coast of Australia, disproving the existence of a southern continent between 40° S and 70° S, and the discovery of South Georgia and the Sandwich Islands in the South Atlantic. Turning to Cook's final voyage, King said it was "distinguished, above all the rest, by the extent and importance of its discoveries." There were "several smaller islands in the Southern Pacific" that Cook had charted for the first time, but they were so inconsequential that King did not list them, quickly shifting to discoveries "north of the equinoctial line." The first of these was Hawaii. This chain's "situation and productions, bid fairer for becoming an object of consequence, in the system of European navigation, than any other discovery in the South Sea." That judgment seems to have been vindicated. Although Cook delineated a host of prominent features in the South Pacific – Easter Island, Tahiti, Tonga, New Zealand, Australia – they had all been explored to some extent by other Europeans. Hawaii was a genuine discovery. Describing the third voyage's legacy more generally, Barry Gough calls the expedition "an achievement in exploration of the first magnitude."[12]

King observed that Cook had "explored what had hitherto remained unknown of the Western coast of America, from the latitude of 43° to 70° North." This survey, 3,500 miles in length, put him in position to ascertain "the proximity of the two great continents of Asia and America." After passing "the straits between them," Cook "surveyed the coast, on each side, to such a height of Northern latitude, as to demonstrate the impracticability of a passage, in that hemisphere, from the Atlantic into the Pacific Ocean, either by an Eastern or Western course." Summing up, King stated: "In short, if we except the sea of Amur, and the Japanese Archipelago, which still remain imperfectly known to Europeans, he has completed the hydrography of the habitable globe." This was the opening La Pérouse seized on, but as it should happen, the fur traders following in Cook's wake and King's text revealed that the Northwest Coast of America between 45° and 55° N was also "imperfectly known," reopening the question of a mid-latitude Northwest Passage – or, at least, an evolved image of the same. In King's estimation, Cook had proved again that "by repeated trials" of exotic foods "voyages might be protracted to the unusual length of three or even four years, in unknown regions, and under every change and rigour of climate, not only without affecting the health, but even without diminishing the probability of life, in the smallest degree."[13] In that light, Cook's experiments with walrus meat and sugar-cane beer were both vindicated.

The unqualifiedly positive tone that Douglas, King, Samwell, and other contemporaries displayed toward Cook's third voyage, even in light of his death, was of a piece with the reception that followed his previous expeditions. Now that time had passed and passions had cooled, Johann Forster could refer to Cook as "my late friend, the great circumnavigator" and "the greatest geographer of the age." He added: "If we consider his extreme abilities, both natural and acquired, the firmness and constancy of his mind, his truly paternal care for the crew entrusted to him ... we must acknowledge him to have been one of the greatest men of his age." The name "*Cook* will never fall into oblivion," Forster concluded, and so far he has been proven right.[14]

⚓

The esteem Cook's generation held for him should not be overweighted by posterity, but neither should it be ignored, especially when we consider that modern authors have mined the record of the third voyage looking for evidence written by Cook's contemporaries that call his thoroughness and competency into question. This has been especially true of anthropological scholarship, the dominant strain in the literature analyzing Cook, which necessarily concentrates on his story in Polynesia. This focus has had three effects: 1) it makes a portion of Cook's life work the sum of it, 2) it minimizes the importance of his encounters in North America and excludes those in Siberia altogether, and 3) it has severely impeded our understanding of his voyaging in the icy latitudes.

Summarizing Cook's legacy, Beaglehole posed him as the St. George of cartography – the "slayer of myth." After a life of accomplishment, he argued: "It was as if some Fate had used him fully for a mighty and appointed purpose, and then immediately thrown him aside."[15] Indeed, by sailing 200,000 miles and covering every degree of global longitude several times over and 142 of 180 degrees of latitude, Cook put himself in position to correct two centuries' worth of geographic error, contradiction, and careless speculation. Beaglehole argued that to differentiate our modern understanding of the world's physical layout and what was known in 1768, when Cook first embarked for the Pacific, "we must obliterate our own geographical knowledge." Given Cook's wide-ranging scope in the basin, accentuated by his death site, it is easy to agree with John Gascoigne's assessment that "Cook had become part of the Pacific" and can never be separated from its history.[16]

But what truly distinguished Cook's voyages was their interdisciplinary nature. The man himself, to use modern phrasing, could multitask across a wide span of practical skills and theoretical interests. The former ranged from effective navigation (for which he had no equal) to the equally rare ability to manage people and shipboard life, including preservation of his crew's well-being. His path-breaking observations on the dispersion of humanity informed the nascent social science of anthropology, and he pioneered the proto-science of polar climatology. To achieve this, Cook was assisted by three teams of naturalists, astronomers, map-makers and visual artists, plus several capable officers, notably Charles Clerke and James King. It is commonly misstated that Cook was the first navigator to have a set of scientists and illustrators with him on a voyage of discovery, an exaggeration borne of the constant privileging of *Endeavour*'s voyage. Nonetheless, the template of Enlightenment discovery was institutionalized by Cook, whose tracks across the world's oceans provided the platforms that allowed naturalists and artists to make their vital contributions. Bougainville may have been first, but Cook made it the norm.

In this fashion, Cook transformed the arc of British exploration from an imperial joust with Spain into a competition with France for Enlightenment triumph. Instead of glorying in the privateering tradition of Francis Drake, British maritime accomplishment came to be measured in the amount of new understanding about the world being generated. The key to this development was what Bernard Smith terms the graphic-arts programme – portraits, cartographic charts, and drawings of flora, fauna, landscapes, and headlands that numbered in the hundreds, many of which were published. It was this visual encyclopedia, created over the course of three extensive voyages, combined with his accounts of his navigational exploits, that distinguishes Cook as the greatest explorer of all time. His reports, Lynne Withey states, created "a new kind of travel book combining a descriptive narrative with drawings of representative men and women from various parts of the world."[17]

The literature of Cook had a profound effect on Western civilization. In Alan Frost's estimation, Cook provided Europe its first glimmer of understanding about some parts of the world, thereby marking "the imagination of his age in the manner of a Newton or a Darwin." Victor Suthren expands on this notion, arguing that the goal of the Enlightenment agenda was not merely unlocking "the logic of the universe" but testing "the capacity of man as a reasoning creature to understand it fully." Over time, this agenda was supplanted at severe cost by other modes of thought and engagement. The Royal Society's code of conduct for explorers, Suthren concludes, projected "a civilized and respectful examination that contrasts somewhat painfully with the rapacious imperialism of the nineteenth century, or the racist horrors of the twentieth."[18]

There was a parallel evolution in cultural expression. The Romantic poets – William Wordsworth, Lord Byron, and Samuel Taylor Coleridge – were weaned on Cook's exploits as children. The Romantic movement, which became Europe's dominant cultural mode in the 1790s, valued emotion and subjectivity in contradistinction to the Enlightenment's emphasis on logic and empirical observation. Romantics seized the dramatic prose and vivid imagery of Cook's era and converted the signifying power of exploration from an external passage to some exotic place "out there" into allegories that internalized the explorer's experience by valorizing personal journeys of self-discovery. The Romantics appropriated the Enlightenment's discovery impulse by turning its reasoning processes inward, thereby giving vent to impassioned expressionism and laying the groundwork for the existentialists who followed. In their excesses, the Romantics occasionally rebelled against the very idea of reason itself, much in the way today's postmodernists view the notion of trustworthy knowledge as just one among many viable social constructions. To personalize this via a South Pacific analogy, Cook, the prototypical hero of the Enlightenment era (an authority figure cool under pressure, laconic in outlook, ascetic in lifestyle, and duty bound) gave way to Fletcher Christian of *Bounty* fame, the prototypical hero of the Romantic age (rebellious by definition, liberated from mission, and passionate in judgment). In our post-truth era where, to use nautical metaphors, everything is fluid and nothing is moored, we can see what has been forfeited by forsaking authoritative sources of information or epistemologies.

During the era when Cook was worshipped, which stretched from the late eighteenth century to the end of the Victorian age, the legend was born that he had formed his shipboard assistants into a generation of nautical supermen. The origin of this fable was a comment by William Windham, a member of Parliament, to James Burney, a lieutenant on *Resolution* during the third voyage. Prompted by William Bligh's extraordinary sail of a launch from Polynesia, where he was set adrift after the mutiny on *Bounty*, to Batavia in the Dutch East Indies, Windham declared: "What officers you are ... You *men of Captain Cook;* you rise upon us in every trial!" Bligh was a relentless perfectionist, but he was no Cook. The officers who served under Cook were generally effective as a cohort, but only George Vancouver became an eminent explorer.[19]

⚓

A more measured assessment of Cook's associates is the least diminution of the great navigator's legend. Frank McLynn notes: "Few things have plummeted more disastrously than Cook's reputation," principally because "he is the object of almost universal execration in all societies that have lived through colonialism." As Glyn Williams explains, the anticolonial revolutions after the Second World War informed the postcolonial out-look that emerged in the 1970s and 1980s. This movement reached a crescendo in 1992 during the public protests that accompanied the Columbus Quincentennial throughout the Western Hemisphere and In-digenous resistance to the First Fleet Bicentennial in Australia that same year. (In a prequel fourteen years earlier, the council of the Mowachaht First Nation refused to allow scholars attending a Simon Fraser University–sponsored conference about Cook to land at Nootka Sound.[20]) Not co-incidentally, the apogee of postcolonial scholarship critical of Cook, Gananath Obeyesekere's *The Apotheosis of Captain Cook,* also appeared in 1992.

Within the framework of the new social history, with its emphases on race, class, and gender, Cook either did nothing exceptional – serving merely as agent of metacultural forces such as the British Empire and Enlightenment science – or he is responsible for the modern plight of Indigenous peoples in Polynesia, Australia, the Northwest Coast of Amer-ica, and Alaska. Is this the final word on James Cook? John Robson charges that modern scholars have rewritten history by "ascribing present-day interpretations, morals and attitudes to eighteenth century events. Cook may not have deserved some of the adulation and praise that earlier gen-erations have lavished upon him, but he certainly does not deserve the negative twists currently being placed on his actions 200 years ago."[21] Pope Francis, in his celebrated remarks before Congress in September 2015, addressed the intellectual fashion Robson criticizes. Francis asserted: "Tragically, the rights of those who were here long before us were not always respected. For those peoples and their nations ... first contacts were often turbulent and violent, but it is difficult to judge the past by the criteria of the present." The important thing going forward, he averred, is not to repeat "the sins and errors of the past. We must resolve now to live as nobly and as justly as possible."[22]

Greg Dening deems the "fatal impact" argument "an unhelpful ana-lytic concept" because the "relationship between Native Polynesians and intruding Euro-American Strangers" was symbiotic. Williams finds the thesis patronizing because it depicts Pacific peoples as helpless victims and concurs with Dening that "acculturation and adaptation rather than catastrophe and extinction" are more appropriate characterizations of the encounter's dynamic. Andrew Lambert argues that Cook has been "made to carry the blame for the actions of those who came after, largely because he was transferred into a nationalist icon by white Australians and New Zealanders, for quite other purposes." In that respect, Cook has become, in Deborah Bird Rose's memorable phrase, "the emblematic invader." Stuart Murray, writing about postcolonial excesses, dismisses the idea that "only Aboriginal commentators can interpret the full meaning of Cook's contacts," terming such a position "the language of cultural primacy." This outlook is self-defeating, Murray argues, because these scholars, "while making crucial and valid points about the frequently blinkered nature of European scholarship regarding culture contact, are nevertheless themselves over-deterministic in their categorisation of Cook as simply an exemplar of empire."[23]

Nicholas Thomas synthesizes this discourse by citing an Indigenous memorial at the mouth of Australia's Endeavour River. The Milbi Wall, a distillation of Indigenous myth and history, asserts that Cook was not responsible for colonial impositions on the Guugu Yimidhirr. His arrival was astonishing, but, quoting from the memorial's text, "Cook gained the confidence of the Aboriginal people without violence." It was the ensuing gold rush that created conditions where "greed left no room for respect." Sacred places were trampled, food and water were stolen, and pathogens were introduced. Guns killed many. These events were followed by the cultural repressions of the missionary period. Within this context, Thomas concludes: "Cook's visit does not define what has happened since. It amounts, rather, to the exception that proves the rule." He was "an anomaly that prefigured nothing."[24] Cook would have wanted no part of colonial barbarism or its evangelical overlay.

An important semantical aspect of the idealization of Cook from his own time to Beaglehole's in the middle of the twentieth century was lack of nuance in the use of the term "discovery." Bernard Smith once trenchantly observed: "The discovery of the world is really a subject for prehistorians. Cook was not a discoverer of new lands in any fundamental sense of the word." His discovery of South Georgia and the Sandwich Islands in the South Atlantic and his intimation of Antarctica are evidence to the contrary, but Smith's larger point holds: wherever Cook went, he usually found people who had lived there for generations. Smith avows that there may have been a preliterate "anti-Cook" sentiment among Indigenous peoples in the Pacific in the immediate aftermath of his voyages, but it only became salient when it surfaced in the writings of Europeans who were disposed to discredit Cook for their own purposes.[25] Nineteenth-century Christian missionaries in Hawaii propagated spurious histories about Cook as retribution for, it was supposed, the sin of allowing himself to be treated as a deity.

This process has been repeated by modern scholars in solidarity with the wronged, distressed, and displaced Indigenous inhabitants of places Cook might have been the first European to visit, but it has come at a price to historiography. In *From Maps to Metaphors*, Robin Fisher and Hugh Johnston decry the "manipulation of Cook's memory to suit current social and political concerns." They add that it is possible to faithfully tell the Indigenous side of the encounter story without diminishing the achievements of Euro-American explorers, but "laundering history for public consumption does not lead to a better understanding of the past." In a related vein, Daniel Clayton asserts that decentring Cook from the encounter story is appropriate but can be overdone by reflexively equating "travel writing and narratives of exploration to a domineering imperial gaze and ineluctable colonial project."[26] The "fatal impact" interpretation is an oversimplification, just like the original idealization of Cook was.

Ironically, no explorer before or after Cook was as aware of, or sensitive to, the cultural consequences of his encounters. During the first voyage, Cook said of Indigenous Australians: "They are far more happier than we Europeans; being wholy unacquainted not only with the superfluous but the necessary Conveniences so much sought after in Europe ... They live in a Tranquillity which is not disturb'd by the Inequality of Condition: The Earth and sea of their own accord furnishes them with all things necessary for life, they covet not Magnificent Houses, Household-stuff etc."[27] Few criticisms of Western materialism have been more incisive than this one, especially given its August 1770 vintage. Cook's biting analysis could today be considered progressive thinking about the political economy of the modern world. The ethos of limitless acquisition remains unassailable outside of a few pockets of people who embrace the limits of growth in the interest of preserving the sustainability of a finite environment. In some ways, Cook comported himself with more sensitivity in regard to the world's poor (if, indeed, they were poor) than many elements of today's globalized economy.

Cook's openness to the worldview of Indigenous people during the *Endeavour* voyage was not an anomaly. During the second expedition's call at Tanna in Vanuatu, Cook's party was surveying the island when they met two dozen islanders. The residents provided the foreigners with food and helpful directions back to their ship, prompting Cook to reflect that for these "Civil and good Natured" people "its impossible for them to know our real design, we enter their Ports without their daring to make opposition, we attempt to land in a peaceable manner, if this succeeds its well, if not we land nevertheless and mentain the footing we thus got by the Superiority of our fire arms, in what other light can they at first look upon us but as invaders of their Country."[28]

By expressing such reservations, Cook wrote up the particulars of his own impeachment or, more properly, an indictment of the culture that sent him on his missions, ready made for future scholars to marshal. He recognized the unintended but nonetheless deleterious effects of his encounters, and he was equally aware of the global pattern of abuse. After describing the cultural damage caused by his crew's interaction with the Māori, Cook stated that the "concequences of a commerce with Europeans" had introduced "among them wants and perhaps diseases which they never before knew and which serves only to disturb that happy tranquillity they and their fore Fathers had injoy'd. If any one denies the truth of this assertion let him tell me what the Natives of the whole extent of America have gained by the commerce they have had with Europeans."[29]

Lynne Withey reminds us that, given the trajectory of global history, no Indigenous population was going to "remain undiscovered forever." A further complication lies in the fact that "some of the Pacific islanders welcomed their involvement with Europeans." In that sense, "discovery was a two-way business, although certainly the advantages in the exchange were heaped on the side of Europeans." John Gascoigne observes that, "on the whole, the voyagers approached these new societies with a reasonably open mind and with little compulsion to remold them into their own image and likeness."[30] As a product of his own time, Cook assuredly felt he came from a superior civilization, but he tried to consider other cultures on their own terms. Accordingly, Cook was among the first Europeans to realize that the previously unfamiliar peoples he met, even the least advanced technologically, maintained societies with autonomous value. He surely thought of them as alien, but they were also quite worthy. It is no exaggeration to state that Cook popularized a relativistic outlook that propelled Western civilization toward greater respect for Indigenous peoples.

The whole point of voyages such as Cook's, Noel Currie explains, was incorporation: bringing "the territories, products, and inhabitants of the non-European world into European knowledge." But the more sensible explorers knew the limits of their compass. James King pointed out the weaknesses inherent in his ethnographic methodology. Writing at Nootka Sound, he opined that cultural comparisons were dubious because "we cannot be said to converse with the people, we can only judge from outward actions, and not knowing all the Causes that give rise to them." Given these limitations, "we must be constantly led into error;

this also inclines us to form conclusions in the narrow confind sphere of our observations, and what has immediately happend to our selves; whence one person will represent these People as Sullen, Obstinate, and Mistrustful, and another will say they are docile, good natured and unsuspicious."[31]

Despite these limits to cultural analysis, the many admirable qualities Cook found in the cultures he met, along with similar observations by Bougainville on behalf of the French Enlightenment, guided Europe to an appreciation of what those distant Pacific civilizations could teach. As Gascoigne conceives it, Europeans began to exhibit a less exalted view of themselves and developed "an increasing recognition of common humanity." Transcultural insights gleaned from explorers by readers of their travel accounts at home became increasingly widespread in the Western mind. Greg Dening states that the Polynesian encounter was the cultural moment when "Europe comprehended the artificiality of its own civilisation before the simplicity of native lives."[32] The valorization of pluralism and multiculturalism that dominates the avowed sensibility of Western civilization in the modern era originated with Enlightenment-era exploration.

There is another context within which to evaluate Cook's encounters and one with great resonance in the twenty-first century – globalization. Gascoigne says Cook's voyages "provided the maps and cultural and sci-entific cargo for a much more thorough-going European penetration of the Pacific," even if he ignored the prospect of Christian evangelization in the Pacific basin and never envisaged the imperial aftermath of his ef-forts. As noted earlier, Cook's attitude about British prospects in the North Pacific fur trade was dismissive; the prescription for what followed was written by James King. Nevertheless, unintentionally or not, Cook was in "the vanguard of a global movement which was to leave no corner of the earth unconnected with another."[33]

Bernard Smith presaged Gascoigne's thesis with the following remark, which ends his path-breaking interdisciplinary study of the great navigator: "It could be said of Cook more than of any other person that he helped to make the world one world; not a harmonious world, as the men of the Enlightenment had so ardently hoped, but an increasingly interdepend-ent one. His ships began the process of making the world a global village." In that sense, as Smith states elsewhere, Cook was "one of the great form-ative agents in the creation of the modern world." While aspects of Cook's voyages reflected the spread of international rivalries, they also under-mined them by emphasizing the reality of global interconnectedness.[34]

Smith pointed to another idiosyncrasy in Cook historiography: discussions of his achievements tend to turn into a debate "as to whether modern industrial society is a curse or a blessing." William Hunt analyzed this quasi-theological rationale in the formulation of his argument that anti-Cook recrimination gained currency because "the Lost Eden theme is ancient in our culture." Explicating this biblical metaphor, Hunt adds: "We are happy when we rediscover our fall from Eden, since it is satisfying to find meaning in one's experience." The danger in seeing exploration merely as "another instance of Eden's destruction" is that it risks becoming simply "an exercise in exposing our ancestor's wrongs": it does nothing to address the consequences of inheriting that past, suggesting that modern society is complicit by virtue of the gain derived from the world our predecessors created. Striving for a reconciliatory equilibrium, Simon Winchester writes: "The benefits of Western modernity are quite obvious and should be sought after by all. But the wise benefits of antiquity should not be discounted either."[35]

Summing up, Smith observed that "the time is now long past when James Cook can be regarded, simple mindedly, as a hero of the Empire and a discoverer of new lands." The challenge with Cook studies going forward is finding interpretive balance. Anne Salmond has addressed this dynamic, noting that "Cook has been portrayed as the great white hero bringing Western civilization to a benighted land. Then it became fashionable to denounce him as the harbinger of white colonisation [and] ... the ills and harms that would follow." She called these explanatory extremes "one-eyed caricatures making propaganda out of past events."[36] No distortion has been more tendentious than the interpretive trope that positions Cook (and all Enlightenment-era explorers for that matter) under the gothic tyranny of Kurtz, Joseph Conrad's fictional character in *Heart of Darkness*. Nor is it fair to liken Cook's worst moments of conflict at Poverty Bay or Moorea with the "battle" at Wounded Knee or to liken his trading practices to duplicitous First Nations treaty councils.

⚓

But moving from historiography to present-day realities, what can be done about dissipating polar ice packs and displaced and disrupted Indigenous peoples? And what about collateral concerns such as the discovery of renewable energy systems, the preservation of clean water, or the development of food supplies in parts of the world struck by chronic famine? The answers may repose in a new Age of Enlightenment. The wisdom necessary to mitigate these and other modern challenges lies in the paradigm of the original Enlightenment – a quest for scientific knowledge and cultural understanding that, by tolerating differences, crosses intellectual and geographic boundaries. Indeed, the very notion of solving problems, overcoming difficulties, and enhancing the human prospect – in a word, progress – is an Enlightenment ideal.

Cook was not an entirely disinterested agent of the British Empire. But it would be unfair to say that his motivating ethos was power or wealth. He was inspired to map and otherwise illustrate the unknown, define the ultimate dispersion of humanity, and describe the principles that guide polar hydrology. Western civilization, at its best moment during the Enlightenment and with Cook as its emblematic figure, was the first culture in history to conduct a systematic study of the Earth's physical processes and its many polities. This analytical model allowed him to discern both differences and commonalities, but, more importantly, it enabled him to draw valuable lessons about the universality of the human experience. We still need explorers who can cross cultural divides, navigate through common problems such as the prevention and treatment of pandemics, and chart pathways toward a healthy global civilization and the planet on which its sustainability depends.

Notes

INTRODUCTION

1 James Cook, *The Journals of Captain James Cook on His Voyages of Discovery*, vol. 2, *The Voyage of the* Resolution *and* Adventure, *1772–1775*, ed. J.C. Beaglehole (Cambridge: Cambridge University Press/Hakluyt Society, 1961), 322. Hereafter, the journals, including vol. 1, *The Voyage of the* Endeavour, *1768–1771* (1955), and vol. 3, *The Voyage of the* Resolution *and* Discovery, *1776–1780* (1967), are cited as *JCJC*, followed by the volume and page number.

2 Tony Horwitz, *Blue Latitudes: Boldly Going Where Captain Cook Has Gone Before* (New York: Picador/Henry Holt, 2002), 219.

3 *JCJC*, 1:74.

4 Horwitz, *Blue Latitudes*, 220.

5 Simon Winchester, *Pacific: Silicon Chips and Surfboards, Coral Reefs and Atom Bombs, Brutal Dictators and Fading Empires* (New York: Harper, 2015), 4.

6 For the saliency of the ethnohistorical perspective in Cook studies, see Greg Dening, *Performances* (Chicago: University of Chicago Press, 1996); Marshall Sahlins, *How "Natives" Think: About Captain Cook, For Example* (Chicago: University of Chicago Press, 1995); Anne Salmond, *The Trial of the Cannibal Dog: The Remarkable Story of Captain Cook's Encounters in the South Seas* (New Haven, CT: Yale University Press, 2003); Gananath Obeyesekere, *The Apotheosis of Captain Cook: European Mythmaking in the Pacific* (Princeton, NJ: Princeton University Press, 1992); and Nicholas Thomas, *Cook: The Extraordinary Voyages of Captain Cook* (New York: Walker, 2003).

7 *JCJC*, 3:ccxxi–ccxxii. Robin Fisher and Hugh Johnston, eds., *Captain James Cook and His Times* (Seattle: University of Washington Press, 1979), 3–4; and Michael E. Hoare, "Two Centuries' Perceptions of James Cook: George Forster to Beaglehole," in Fisher and Johnston, *Captain Cook and His Times*, 214.

8 *JCJC*, 3:ccxxi–ccxxii.

9 *JCJC*, 3:cvii–cviii.

10 Thomas, *Cook*, 332.

11 *JCJC*, 3:vii, viii. Authors quoted, in order: Anne Salmond, "Tute: The Impact of Polynesia on Captain Cook," in *Captain Cook: Explorations and Reassessments*, ed. Glyndwyr Williams (Woodbridge, UK: Boydell, 2004), 78 ["behaviour had shifted"]; Glyn Williams, *Naturalists at Sea: Scientific Travelers from Dampier to Darwin* (New Haven, CT: Yale University Press, 2013), 122 ["out of character"]; and John Robson, "A Comparison of the Charts Produced during the Voyages of Louis-Antoine de Bougainville and James Cook," in Williams, *Captain Cook*, 144 ["tired and sick man"].

12 *JCJC*, 2:587.

CHAPTER 1: THE NORTH SEA AND CANADA

1 Sophie Forgan, "James Cook from Yorkshire" (paper presented at the "Imagining Anchorage Symposium," Anchorage, Alaska, June 18, 2015), 3. Copy in possession of the author.

2 Ibid., 5, 14–15; Richard C. Allen, "'Remember Me to My Good Friend Captain Walker': James Cook and the North Yorkshire Quakers," in Williams, *Captain Cook*, 21–36.

3 Forgan, "James Cook from Yorkshire," 19.

4 Victor Suthren, *To Go upon Discovery: James Cook and Canada, from 1758 to 1779* (Toronto: Dundurn, 2000), 7.

5 Ibid., 60.

6 D. Peter MacLeod, *Northern Armageddon: The Battle of the Plains of Abraham and the Making of the American Revolution* (New York: Knopf, 2016), 15.

7 J.C. Beaglehole, *The Life of Captain James Cook* (Palo Alto, CA: Stanford University Press, 1974), 41.

8 C.P. Stacey, *Quebec, 1759: The Siege and the Battle* (London: Macmillan, 1973), 51.

9 Suthren, *To Go upon Discovery*, 79.

10 Ibid., 85–86.

11 Ibid., 88.

12 Ibid., 93.

13 Stacey, *Quebec*, 119, 146, 172.

14 Suthren, *To Go upon Discovery*, 100.

15 John Robson, *Captain Cook's War and Peace: The Royal Navy Years, 1755–1768* (Barnsley, UK: Seaforth, 2009), 101 ["fast in the ice"], 103 ["thick fogs"].

16 Ibid., 106.

17 Ibid., 109.

18 Suthren, *To Go upon Discovery*, 119.

19 Beaglehole, *The Life of Captain James Cook*, 59.

20 Suthren, *To Go upon Discovery*, 131.

21 Ibid., 133.

22 Ibid., 140.

23 Ibid., 147.

24 Ibid., 148.

25 Wain Fimeri, Paul Rudd, and Matthew Thomason, dirs., *Captain Cook: Obsession and Discovery*, written by Vanessa Collingridge and Cam Eason, produced by Andrew Ferns (Sydney: Australian Broadcasting Corporation with Cook Films, Ferns Productions, South Pacific Pictures, and December Films Productions, 2007). This documentary was the outgrowth of Collingridge's *Captain Cook: A Legacy under Fire* (Guilford, CT: Lyons, 2002).

26 Felipe Fernández-Armesto, *Pathfinders: A Global History of Exploration* (New York: Norton, 2006), 349.

27 *JCJC*, 2:658.

CHAPTER 2: THE REPUBLIC OF LETTERS

1 Joseph Banks, *The Endeavour Journal of Joseph Banks, 1768–1771*, ed. J.C. Beaglehole (Sydney: Angus and Robertson, 1962), 1:103.

2 *JCJC*, 1:xxii, cii.

3 Beaglehole, *The Life of Captain James Cook*, 131–32; John Hawkesworth, *An Account of the Voyages Undertaken ... for Making Discoveries in the Southern Hemisphere ...* (London:

W. Strahan and T. Cadell, 1773), 1:438, available online at southseas.journals/nla.gov.
au; Frank McLynn, *Captain Cook: Master of the Seas* (New Haven, CT: Yale University
Press, 2011), 455.

4 Simon Baker, *The Ship: Retracing Cook's* Endeavour *Voyage* (London: BBC, 2002), 36.

5 McLynn, *Captain Cook*, 16 [meritocratic avenue]; and Martin Dugard, *Farther Than Any
Man: The Rise and Fall of Captain James Cook* (New York: Washington Square, 2002), 11.

6 For Cook's primacy over Bougainville, see Robson, "A Comparison of the Charts," 144,
where he cites John Noble Wilford for crediting *Endeavour* as "the first scientific voyage
of discovery."

7 Baker, *The Ship*, 15, 20.

8 Ibid., 107–8.

9 Hawkesworth, *An Account*, 23:v.

10 Banks, *The* Endeavour *Journal*, 1:36.

11 Fernández-Armesto, *Pathfinders*, 145–46, 208, 223.

12 Barbara Belyea, "Just the Facts: Vancouver and Theoretical Geography" (paper presented at the "Vancouver Conference on Exploration and Discovery," Simon Fraser
University, April 23–26, 1992). Copy in possession of the author.

13 Howard T. Fry, *Alexander Dalrymple (1737–1808) and the Expansion of British Trade*
(Toronto: University of Toronto Press, 1970), 35, 267.

14 Ibid., 59; Beaglehole's *The Life of Captain James Cook*, 125, poses a foundational element
of the caricature of the naive Dalrymple candidacy by stipulating that the geographer's longest sail ever was a mere nineteen days, a view reiterated in Collingridge, *Captain
Cook*, 86, and McLynn, *Captain Cook*, 70. Suthren, *To Go upon Discovery*, 160, mocks
Dalrymple as a "quasi-passenger" aboard *Cuddalore*.

15 Fry, *Alexander Dalrymple*, 98, 17. *JCJC*, 1:ciii.

16 *JCJC*, 1:lxxix; Fernández-Armesto, *Pathfinders*, 291.

17 *JCJC*, 1:cclxxxi.

18 *JCJC*, 1:cclxxxiii.

19 Banks, *The* Endeavour *Journal*, 1:29; and Robson, "A Comparison of the Charts," 142–
44, 160.

20 *JCJC*, 1:514.

21 *JCJC*, 1:514.

22 Robert J. Miller, *Native America, Discovered and Conquered: Thomas Jefferson, Lewis and
Clark, and Manifest Destiny* (Westport, CT: Praeger, 2006).

23 *JCJC*, 1:514–15.

24 *JCJC*, 1:517–19.

25 John Robson, *Captain Cook's World: Maps of the Life and Voyages of James Cook R.N.* (Seattle:
University of Washington Press, 2000), 12; and Lynne Withey, *Voyages of Discovery:
Captain Cook and the Exploration of the Pacific* (Berkeley: University of California Press,
1989), 140.

CHAPTER 3: THE SOUTH PACIFIC

1 *JCJC*, 1:59.

2 Banks, *The* Endeavour *Journal*, 1:221.

3 *JCJC*, 1:74.

4 *JCJC*, 1:62.

5 Banks, *The* Endeavour *Journal*, 1:239–40.

6 Ibid., 1:283.

7 Hawkesworth, *An Account*, 23:111; and Banks, *The* Endeavour *Journal*, 1:283.

8 Banks, *The* Endeavour *Journal*, 1:286.

9 Ibid., 1:308.

10 *JCJC*, 1:161.

11 Hawkesworth, *An Account*, 23:286; and Banks, *The* Endeavour *Journal*, 1:399.

12 *JCJC*, 1:170.

13 Thomas, *Cook*, 91; *JCJC*, 1:171; and Banks, *The* Endeavour *Journal*, 1:403.

14 *JCJC*, 1:193.

15 Banks, *The* Endeavour *Journal*, 1:468; and *JCJC*, 1:254.

16 *JCJC*, 1:254; and Banks, *The* Endeavour *Journal*, 1:468.

17 Banks, *The* Endeavour *Journal*, 1:424, 440, 442 ["go hard"], 469, 471–72.

18 *JCJC*, 1:262, 290.

19 *JCJC*, 1:290–91.

20 *JCJC*, 1:272–73.

21 Banks, *The* Endeavour *Journal*, 2:38–39.

22 Ibid., 2:41.

23 Ibid., 2:79, 81.

24 Sydney Parkinson, *A Journal of a Voyage to the South Seas in His Majesty's Ship the Endeavour* (1784), 186, southseas.nla.gov.au.

25 *JCJC*, 1:380.

26 For examples of Cook as a supposedly fatigued explorer along the Northwest Coast, see Stephen R. Bown, *Madness, Betrayal and the Lash: The Epic Voyage of Captain George Vancouver* (Vancouver: Douglas and McIntyre, 2008), 26–27; Barry M. Gough, The *Northwest Coast: British Navigation, Trade, and Discoveries to 1812* (Vancouver: UBC Press, 1992), 9, 42; Peter Stark, *Astoria: John Jacob Astor and Thomas Jefferson's Lost Pacific Empire* (New York: HarperCollins, 2014), 70–71; Noel E. Currie, *Constructing Colonial Discourse* (Montreal and Kingston: McGill-Queen's University Press, 2005), 151; and Glyndwr Williams, *Voyages of Delusion: The Search for the Northwest Passage in the Age of Reason* (London: HarperCollins, 2002), 309.

27 *JCJC*, 1:387.

28 *JCJC*, 1:399.

29 Fernández-Armesto, *Pathfinders*, 290; and *JCJC*, 1:cxcii.

30 Banks, *The* Endeavour *Journal*, 2:145.

31 Ibid., 1:45.

32 *JCJC*, 1:404. The trope of Cook as the diminished, erratic, or slightly demented explorer during the third voyage is so common as to be axiomatic. See, for example, Thomas, *Cook*, 332–33; McLynn, *Captain Cook*, chs. 13–17, especially 282–84, 325–30, 343, 365–71, 398–400; Collingridge, *Captain Cook*, 292–98, 308–11, 320–22, 328–30; Dugard, *Farther Than Any Man*, 255–56; John Gascoigne, *Captain Cook: Voyager between Worlds* (London: Hambledon Continuum, 2008); 58–59; Horwitz, *Blue Latitudes*, 330–32, 346, 381; Gough, *The Northwest Coast*, 54; Withey, *Voyages of Discovery*, 334–35, 338, 378; Williams, *Voyages of Delusion*, 307; and Glyndwr Williams, *Arctic Labyrinth: The Quest for the Northwest Passage* (London: Penguin, 2009), 148.

33 *JCJC*, 1:408.

34 Hawkesworth, *An Account*, 23:659; and *JCJC*, 1:408, 410.

35 Banks, *The* Endeavour *Journal*, 2:148; and *JCJC*, 1:413.

36 See Beaglehole, *The Life of Captain James Cook*, 605, 628, 633; Gough, *The Northwest Coast*, 53; McLynn, *Captain Cook*, 274, 352, 360, 362; Williams, *Voyages of Delusion*, 326;

Williams, *Arctic Labyrinth*, 146; Glyndwr Williams, "Myth and Reality: James Cook and the Theoretical Geography of Northwest America," in Fisher and Johnston, *Captain Cook and His Times*, 69–79.

37 *JCJC*, 1:418.

38 See Collingridge, *Captain Cook*, 328–30; and McLynn, *Captain Cook*, 365–71.

39 *JCJC*, 1:426.

40 Parkinson, *Journal*, 213; and Banks, *The* Endeavour *Journal*, 2:180.

41 Beaglehole, *The Life of Captain James Cook*, 264.

42 Banks, *The* Endeavour *Journal*, 2:249.

43 *JCJC*, 1:467.

44 *JCJC*, 1:505–6.

CHAPTER 4: TOWARD THE SOUTH POLE

1 Banks, *The* Endeavour *Journal*, 2:347; and Peter Moore, *Endeavour: The Ship and the Attitude That Changed the World* (London: Vintage, 2019), 259.

2 Beaglehole, *The Life of Captain James Cook*, 276; and Banks, *The* Endeavour *Journal*, 2:329, 335.

3 *JCJC*, 1:xli–xlii.

4 *JCJC*, 2:941.

5 Banks, *The* Endeavour *Journal*, 2:343.

6 Bernard Smith, *Imagining the Pacific: In the Wake of Cook's Voyages* (New Haven, CT: Yale University Press, 1992), 45; and George Forster, *A Voyage Round the World*, ed. Nicholas Thomas and Oliver Berghof, 2 vols. (Honolulu: University of Hawai'i Press, 2000), 2:697.

7 Michael E. Hoare, ed., *The* Resolution *Journal of Johann Reinhold Forster, 1772–1775* (London: Hakluyt Society, 1982), 1:106. As an example of the near invisibity of sea ice science in the historiography of the second voyage, Hoare's introduction to Forster's journal devotes forty-six pages to the expedition's "Science and Its Records" but only part of a single paragraph to glaciology. Hoare admits, though, that "too little consideration has been given" to that subject (ibid.).

8 Forster, *A Voyage Round the World*, xxi. See Introduction, note 6, for citations documenting the anthropological orientation of Cook studies.

9 Beaglehole, *The Life of Captain James Cook*, 300.

10 James Cook, *A Voyage towards the South Pole and Round the World* ... (London, 1777), vol. 1, bk. 1, ch. 1, available at https://en.wikisource.org/wiki/A_Voyage_Towards_the_South_pole_and_Around_the_World.

11 Ibid., vol. 1, bk. 1, Intro.

12 *JCJC*, 2:xxi.

13 Cook, *A Voyage towards the South Pole*, vol. 1, bk. 1, Intro.

14 Forster, *A Voyage Round the World*, 62, 697; *JCJC*, 2:55, 57.

15 Forster, *A Voyage Round the World*, 65–67; and Hoare, *The* Resolution *Journal*, 2:193.

16 Forster, *A Voyage Round the World*, 68, 697; Hoare, *The* Resolution *Journal*, 2:196; and Cook, *A Voyage towards the South Pole*, vol. 1, bk. 1, ch. 2.

17 *JCJC*, 2:59, 62–63; and Cook, *A Voyage towards the South Pole*, vol. 1, bk. 1, ch. 2.

18 Barry Lopez, *Arctic Dreams: Imagination and Desire in a Northern Landscape* (New York: Vintage, 1986), 237.

19 *JCJC*, 2:63; and Hoare, *The* Resolution *Journal*, 2:197.

20 Cook, *A Voyage towards the South Pole*, vol. 1, bk. 1, ch. 2; and Hoare, *The Resolution Journal*, 2:193, 200.

21 *JCJC*, 2:63; Hoare, *The* Resolution *Journal*, 2:199–200. Johann Forster also adopted the terms packed ice and field ice from the "Greenlandmen" but with interpolated meanings. Cook's denominations made more practical sense and have endured as conventional usage, which speaks to his influence as a polar explorer.

22 *JCJC*, 2:63.

23 *JCJC*, 2:64, 66; and Cook, *A Voyage towards the South Pole*, vol. 1, bk. 1, ch. 2.

24 Cook, *A Voyage towards the South Pole*, vol. 1, bk. 1, ch. 2.

25 *JCJC*, 2:70–72.

26 *JCJC*, 2:74, 77; Johann Reinhold Forster, *Observations Made during a Voyage Round the World*, ed. Nicholas Thomas, Harriet Guest, and Michael Dettelbach (Honolulu: University of Hawai'i Press, 1996), 372; Forster, *A Voyage Round the World*, 71; and Hoare, *The* Resolution *Journal*, 200. Sir John Pringle, president of the Royal Society, alleged that Cook was the first explorer to melt icebergs to get water. Johann Forster debunked this in his *History of the Voyages and Discoveries Made in the North* (Dublin: Luke White and Pat Byrne, 1786), arguing that Cook had "merit sufficient of his own without the addition of this circumstance" (284).

27 *JCJC*, 2:72. On historians vis-à-vis Cook and Russian cartography, see Chapter 3, this book, note 36.

28 *JCJC*, 2:80.

29 *JCJC*, 2:81.

30 *JCJC*, 2:85–86.

31 Forster, *A Voyage Round the World*, 75.

32 *JCJC*, 2:98.

33 Cook, *A Voyage towards the South Pole*, vol. 1, bk. 1, ch. 3.

34 Forster, *A Voyage Round the World*, 76; and Forster, *Observations*, 82.

35 *JCJC*, 2:lxvi.

36 *JCJC*, 2:105–6.

37 Beaglehole, *The Life of Captain James Cook*, 641–43; Collingridge, *Captain Cook*, 328–29; Dugard, *Farther Than Any Man*, 270; Horwitz, *Blue Latitudes*, 380–81; and Withey, *Voyages of Discovery*, 378–79.

38 Beaglehole, *The Life of Captain James Cook*, 324.

39 David L. Nicandri, "The Rhyme of the Great Navigator: The Literature of Captain Cook and Its Influence on the Journals of Lewis and Clark; Part 2: The Grandest Sight," *We Proceeded On* 42, 2 (2016): 17–23.

40 *JCJC*, 2:175.

41 Collingridge, *Captain Cook*, 240–41.

42 Forster, *A Voyage Round the World*, 121.

43 *JCJC*, 2:173.

44 Cook, *A Voyage towards the South Pole*, vol. 1, bk. 1, ch. 8.

45 *JCJC*, 2:189.

46 *JCJC*, 2:189.

47 Forster, *A Voyage Round the World*, 143, 200.

48 Popularized by Alan Moorehead's *The Fatal Impact* (1966) and replicated by others, including Ian Cameron's *Lost Paradise: The Exploration of the Pacific* (1987).

49 Forster, *A Voyage Round the World*, 209; *JCJC*, 2:219; and Hoare, *The* Resolution *Journal*, 2:354–55.

50 *JCJC*, 2:290; and Hoare, *The* Resolution *Journal*, 3:423–25.

51 Beaglehole, *The Life of Captain James Cook*, 447.

52 Beaglehole, *The Life of Captain James Cook*, 446; Currie, *Constructing Colonial Discourse*, 87–126.
53 *JCJC*, 2:299.
54 *JCJC*, 2:299–300.

CHAPTER 5: THE LIMIT OF AMBITION

1 Forster, *A Voyage Round the World*, 286; and Hoare, *The* Resolution *Journal*, 3:432.
2 Cook, *A Voyage towards the South Pole*, vol. 1, bk. 2, ch. 6; and *JCJC*, 2:305.
3 *JCJC*, 2:304–5; Cook, *A Voyage towards the South Pole*, vol. 1, bk. 2, ch. 6; and Hoare, *The* Resolution *Journal*, 3:435.
4 Cook, *A Voyage towards the South Pole*, vol. 1, bk. 2, ch. 6; Forster, *Observations*, 61; Hoare, *The* Resolution *Journal*, 3:436–38.
5 *JCJC*, 2:308; and Cook, *A Voyage towards the South Pole*, vol. 1, bk. 2, ch. 6.
6 *JCJC*, 2:309.
7 Dugard, *Farther Than Any Man*, 167.
8 Hoare, *The* Resolution *Journal*, 3:441; and Cook, *A Voyage towards the South Pole*, vol. 1, bk. 2, ch. 6.
9 *JCJC*, 2:312; Forster, *A Voyage Round the World*, 291; and Hoare, *The* Resolution *Journal*, 3:438–39.
10 *JCJC*, 2:314.
11 Forster, *A Voyage Round the World*, 292; and Hoare, *The* Resolution *Journal*, 3:443.
12 Forster, *A Voyage Round the World*, 292; and Hoare, *The* Resolution *Journal*, 3:445.
13 Hoare, *The* Resolution *Journal*, 3:443–44.
14 Forster, *A Voyage Round the World*, 293.
15 Hoare, *The* Resolution *Journal*, 3:448; Cook, *A Voyage towards the South Pole*, vol. 1, bk. 2, ch. 6; and *JCJC*, 2:318.
16 *JCJC*, 2:319; and Hoare, *The* Resolution *Journal*, 3:450.
17 Cook, *A Voyage towards the South Pole*, vol. 1, ch. 6; *JCJC*, 2:320; and Forster, *A Voyage Round the World*, 294.
18 *JCJC*, 2:321.
19 Forster, *A Voyage Round the World*, 68; and Forster, *Observations*, 63.
20 *JCJC*, 2:321.
21 *JCJC*, 2:322.
22 *JCJC*, 2:323.
23 Forster, *A Voyage Round the World*, 294–95.
24 Cook, *A Voyage towards the South Pole*, vol. 1, bk. 2, ch. 6; and Thomas, *Cook*, 228.
25 *JCJC*, 2:lxxxvi–lxxxvii; and Forster, *A Voyage Round the World*, 294.
26 *JCJC*, 2:324.
27 *JCJC*, 2:325.
28 Cook, *A Voyage towards the South Pole*, vol. 1, bk. 2, ch. 6.
29 *JCJC*, 2:325, 328.
30 Forster, *A Voyage Round the World*, 295.

CHAPTER 6: TEMPORIZING IN THE TROPICS

1 Cook, *A Voyage towards the South Pole*, vol. 1, bk. 2, ch. 6; and *JCJC*, 2:331–32.
2 Forster, *A Voyage Round the World*, 298.
3 McLynn, *Captain Cook*, 232; Sir James Watt, "Medical Aspects and Consequences of Cook's Voyages," in Fisher and Johnston, *Captain Cook and His Times*, 156; and Hoare, *The* Resolution *Journal*, 3:455.

4 Forster, *Observations*, 372; Cook, *A Voyage towards the South Pole*, vol. 1, bk. 2, ch. 6; Forster, *A Voyage Round the World*, 296; and Hoare, *The* Resolution *Journal*, 3:457.

5 *JCJC*, 2:348–49, 351, 354–55.

6 *JCJC*, 369.

7 *JCJC*, 383, 392.

8 Cook, *A Voyage towards the South Pole*, vol. 1, bk. 2, ch. 15.

9 *JCJC*, 2:433–34.

10 Forster, *A Voyage Round the World*, 407; and *JCJC*, 2:436.

11 Cook, *A Voyage towards the South Pole*, vol. 1, bk. 2, ch. 3; Hoare, *The* Resolution *Journal*, 3:550.

12 *JCJC*, 2:457; and Cook, *A Voyage towards the South Pole*, vol. 2, bk. 3, ch. 7.

13 Hoare, *The* Resolution *Journal*, 4:555; and Forster, *A Voyage Round the World*, 422–23.

14 Forster, *A Voyage Round the World*, 560.

15 *JCJC*, 2:478.

16 *JCJC*, 2:478–79.

17 See *JCJC*, 3:132, 149, 151, 159, 231–32.

18 Forster, *A Voyage Round the World*, 505.

19 *JCJC*, 2:493.

20 *JCJC*, 2:498, 500. The Tannese account is drawn from Glyn Williams, *The Death of Captain Cook: A Hero Made and Unmade* (London: Profile Books, 2008), 128.

21 *JCJC*, 2:519–20.

22 Cook, *A Voyage towards the South Pole*, vol. 2, bk. 3, ch. 10 (emphasis added).

23 *JCJC*, 2:528.

24 Hoare, *The* Resolution *Journal*, 4:647; and Forster, *A Voyage Round the World*, 598.

25 *JCJC*, 2:554, 562, 565–66, 568.

26 *JCJC*, 2:573.

27 *JCJC*, 2:574–76.

28 Forster, *A Voyage Round the World*, 612–13.

29 *JCJC*, 2:617.

30 *JCJC*, 2:580–83.

31 Hoare, *The* Resolution *Journal*, 4:689; and Cook, *A Voyage towards the South Pole*, vol. 2, bk. 4, ch. 1.

32 *JCJC*, 2:587.

CHAPTER 7: COOK AND FORSTER, ON ICE

1 *JCJC*, 2:615.

2 *JCJC*, 617–19; and Hoare, *The* Resolution *Journal*, 4:710.

3 *JCJC*, 2:620.

4 *JCJC*, 2:620–21; Cook, *A Voyage towards the South Pole*, vol. 2, bk. 4, ch. 5; and Beaglehole, *The Life of Captain James Cook*, 428.

5 Dugard, *Farther Than Any Man*, 210; Hoare, *The* Resolution *Journal*, 4:713; and *JCJC*, 2:622.

6 *JCJC*, 2:623–25 (emphasis added).

7 *JCJC*, 2:625.

8 *JCJC*, 2:625.

9 *JCJC*, 2:625; and Withey, *Voyages of Discovery*, 298.

10 *JCJC*, 2:626.

11 See Chapter 3, this book, note 36.

12 *JCJC*, 2:628–29.

13 *JCJC*, 2:ci.

14 Forster, *A Voyage Round the World*, 649.

15 Ibid., 650.

16 *JCJC*, 2:632–33; Cook, *A Voyage towards the South Pole*, vol. 2, bk. 4, ch. 5; Forster, *Observations*, 119; and Forster, *A Voyage Round the World*, 649.

17 *JCJC*, 2:632; Hoare, *The* Resolution *Journal*, 4:721.

18 Cook, *A Voyage towards the South Pole*, vol. 2, bk. 4, ch. 5; and *JCJC*, 2:634, 636.

19 Cook, *A Voyage towards the South Pole*, vol. 2, bk. 4, ch. 6; and *JCJC*, 2:636.

20 Cook, *A Voyage towards the South Pole*, vol. 2, bk. 4, ch. 6.

21 *JCJC*, 2:637–38.

22 Forster, *A Voyage Round the World*, 651.

23 Cook, *A Voyage towards the South Pole*, vol. 2, bk. 4, ch. 6.

24 Ibid.

25 *JCJC*, 2:638.

26 *JCJC*, 2:640.

27 *JCJC*, 2:642.

28 *JCJC*, 2:642; and Cook, *A Voyage towards the South Pole*, vol. 2, bk. 4, ch. 7.

29 *JCJC*, 2:643.

30 Forster, *A Voyage Round the World*, 651.

31 *JCJC*, 2:643.

32 *JCJC*, 2:644.

33 *JCJC*, 2:644.

34 *JCJC*, 2:644–45.

35 *JCJC*, 2:766.

36 *JCJC*, 2:645–46.

37 Forster, *A Voyage Round the World*, 750, 782.

38 W.F. Weeks, *On Sea Ice* (Fairbanks: University of Alaska Press, 2010), 295. On polynyas, see 283–90.

39 *JCJC*, 2:646; and Hoare, *The* Resolution *Journal*, 2:201–2, 4:725.

40 *JCJC*, 2:646.

41 Weeks, *On Sea Ice*, 2–3, 43, 47, 107.

42 Ibid., 5, 12.

43 Ibid., 50, 60–64, 141, 146, 148, 153, 171.

44 Ibid., 78–79, 86–87.

45 Ibid., 79, 90. See 144–90 for a more technical description of this desalination process.

46 Ibid., 126, 132, 144.

47 Hoare, *The* Resolution *Journal*, 1:106; and Forster, *Observations*, 78.

48 Forster, *Observations*, 61.

49 Ibid., 62, 64.

50 Ibid., 62, 65.

51 Ibid., 62, 64–65.

52 Ibid., 60, 65.

53 Ibid., 66–67, 69.

54 Ibid., 70–71.

55 Ibid., 71–73.

56 Ibid., 77.

57 Ibid., 77–78; Sarah M. Kang and Richard Seager, "Croll Revisited: Why Is the Northern Hemisphere Warmer Than the Southern Hemisphere?," https://ocp.ldeo.columbia.edu/res/div/ocp/pub/seager/Kang_Seager_subm.pdf.

58 Forster, *Observations*, 82, 97, 186, 359, 363, 376, 402 ["fractious personality"], 414; *JCJC*, 2:886; and Hoare, *The* Resolution *Journal*, 1:75.
59 Forster, *Observations*, 383, 402; Michael E. Hoare, "Two Centuries' Perception of James Cook: George Forster to Beaglehole," in Fisher and Johnston, *Captain Cook and His Times*, 215.
60 *JCJC*, 2:647.
61 *JCJC*, 2:651–52.
62 *JCJC*, 2:652–54.
63 *JCJC*, 2:654–55.
64 *JCJC*, 2:660–62.
65 *JCJC*, 2:669, 671.
66 *JCJC*, 2:678, 682, 953.
67 Cook, *A Voyage towards the South Pole*, vol. 2, bk. 4, ch. 11; and *JCJC*, 2:cxi.
68 *JCJC*, 2:692.
69 Banks, *The* Endeavour *Journal*, 1:105.
70 Thomas, *Cook*, xix–xx, 250.
71 Horwitz, *Blue Latitudes*, 134.
72 Cook, *A Voyage towards the South Pole*, vol. 1., Intro.; and *JCJC*, 2:cxliii.
73 Horwitz, *Blue Latitudes*, 73; and McLynn, *Captain Cook*, 51, 53.
74 Forster, *A Voyage Round the World*, 792; *JCJC*, 2:cxliii; and Thomas, *Cook*, 257.
75 Forster, *A Voyage Round the World*, 694, 790.
76 Smith, *Imagining the Pacific*, 151–52.

CHAPTER 8: AN ANCIENT QUEST, A NEW MISSION
1 Robson, *Captain Cook's World*, 115; *JCJC*, 2:957; Collingridge, *Captain Cook*, 231; and Hoare, *The* Resolution *Journal*, 4:728.
2 Thomas, *Cook*, 265.
3 *JCJC*, 2:958–59.
4 *JCJC*, 2:960.
5 Beaglehole, *The Life of Captain James Cook*, 484.
6 *JCJC*, 3:xxxii, xli.
7 Except as otherwise noted, the sources for the discussion on the early history of the Northwest Passage are Williams, *Arctic Labyrinth*, xviii, 34–37, 42, 46, 74, 86–87, 108–9, 116–19, 122–23, 125–27, 131, 379–86; Williams, *Voyages of Delusion*, xviii, xix, 13, 20–21, 130–33, 150, 241–42, 246, 250–53, 260–63; and Fernández-Armesto, *Pathfinders*, 219–20, 254, 270, 273.
8 Gough, *The Northwest Coast*, 28.
9 Forster, *History of the Voyages and Discoveries Made in the North*, 454–55; and *JCJC*, 3:xlvi.
10 Williams, *Arctic Labyrinth*, 119.
11 Barry Gough, *Juan de Fuca's Strait: Voyages in the Waterway of Forgotten Dreams* (Madeira Park, BC: Harbour Publishing, 2012), 194, 218.
12 *JCJC*, 3:lxi; Williams, *Arctic Labyrinth*, 120; and Williams, *Voyages of Delusion*, 263.
13 Samuel Hearne, *A Journey from Prince of Wales's Fort in Hudson's Bay to the Northern Ocean*, ed. Richard Glover (Toronto: Macmillan, 1958), lxviii, lxix.
14 I.S. MacLaren, "Notes on Samuel Hearne's *Journey* from a Bibliographical Perspective," *Papers of the Bibliographical Society of Canada* 31, 2 (Fall 1993): 21–45, and "Samuel Hearne's Accounts of the Massacre at Bloody Fall, 17 July 1771," *ARIEL: A Review of English Literature* 22, 1 (January 1991): 25–51.
15 Hearne, *A Journey*, 93, 95.

16 Ibid., lxix–lxx.
17 Ibid., lxix.
18 Daines Barrington, *The Probability of Reaching the North Pole, Discussed* (London: C. Heydinger, 1775), 82–83, 86.
19 Ibid., 7, 11, 52, 81.
20 Ibid., 17–18, 23.
21 Ibid., 29.
22 Ibid., 42, 54–55.
23 Ibid., 69, 87–89.
24 *JCJC*, 3:xlvii.
25 John Norris, "The Strait of Anian and British America: Cook's Third Voyage in Perspective," *BC Studies* 36 (Winter 1977–78): 18.
26 James Cook and James King, *A Voyage to the Pacific Ocean in the Years 1776, 1777, 1778, 1779 and 1780 ... Vol. I and II written by Captain Cook, Vol. III by Captain King*, ed. John Douglas (London: G. Nicol and T. Cadell, 1784), 1:xxx–xxxi.
27 *JCJC*, 3:1497. Beaglehole professed that if Cook's northern voyage had not been set in motion as early as it had, 1775, the war with the colonies may have pre-empted it. See *JCJC*, 3:lxvii.
28 Beaglehole, *The Life of Captain James Cook*, 688.
29 Williams, *Voyages of Delusion*, 288.
30 Robin Inglis, "Successors and Rivals to Cook: The French and the Spaniards," in Williams, *Captain Cook*, 173.
31 *JCJC*, 3:ccxxi; Cook, *A Voyage to the Pacific*, 1:xli.
32 *JCJC*, 3:ccxxi–ccxxii.
33 Cook, *A Voyage to the Pacific*, 1:xlix.
34 Ibid., 1:xlix–lii.
35 *JCJC*, 3:ccxx–ccxxii; and Cook, *A Voyage to the Pacific*, 1:xiv, xxvii.
36 *JCJC*, 3:ccxxi.
37 *JCJC*, 3:ccxxii.
38 Cook, *A Voyage to the Pacific*, 1:lxxxviii.
39 Williams, *Voyages of Delusion*, 239.
40 Cook, *A Voyage to the Pacific*, 1:xxviii–xxix.
41 Ibid., 1:xxx.

CHAPTER 9: SOUTHERN STAGING GROUNDS
1 Williams, "Myth and Reality: James Cook and the Theoretical Geography of Northwest America," 70; Beaglehole, *The Life of Captain James Cook*, 445; and *Cook's Log* 42, 2 (April–June 2019): 3.
2 Andrew S. Cook, "James Cook and the Royal Society," in Williams, *James Cook*, 52.
3 Beaglehole, *The Life of Captain James Cook*, 493.
4 Richard Henry Dana Jr., *Two Years before the Mast* (New York: Signet, 2000), 336; and *JCJC*, 3:1488.
5 *JCJC*, 1:505.
6 Richard A. Van Orman, *The Explorers: Nineteenth-Century Expeditions in Africa and the American West* (Albuquerque: University of New Mexico Press, 1984), 134; and Fernández-Armesto, *Pathfinders*, 334.
7 James Trevenen, *A Memoir of James Trevenen*, ed. Christopher Lloyd and R.C. Anderson (London: Navy Records Society, 1959), 74; and James Kenneth Munford, ed., *John*

Ledyard's Journal of Captain Cook's Last Voyage (Corvallis: Oregon State University Press), 135.

8 Smith, *Imagining the Pacific*, 46; and Beaglehole, *The Life of Captain James Cook*, 502.
9 *JCJC*, 3:743, 1514, 1516.
10 *JCJC*, 1:cxciii; and Smith, *Imagining the Pacific*, 181.
11 *JCJC*, 3:28–29, 990, 769.
12 *JCJC*, 3:42–43.
13 *JCJC*, 3:48, 779.
14 Robson, *Captain Cook's World*, 151; and Collingridge, *Captain Cook*, 294.
15 *JCJC*, 3:803, 995.
16 Dugard, *Farther Than Any Man*, 249; and McLynn, *Captain Cook*, 292.
17 Sinclair H. Hitchings, "Introduction," Munford, *John Ledyard's Journal*, xxviii; and McLynn, *Captain Cook*, 293.
18 Van Orman, *The Explorers*, 131.
19 *JCJC*, 3:824.
20 *JCJC*, 3:78; and Cook, *A Voyage to the Pacific*, 1:206.
21 *JCJC*, 3:91.
22 McLynn, *Captain Cook*, 313.
23 Gascoigne, *Captain Cook*, 86; and *JCJC*, 3:101.
24 Cook, *A Voyage to the Pacific*, 1:234, 260.
25 *JCJC*, 3:1044.
26 Christine Holmes, ed., *Captain Cook's Final Voyage: The Journal of Midshipman George Gilbert* (Honolulu: University of Hawai'i Press, 1982), 33; and *JCJC*, 3:918.
27 *JCJC*, 3:160; and Cook, *A Voyage to the Pacific*, 1:364.
28 Beaglehole, *The Life of Captain James Cook*, 548; and Thomas, *Cook*, 332.
29 McLynn, *Captain Cook*, 313; Collingridge, *Captain Cook*, 292, 295, 297; and Thomas, *Cook*, 333.
30 Holmes, *Captain Cook's Final Voyage*, 35; and Withey, *Voyages of Discovery*, 338.
31 Alan Frost, "New Geographical Perspectives and the Emergence of the Romantic Imagination," in Fisher and Johnston, *Cook and His Times*, 19; and Hoare, *The Resolution Journal*, 3:551.
32 Beaglehole, *The Life of Captain James Cook*, 453–54.
33 *JCJC*, 3:974, 1056.
34 *JCJC*, 3:194.
35 Cook, *A Voyage to the Pacific*, 2:80.
36 Smith, *Imagining the Pacific*, 193.
37 Collingridge, *Captain Cook*, 309.
38 Holmes, *Captain Cook's Final Voyage*, 46–47, 50; and *JCJC*, 3:1383.
39 *JCJC*, 3:232; and Thomas, *Cook*, 347.
40 McLynn, *Master of the Seas*, 282–83; and Hoare, "George Forster to Beaglehole," 226.
41 Dana, *Two Years before the Mast*, 376, 381; and Salmond, "Tute," 87.
42 Obeysekere, *The Apotheosis*, 133; and Gascoigne, *Captain Cook*, 40, 42.
43 Salmond, "Tute," 85. The roster of crewmembers is found at *JCJC*, 3:1458–78.
44 Van Orman, *The Explorers*, 18.
45 Smith, *Imagining the Pacific*, 208–9.
46 *JCJC*, 3:240.
47 *JCJC*, 3:241–42, 1387.
48 McLynn, *Captain Cook*, 328.

49 *JCJC*, 3:245.
50 Collingridge, *Captain Cook*, 313; and *JCJC*, 3:252.
51 *JCJC*, 3:253, 256.

CHAPTER 10: TERRA BOREALIS

1   Holmes, *Captain Cook's Final Voyage*, 60; and *JCJC*, 3:1074.
2   *JCJC*, 3:256, 260, 1079, 1346–47.
3   *JCJC*, 3:263, 1081.
4   *JCJC*, 3:263–66, 269, 1084.
5   *JCJC*, 3:269; and Fernández-Armesto, *Pathfinders*, 25.
6   *JCJC*, 3:277; and Cook, *A Voyage to the Pacific*, 2:223.
7   *JCJC*, 3:277, 279, 1392–93.
8   Collingridge, *Captain Cook*, 230; Dugard, *Farther Than Any Man*, 258; and *JCJC*, 3:286.
9   *JCJC*, 3:ccxxi.
10 *JCJC*, 3:288.
11 Holmes, *Captain Cook's Final Voyage*, 68; and *JCJC*, 3:289.
12 *JCJC*, 3:289.
13 *JCJC*, 3:293.
14 Gough, *Northwest Coast*, 54; Bown, *Madness, Betrayal and the Lash*, 27; Stark, *Astoria*, 70; and *JCJC*, 3:293–94.
15 *JCJC*, 3:294; and Williams, *Voyages of Delusion*, 309.
16 *JCJC*, 3:295–96.
17 Holmes, *Captain Cook's Final Voyage*, 72; and *JCJC*, 3:1088–89.
18 *JCJC*, 3:302–3; and Gascoigne, *Captain Cook*, 92.
19 *JCJC*, 3:297, 1096.
20 *JCJC*, 3:303.
21 *JCJC*, 3:306; and James Zug, *American Traveler: The Life and Adventures of John Ledyard* (New York: Basic Books, 2005), 79.
22 Munford, *John Ledyard's Journal*, 72.
23 *JCJC*, 3:306; and Gascoigne, *Captain Cook*, 89.
24 *JCJC*, 3:307–8.
25 *JCJC*, 3:309, 1403.
26 *JCJC*, 3:334–35.
27 *JCJC*, 3:336.
28 *JCJC*, 3:337–38.
29 *JCJC*, 3:338.
30 *JCJC*, 3:340, 1105.
31 *JCJC*, 3:342–43.
32 *JCJC*, 3:342–43.
33 *JCJC*, 3:1106; and Munford, *John Ledyard's Journal*, 80.
34 *JCJC*, 3:1106, 1108.
35 *JCJC*, 3:345.
36 *JCJC*, 3:348–49, 351.
37 *JCJC*, 3:351.
38 *JCJC*, 3:ccxxii.
39 *JCJC*, 3:352.
40 *JCJC*, 3:352–53, 1420.
41 *JCJC*, 3:353.
42 *JCJC*, 3:356–57.

43 *JCJC*, 3:358–59.
44 *JCJC*, 3:358–59.
45 *JCJC*, 3:360.
46 Munford, *John Ledyard's Journal*, 81; and *JCJC*, 3:361.
47 *JCJC*, 3:361.
48 *JCJC*, 3:362–63.
49 *JCJC*, 3:364.
50 *JCJC*, 3:364.
51 *JCJC*, 3:365, 469.
52 *JCJC*, 3:366–67.
53 *JCJC*, 3:367, 421.
54 *JCJC*, 3:367; ccxxi; and McLynn, *Captain Cook*, 354.
55 *JCJC*, 3:368.
56 *JCJC*, 3:368.
57 Williams, *Naturalists*, 122; Williams, *Voyages of Delusion*, 318; and McLynn, *Captain Cook*, 353.
58 *JCJC*, 3:368; and Cook, *A Voyage to the Pacific*, 2:opposite 353.
59 *JCJC*, 3:371.
60 *JCJC*, 3:371
61 Collingridge, *Captain Cook*, 320; and *JCJC*, 3:1424.
62 Thomas, *Cook*, 368.

CHAPTER 11: BLINK

1 See, for example, Collingridge, *Captain Cook*, 316–25; Dugard, *Farther Than Any Man*, 261–68; Withey, *Voyages of Discovery*, 373–75; Beaglehole, *The Life of Captain James Cook*, 571–636; and Thomas, *Cook*, 361–77.
2 *JCJC*, 3:1117, 1423–24.
3 *JCJC*, 3:374–76.
4 *JCJC*, 3:380, 382.
5 *JCJC*, 3:383–84.
6 *JCJC*, 3:384, 672; and Munford, *John Ledyard's Journal*, 83.
7 *JCJC*, 3:384–86.
8 *JCJC*, 3:386, 388; Beaglehole, *The Life of Captain James Cook*, 609; and McLynn, *Captain Cook*, 355.
9 *JCJC*, 3:388–89.
10 See McLynn, *Captain Cook*, 355, regarding Cook's competence; and Trevenen, *A Memoir*, 26–27.
11 *JCJC*, 3:389–90.
12 *JCJC*, 3:391, 1424.
13 *JCJC*, 3:392–93.
14 *JCJC*, 3:394.
15 *JCJC*, 3:395–96.
16 *JCJC*, 3:397–99.
17 *JCJC*, 3:400, 402.
18 *JCJC*, 3:403.
19 *JCJC*, 3:1428–29.
20 *JCJC*, 3:405.
21 *JCJC*, 3:406.
22 *JCJC*, 3:406.

23 *JCJC*, 3:407, 1431.
24 *JCJC*, 3:409.
25 *JCJC*, 3:410–14, 1134; and J.C. Beaglehole, ed., *Cook and the Russians* (London: Hakluyt Society, 1973), 6.
26 *JCJC*, 3:414.
27 *JCJC*, 3:414.
28 *JCJC*, 3:1134.
29 *JCJC*, 3:414–15.
30 *JCJC*, 3:416 (emphasis added).
31 *JCJC*, 3:417.
32 *JCJC*, 3:417–18; Weeks, *On Sea Ice*, 330–35, 346–57, 366.
33 *JCJC*, 3:418.
34 *JCJC*, 3:418.
35 Collingridge, *Captain Cook*, 325.
36 *JCJC*, 3:419.
37 Munford, *John Ledyard's Journal*, 87–88; and *JCJC*, 3:419–20.
38 *JCJC*, 3:420, 1453–54; and Munford, *John Ledyard's Journal*, 87.
39 *JCJC*, 3:420.
40 *JCJC*, 3:422–23. Subsequent explorers with more time to study the matter determined that a surface current indeed runs northward through Bering Strait on its eastern side and then follows the coast of Alaska past Cape Lisburne.
41 *JCJC*, 3:423.
42 *JCJC*, 3:1455; and King, *Voyage to the Pacific*, 3:274.
43 Gough, *Juan de Fuca's Strait*, 78; Trevenen, *A Memoir*, 27; Dugard, *Farther Than Any Man*, 267; and Andrew Lambert, "Retracing the Captain: 'Extreme History,' Hard Tack and Scurvy," in Williams, *Captain Cook*, 252.
44 *JCJC*, 3:424.
45 Collingridge, *Captain Cook*, 325; McLynn, *Captain Cook*, 359–60; and *JCJC*, 3:ccxxii.
46 *JCJC*, 3:424.
47 *JCJC*, 3:424.
48 *JCJC*, 3:424; and Weeks, *On Sea Ice*, 44, 57–63.
49 *JCJC*, 3:424, 1455; and Weeks, *On Sea Ice*, 126, 135, 500–2.
50 *JCJC*, 3; 424–25.
51 *JCJC*, 3:425.
52 Thomas, *Cook*, 374.
53 *JCJC*, 3:425–26.
54 *JCJC*, 3:426–27.
55 *JCJC*, 3:427.
56 *JCJC*, 3:427.
57 Collingridge, *Captain Cook*, 325.
58 Williams, *Voyages of Delusion*, 328.
59 Harry Stern, "Sea Ice in the Western Portal of the Northwest Passage from 1778 to the Twenty-First Century," in *Arctic Ambitions: Captain Cook and the Northwest Passage*, ed. James K. Barnett and David L. Nicandri (Seattle: University of Washington Press, 2015), 350, 353; and *JCJC*, 3:cxxxiv.
60 *JCJC*, 3:427; and Beaglehole, *The Life of Captain James Cook*, 622.
61 *JCJC*, 3:427.
62 *JCJC*, 3:428–29.
63 *JCJC*, 3:429.

64 *JCJC*, 3:430–31; and Munford, *John Ledyard's Journal*, 88.
65 *JCJC*, 3:432.
66 *JCJC*, 3:432–33; and Munford, *John Ledyard's Journal*, 99–100.
67 *JCJC*, 3:433.
68 Beaglehole, *The Life of Captain James Cook*, 625; and *JCJC*, 3:434.

CHAPTER 12: NORTHERN INTERLUDE
1 *JCJC*, 3:xxxiii.
2 *JCJC*, 3:1433.
3 *JCJC*, 3:434–35.
4 *JCJC*, 3:435.
5 *JCJC*, 3:435.
6 *JCJC*, 3:435–36.
7 *JCJC*, 3:437.
8 *JCJC*, 3:437–38.
9 *JCJC*, 3:438.
10 *JCJC*, 3:438.
11 *JCJC*, 3:439.
12 *JCJC*, 3:439, 1433.
13 *JCJC*, 3:440–41.
14 Brian Fagan, *The Little Ice Age: How Climate Made History, 1300–1850* (New York: Basic Books, 2000), 47–49, 103, 113.
15 Hearne, *A Journey*, 65, 86.
16 W. Kaye Lamb, ed., *The Journals and Letters of Sir Alexander Mackenzie* (London: Cambridge University Press, 1970), 255–56; and Fagan, *Little Ice Age*, xiii.
17 Fry, *Alexander Dalrymple*, 220.
18 *JCJC*, 3:441, 1434–35.
19 *JCJC*, 3:441.
20 *JCJC*, 3:ccxxii, 441–42.
21 *JCJC*, 3:441–42, 444.
22 *JCJC*, 3:444.
23 *JCJC*, 3:1437, 1454–55.
24 Beaglehole, *The Life of Captain James Cook*, 628.
25 *JCJC*, 3:444–45, 1455.
26 *JCJC*, 3:445–48.
27 *JCJC*, 3:448–49.
28 *JCJC*, 3:1445; Williams, *Naturalists at Sea*, 122; and Evguenia Anichtchenko, "From Russia with Charts: Cook and Russians in the North Pacific," in Barnett and Nicandri, *Arctic Ambitions*, 63.
29 *JCJC*, 3:449, 1445.
30 *JCJC*, 3:450, 1446; and Beaglehole, *Cook and the Russians*, 3.
31 *JCJC*, 3:451, 1448.
32 *JCJC*, 3:451, 1446, 1448–49.
33 *JCJC*, 3:451–52, 654.
34 *JCJC*, 3:452–53.
35 *JCJC*, 3:453–54.
36 *JCJC*, 3:454.
37 *JCJC*, 3:456.
38 *JCJC*, 3:456.

39　Norris, "The Strait of Anian," 18; I.S. MacLaren, "In Consideration of the Evolution of Explorers and Travelers into Authors: A Model," *Studies in Travel Writing* 15, 3 (September 2011): 232; and Belyea, "Just the Facts," 5–6.
40　Williams, *Naturalists at Sea*, 36.
41　*JCJC*, 3:456.
42　Anichtchenko, "From Russia with Charts," 69; Claudio Saunt, *West of the Revolution: An Uncommon History of 1776* (New York: Norton, 2014), 51; and Zug, *American Traveler*, 77.
43　*JCJC*, 3:457, 672, 676, 1533; and Anichtchenko, "From Russia with Charts," 78.
44　*JCJC*, 3:1531.
45　*JCJC*, 3:1531–32.
46　*JCJC*, 3:1532; and Beaglehole, *The Life of Captain James Cook*, 711.
47　*JCJC*, 3:457.
48　*JCJC*, 3:457.
49　*JCJC*, 3:464–65.
50　*JCJC*, 3:468.
51　*JCJC*, 3:351, 411–12.
52　Williams, *Naturalists at Sea*, 43, 50; and Saunt, *West of the Revolution*, 74.
53　The preceding discussion is based primarily on Glenn Hodges, "Tracking the First Americans," *National Geographic* 227, 1 (January 2015): 124–37; Matt Ridley, "The Prehistory of Us," *Wall Street Journal*, May 2–3, 2015, C1–C2; and Howard Schneider, "Native Informant," *Wall Street Journal*, January 9–10, 2016, C9, a review of *White Eskimo* by Stephen R. Bown.
54　Fernández-Armesto, *Pathfinders*, 14.
55　Ibid., 242.
56　*JCJC*, 3:470.
57　Obeyesekere, *Apotheosis*, 42; and *JCJC*, 3:1454.
58　Munford, *John Ledyard's Journal*, 90.

CHAPTER 13: INTIMATIONS OF MORTALITY

1　Bown, *Madness, Betrayal and the Lash*, 29; and McLynn, *Captain Cook*, 398.
2　Thomas, *Cook*, 376.
3　King, *Voyage to the Pacific*, 3:49.
4　Holmes, *Captain Cook's Final Voyage*, 99–100; and Munford, *John Ledyard's Journal*, 102.
5　*JCJC*, 3:474, 477.
6　*JCJC*, 3:cxli; McLynn, *Captain Cook*, 369; and Collingridge, *Captain Cook*, 329.
7　Holmes, *Captain Cook's Final Voyage*, 100; George Vancouver, *A Voyage of Discovery to the North Pacific and Round the World, 1791–95*, ed. W. Kaye Lamb (Cambridge: Cambridge University Press, 1984), 3:842–43.
8　*JCJC*, 3:478–79.
9　*JCJC*, 3:478–79.
10　*JCJC*, 3:479.
11　McLynn, *Captain Cook*, 369; Collingridge, *Captain Cook*, 329; and Withey, *Voyages of Discovery*, 378.
12　*JCJC*, 3:479.
13　*JCJC*, 3:479–80.
14　*JCJC*, 3:479–80.
15　*JCJC*, 3:480.

16 Caroline Alexander, *The Bounty: The True Story of the Mutiny on the Bounty* (New York: Viking, 2003), 116.

17 Barbara Tuchman, *The First Salute: A View of the American Revolution* (New York: Knopf, 1988), 109; Bown, *Madness, Betrayal and the Lash*, 114; and Alexander, *The Bounty*, 138.

18 Tuchman, *The First Salute*, 5–6; King, *Voyage to the Pacific*, 3:418; and Forster, *History of the Voyages and Discoveries Made in the North*, 404.

19 Williams, *The Death of Captain Cook*, 91; Munford, *John Ledyard's Journal*, 102; Withey, *Voyages of Discovery*, 378, 390; and McLynn, *Captain Cook*, 466.

20 Thomas, *Cook*, 442.

21 Robson, *Captain Cook's World*, 159, plate 3.23; *JCJC*, 3:480–81; and King, *Voyage to the Pacific*, 3:103, 105.

22 Munford, *John Ledyard's Journal*, 221.

23 McLynn, *Captain Cook*, 373.

24 *JCJC*, 3:483.

25 *JCJC*, 3:485.

26 *JCJC*, 3:486–87.

27 Holmes, *Captain Cook's Final Voyage*, 100; and *JCJC*, 3:487.

28 Holmes, *Captain Cook's Final Voyage*, 101; and *JCJC*, 3:489–90.

29 *JCJC*, 3:490–91.

30 *JCJC*, 3:clxxi–clxxii; Williams, *The Death of Captain Cook*, 49; I.S. MacLaren, "Bones of Empire: Cook and Franklin Reaching to Alaska for a Northwest Passage," in *Imagining Anchorage: The Making of America's Northernmost Metropolis*, ed. James K. Barnett and Ian Harman (Fairbanks: University of Alaska Press, 2018), 109; and Obeyesekere, *Apotheosis*, 216.

31 MacLaren, "In Consideration of the Evolution of Explorers and Travelers into Authors," 227–28. MacLaren's foundational work on this topic is "Exploration/Travel Literature and the Evolution of the Author," *Journal of Canadian Studies* 5 (1992): 39–68.

32 *JCJC*, 3:267.

33 *JCJC*, 3:503.

34 *JCJC*, 3:503–4; and Dana, *Two Years before the Mast*, 172.

35 King, *Voyage to the Pacific*, 3:11.

36 *JCJC*, 3:118, 151.

37 *JCJC*, 3:508, 510, 517; and Salmond, "Tute," 84.

38 Munford, *John Ledyard's Journal*, 136–37; Thomas, *Cook*, 388; and *JCJC*, 3:516, 525.

39 Munford, *John Ledyard's Journal*, 141; and *JCJC*, 3:1190.

40 McLynn, *Captain Cook*, 390; *JCJC*, 3:clvi–clvii; and Williams, *The Death of Captain Cook*, 82.

41 *JCJC*, 3:1547–48.

42 Beaglehole, *Cook and the Russians*, 5.

43 Arthur Kitson, *Captain James Cook, R.N., F.R.S.: "The Circumnavigator"* (New York: E.P. Dutton, 1907), 490; and *JCJC*, 3:1535.

44 *JCJC*, 3:1552–53.

45 Kitson, *Captain James Cook*, 491–92; and Beaglehole, *The Life of Captain James Cook*, 689.

46 Kitson, *Captain James Cook*, 492–93 (emphasis added); and King, *Voyage to the Pacific*, 3:55.

47 Baker, *The Ship*, 12.

48 Thomas, *Cook*, 396; and Holmes, *Captain Cook's Final Voyage*, 105.

49 Kitson, *Captain James Cook*, 493.

50 Ibid., 494–96.
51 *JCJC*, 3:540.
52 *JCJC*, 3:543–44; and King, *Voyage to the Pacific*, 3:71.
53 *JCJC*, 3:562.
54 *JCJC*, 3:548–49, 1216.
55 Gascoigne, *Captain Cook*, 59; and *JCJC*, 3:550–51, 1186.
56 Dening, *Performances*, 73; Glyndwr Williams, "'As Befits Our Age, There Are No More Heroes': Reassessing Captain Cook," in Williams, *Captain Cook*, 230; and *JCJC*, 3:551.
57 *JCJC*, 3:558, 567–68, 1236.
58 Dening, *Performances*, 72.

CHAPTER 14: SPRINGTIME IN KAMCHATKA

1 King, *Voyage to the Pacific*, 3:115.
2 *JCJC*, 3:578, 587.
3 *JCJC*, 3:582, 1235–36; Holmes, *Captain Cook's Final Voyage*, 127–28; and Beaglehole, *The Life of Captain James Cook*, 678.
4 *JCJC*, 3:633, 636.
5 *JCJC*, 3:636–37; and King, *Voyage to the Pacific*, 3:173.
6 *JCJC*, 3:638, 640.
7 *JCJC*, 3:642.
8 *JCJC*, 3:642–43; and King, *Voyage to the Pacific*, 3:183.
9 *JCJC*, 3:643–44.
10 *JCJC*, 3:645; and Beaglehole, *Cook and the Russians*, 4.
11 *JCJC*, 3:650; King, *Voyage to the Pacific*, 3:184; and Trevenen, *A Memoir*, 27.
12 *JCJC*, 3:651.
13 *JCJC*, 3:651–52.
14 *JCJC*, 3:652.
15 *JCJC*, 3:648.
16 *JCJC*, 3:649–50, 653–54, 1242.
17 *JCJC*, 3:660, 663.
18 *JCJC*, 3:663.
19 *JCJC*, 3:664–65.
20 *JCJC*, 3:665–66.
21 *JCJC*, 3:666–67.
22 *JCJC*, 3:666–67; and King, *Voyage to the Pacific* 3:213–14.
23 King, *Voyage to the Pacific*, 3:217–18; and Anichtchenko, "From Russia with Charts," 77.
24 *JCJC*, 3:668–69.
25 *JCJC*, 3:669–70.
26 *JCJC*, 3:655–57.
27 *JCJC*, 3:670–71, 1246–47.
28 *JCJC*, 3:671; and King, *Voyage to the Pacific*, 3:224.
29 *JCJC*, 3:671, 1248.
30 *JCJC*, 3:672.
31 *JCJC*, 3:672–73; and Beaglehole, *Cook and the Russians*, 6.
32 *JCJC*, 3:674.
33 *JCJC*, 3:673.
34 *JCJC*, 3:676.
35 *JCJC*, 3:540, 541.

36 *JCJC*, 3:677.
37 *JCJC*, 3:677–78, 1257.

CHAPTER 15: DIMINISHING RETURNS

1 *JCJC*, 3:678.
2 *JCJC*, 3:679, 683, 686–87.
3 King, *Voyage to the Pacific*, 3:244; and *JCJC*, 3:687–89, 1263.
4 *JCJC*, 3:689; and King, *Voyage to the Pacific*, 3:245.
5 *JCJC*, 3:689; and King, *Voyage to the Pacific*, 3:245–46.
6 *JCJC*, 3:690, 1264; and King, *Voyage to the Pacific*, 3:246.
7 King, *Voyage to the Pacific*, 3:247; and *JCJC*, 3:690–91.
8 King, *Voyage to the Pacific*, 3:247.
9 Ibid., 3:248.
10 *JCJC*, 3:692–93; and King, *Voyage to the Pacific*, 3:249.
11 *JCJC*, 3:692; and King, *Voyage to the Pacific*, 3:249.
12 *JCJC*, 3:694; and King, *Voyage to the Pacific*, 3:250.
13 *JCJC*, 3:695.
14 King, *Voyage to the Pacific*, 3:251; and *JCJC*, 3:1265.
15 *JCJC*, 3:695; and King, *Voyage to the Pacific*, 3:251–52.
16 *JCJC*, 3:1266; and King, *Voyage to the Pacific*, 3:253.
17 *JCJC*, 3:696–97.
18 *JCJC*, 3:696, 1266.
19 King, *Voyage to the Pacific*, 3:256.
20 *JCJC*, 3:1266–67.
21 King, *Voyage to the Pacific*, 3:258; and *JCJC*, 3:1267.
22 King, *Voyage to the Pacific*, 3:258–60; and *JCJC*, 3:1267–68.
23 *JCJC*, 3:695, 1268.
24 *JCJC*, 3:1269.
25 *JCJC*, 3:698, 1545.
26 *JCJC*, 3:698; and Beaglehole, *Cook and the Russians*, 8.
27 King, *Voyage to the Pacific*, 3:260.
28 Ibid. 260–61; *JCJC*, 3:698.
29 King, *Voyage to the Pacific*, 3:263, 268.
30 Ibid., 3:270.
31 Ibid., 3:271–72.
32 Ibid., 3:274–75.
33 Ibid., 3:276.
34 *JCJC*, 3:699, 1542–44; and Banks, *The Endeavour Journal*, 1:112.
35 *JCJC*, 3:700–1.
36 *JCJC*, 3:1271–72.
37 *JCJC*, 3:701, 1273.
38 King, *Voyage to the Pacific*, 3:285.
39 Ibid., 3:288–89.
40 Ibid., 3:307.
41 Ibid., 3:345.
42 Trevenen, *A Memoir*, 35; and *JCJC*, 3:707, 1277.
43 *JCJC*, 3:707, 1281.
44 *JCJC*, 3:ccxxii.
45 King, *Voyage to the Pacific*, 3:291, 311; and *JCJC*, 3:709.

CHAPTER 16: SEEDING THE FUR TRADE ON THE VOYAGE HOME

1 See, for example, Bown, *Madness, Betrayal and the Lash,* 37; and Gough, *Northwest Coast,* i.
2 King, *Voyage to the Pacific,* 3:347, 369.
3 Ibid., 3:370.
4 Ibid., 3:371–73.
5 Ibid., 3:385.
6 Ibid., 3:379; Inglis, "Successors and Rivals," 167–68.
7 King, *Voyage to the Pacific,* 3:404, 406; and *JCJC,* 3:1287.
8 *JCJC,* 3:709.
9 King, *Voyage to the Pacific,* 3:408–10.
10 Ibid., 3:415–16.
11 *JCJC,* 3:712, 1295; and King, *Voyage to the Pacific,* 3:417.
12 King, *Voyage to the Pacific,* 3:417–18.
13 *JCJC,* 3:712; and King, *Voyage to the Pacific,* 3:422.
14 *JCJC,* 3:713; and King, *Voyage to the Pacific,* 3:422–23.
15 *JCJC,* 3:713.
16 King, *Voyage to the Pacific,* 3:430–31, 436.
17 Ibid., 3:437.
18 Ibid.
19 Ibid., 3:437–38.
20 Ibid., 3:438–39.
21 Ibid., 3:440.
22 Ibid., 3:447–48.
23 Ibid., 3:441, 470.
24 Ibid., 3:473–74, 476–77; and *JCJC,* 3:714.
25 King, *Voyage to the Pacific,* 3:478.
26 Ibid., 3:478–79.
27 Ibid., 3:481.
28 Ibid., 3:483.
29 Ibid.; and Inglis, "The French and the Spaniards," 173.
30 King, *Voyage to the Pacific,* 3:484.
31 *JCJC,* 3:716–17.
32 *JCJC,* 3:718; and King, *Voyage to the Pacific,* 3:488.
33 King, *Voyage to the Pacific,* 3:489.

CONCLUSION

1 *JCJC,* 3:1561–62.
2 *JCJC,* 3:1562.
3 John Robson, "Henry Roberts: 1757–1796," *Cook's Log* 39, 1 (2016): 6; and *JCJC,* 3:cxxxix.
4 Gascoigne, *Captain Cook,* 70.
5 *JCJC,* 3:cciv; and MacLaren, "Bones of Empire," 109.
6 Withey, *Voyages of Discovery,* 406.
7 *JCJC,* 3:cxcix; I.S. MacLaren, "Narrating an Alaskan Cruise: Cook's Journal (1778) and Douglas's Edition of *A Voyage to the Pacific Ocean* (1784)," in Barnett and Nicandri, *Arctic Ambitions,* 231–61; and Currie, *Constructing Colonial Discourse,* 29, 33, 35, 132.
8 Cook, *A Voyage to the Pacific,* 1:liv–lv, lxxviii.
9 *JCJC,* 3:1436; and Cook, *A Voyage to the Pacific,* 1:lxxv.
10 Cook, *A Voyage to the Pacific,* 1:lv.
11 Ibid., 1:liv, lvi–lviii.

12 King, *Voyage to the Pacific*, 3:50; and Gough, *Northwest Coast*, 54.

13 King, *Voyage to the Pacific*, 3:50–51.

14 Forster, *History of the Voyages and Discoveries Made in the North*, 298, 390, 404, 472.

15 *JCJC*, 1:cxx.

16 Ibid., 1:xxxix; and Gascoigne, *Captain Cook*, 48.

17 Smith, *Imagining the Pacific*, 51; and Withey, *Voyages of Discovery*, 461.

18 Alan Frost, "Geographical Perspectives," 19; and Suthren, *To Go upon Discovery*, 165–66.

19 Alexander, *The Bounty*, 165.

20 McLynn, *Captain Cook*, 416; and Williams, "As Befits Our Age," 238.

21 Robson, *Captain Cook's World*, 12.

22 Address of the Holy Father Pope Francis, United States Capitol, Washington, DC, Thursday, September 24, 2015, http://www.vatican.va/content/francesco/en/speeches/2015/september/documents/papa-francesco_20150924_usa-us-congress.html.

23 Dening, *Performances*, 59; Williams, "As Befits Our Age," 242, 244; Lambert, "Retracing the Captain," 255; Stuart Murray, "'Notwithstanding Our Signs to the Contrary': Textuality and Authority at the Endeavour River, June to August 1771," in Williams, *Captain Cook*, 59–60.

24 Thomas, *Cook*, 413.

25 Smith, *Imagining the Pacific*, 239–40.

26 Fisher and Johnston, *Cook and His Times*, 4; Robin Fisher and Hugh Johnston, *From Maps to Metaphors: The Pacific World of George Vancouver* (Vancouver: UBC Press, 1993), 13–14; and Daniel Clayton, "Captain Cook's Command of Knowledge and Space: Chronicles from Nootka Sound," in Williams, *Captain Cook*, 132.

27 *JCJC*, 1:399.

28 *JCJC*, 2:493.

29 *JCJC*, 2:175.

30 Withey, *Voyages of Discovery*, 11; and Gascoigne, *Captain Cook*, 146.

31 Currie, *Constructing Colonial Discourse*, 123; and *JCJC*, 3:1406–7.

32 Gascoigne, *Captain Cook*, 174; and Dening, *Performances*, 215.

33 Gascoigne, *Captain Cook*, 146.

34 Smith, *Imagining the Pacific*, 240; John Robson, "Cook's Place in History," *Cook's Log* 37, 4 (2014): 6; and Gascoigne, *Captain Cook*, 146–47.

35 Smith is quoted in Robson, "Cook's Place," 6; William R. Hunt, "A Framework for James Cook: The Meaning of Exploration in the 18th Century and Now," in *Exploration in Alaska*, ed. Antoinette Shalkop (Anchorage: Cook Inlet Historical Society, 1980), 95; and Winchester, *Pacific*, 444.

36 Smith is quoted in Robson, "Cook's Place," 6; and Salmond is quoted in Williams, "As Befits Our Age," 239.

Bibliography

Alexander, Caroline. *The Bounty: The True Story of the Mutiny on the Bounty.* New York: Viking, 2003.

Allen, Richard C. "'Remember Me to My Good Friend Captain Walker': James Cook and the North Yorkshire Quakers." In *Captain Cook: Explorations and Reassessments,* edited by Glyndwr Williams, 21–36. Woodbridge, UK: Boydell, 2004.

Anichtchenko, Evguenia. "From Russia with Charts: Cook and the Russians in the North Pacific." In *Arctic Ambitions: Captain Cook and the Northwest Passage,* edited by James K. Barnett and David L. Nicandri, 63–88. Seattle: University of Washington Press, 2015.

Baker, Simon. *The Ship: Retracing Cook's Endeavour Voyage.* London: BBC, 2002.

Banks, Joseph. *The Endeavour Journal of Joseph Banks, 1768–1771.* Edited by J.C. Beaglehole. 2 vols. Sydney: Angus and Robertson, 1962.

Barnett, James K., and David L. Nicandri, eds. *Arctic Ambitions: Captain Cook and the Northwest Passage.* Seattle: University of Washington Press, 2015.

Barrington, Daines. *The Probability of Reaching the North Pole, Discussed.* London: C. Heydinger, 1775.

Beaglehole, J.C., ed. *Cook and the Russians.* London: Hakluyt Society, 1973.

–. *The Life of Captain James Cook.* Palo Alto, CA: Stanford University Press, 1974.

Belyea, Barbara. "Just the Facts: Vancouver and Theoretical Geography." Paper presented at "Vancouver Conference on Exploration and Discovery," Simon Fraser University, April 23–26, 1992.

Bown, Stephen R. *Madness, Betrayal and the Lash: The Epic Voyage of Captain George Vancouver.* Vancouver: Douglas and McIntyre, 2008.

Clayton, Daniel. "Captain Cook's Command of Knowledge and Space: Chronicles from Nootka Sound." In *Captain Cook: Explorations and Reassessments,* edited by Glyndwr Williams, 110–36. Woodbridge, UK: Boydell, 2004.

Collingridge, Vanessa. *Captain Cook: A Legacy under Fire.* Guilford, CT: Lyons, 2002.

Cook, Andrew S. "James Cook and the Royal Society." In *Captain Cook: Explorations and Reassessments,* edited by Glyndwyr Williams, 37–58. Woodbridge, UK, Boydell, 2004.

Cook, James. *The Journals of Captain James Cook on His Voyages of Discovery.* Edited by J.C. Beaglehole. 3 vols. Cambridge: Cambridge University Press/Hakluyt Society, 1955–67.

–. *A Voyage towards the South Pole and Round the World ...* 2 vols. London, 1777.

Cook, James, and James King. *A Voyage to the Pacific Ocean in the Years 1776, 1777, 1778, 1779 and 1780 ... Vol. I and II written by Captain Cook, Vol. III by Captain King.* Edited by John Douglas. 3 vols. London: G. Nicol and T. Cadell, 1784.

Currie, Noel E. *Constructing Colonial Discourse: Captain Cook at Nootka Sound.* Montreal and Kingston: McGill-Queen's University Press, 2005.

Dana, Richard Henry Jr. *Two Years before the Mast.* New York: Signet, 2009.

Dening, Greg. *Performances*. Chicago: Chicago University Press, 1996.

Dugard, Martin. *Farther Than Any Man: The Rise and Fall of Captain James Cook*. New York: Washington Square, 2002.

Fagan, Brian. *The Little Ice Age: How Climate Made History, 1300–1850*. New York: Basic Books, 2000.

Fernández-Armesto, Felipe. *Pathfinders: A Global History of Exploration*. New York: Norton, 2006.

Fisher, Robin, and Hugh Johnston, eds. *Captain James Cook and His Times*. Seattle: University of Washington Press, 1979.

–. *From Maps to Metaphors: The Pacific World of George Vancouver*. Vancouver: UBC Press, 1993.

Forgan, Sophie. "James Cook from Yorkshire." Paper presented at the "Imagining Anchorage Symposium," Anchorage, Alaska, June 18, 2015.

Forster, George. *A Voyage Round the World*. Edited by Nicholas Thomas and Oliver Berghof. 2 vols. Honolulu: University of Hawai'i Press, 2000.

Forster, Johann Reinhold. *History of the Voyages and Discoveries Made in the North*. Dublin: Luke White and Pat. Byrne, 1786.

–. *Observations Made during a Voyage Round the World*. Edited by Nicholas Thomas, Harriet Guest, and Michael Dettelbach. Honolulu: University of Hawai'i Press, 1996.

Frost, Alan. "New Geographical Perspectives and the Emergence of the Romantic Imagination." In *Captain James Cook and His Times*, edited by Robin Fisher and Hugh Johnston, 5–19. Seattle: University of Washington Press, 1979.

Fry, Howard T. *Alexander Dalrymple (1737–1808) and the Expansion of British Trade*. Toronto: University of Toronto Press, 1970.

Gascoigne, John. *Captain Cook: Voyager between Worlds*. London: Hambledon Continuum, 2008.

Gough, Barry M. *Juan de Fuca's Strait: Voyages in the Waterway of Forgotten Dreams*. Madeira Park, BC: Harbour, 2012.

–. *The Northwest Coast: British Navigation, Trade, and Discoveries to 1812*. Vancouver: UBC Press, 1992.

Hawkesworth, John. *An Account of the Voyages Undertaken ... for Making Discoveries in the Southern Hemisphere ...* 3 vols. London: W. Strahan and T. Cadell, 1773.

Hearne, Samuel. *A Journey from Prince of Wales Fort in Hudson's Bay to the Northern Ocean*. Edited and introduced by Richard Glover. Toronto: Macmillan, 1958.

Hoare, Michael E., ed. *The* Resolution *Journal of Johann Reinhold Forster, 1772–1775*. 4 vols. London: Hakluyt Society, 1982.

–. "Two Centuries' Perceptions of James Cook: George Forster to Beaglehole." In *Captain Cook and His Times*, edited by Robin Fisher and Hugh Johnston, 211–28. Seattle: University of Washington Press, 1979.

Holmes, Christine, ed. *Captain Cook's Final Voyage: The Journal of Midshipman George Gilbert*. Honolulu: University of Hawai'i Press, 1982.

Horwitz, Tony. *Blue Latitudes: Boldly Going Where Captain Cook Has Gone Before*. New York: Picador/Henry Holt, 2002.

Hunt, William R. "A Framework for James Cook: The Meaning of Exploration in the 18th Century and Now." In *Exploration in Alaska*, edited by Antoinette Shalkop, 89–97. Anchorage: Cook Inlet Historical Society, 1980.

Inglis, Robin. "Successors and Rivals to Cook: The French and the Spaniards." In *Captain Cook: Explorations and Reassessments*, edited by Glyndwr Williams, 161–78. Woodbridge, UK: Boydell, 2004.

Kang, Sarah M., and Richard Seager. "Croll Revisited: Why Is the Northern Hemisphere Warmer Than the Southern Hemisphere?" https://ocp.ldeo.columbia.edu/res/div/ocp/pub/seager/Kang_Seager_subm.pdf.

Kitson, Arthur. *Captain James Cook, R.N., F.R.S.: "The Circumnavigator."* New York: E.P. Dutton, 1907.

Lamb, W. Kaye, ed. *The Journals and Letters of Sir Alexander Mackenzie.* London: Cambridge University Press, 1970.

Lambert, Andrew. "Retracing the Captain: 'Extreme History,' Hard Tack and Scurvy." In *Captain Cook: Explorations and Reassessments,* edited by Glyndwr Williams, 246–55. Woodbridge, UK: Boydell, 2004.

Lopez, Barry. *Arctic Dreams: Imagination and Desire in a Northern Landscape.* New York: Vintage, 1986.

MacLaren, I.S. "Bones of Empire: Cook and Franklin Reaching to Alaska for a Northwest Passage." In *Imagining Anchorage: The Making of America's Northernmost Metropolis,* edited by James K. Barnett and Ian Hartman, 105–25. Fairbanks: University of Alaska Press, 2018.

–. "In Consideration of the Evolution of Explorers and Travelers into Authors: A Model." *Studies in Travel Writing* 15, 3 (September 2011): 221–41.

–. "Exploration/Travel Literature and the Evolution of an Author." *Journal of Canadian Studies* 5 (1992): 39–68.

–. "Narrating an Alaskan Cruise: Cook's Journal (1778) and Douglas's Edition of *A Voyage to the Pacific Ocean* (1784)." In *Arctic Ambitions: Captain Cook and the Northwest Passage,* edited by James K. Barnett and David L. Nicandri, 231–62. Seattle: University of Washington Press, 2015.

–. "Notes on Samuel Hearne's *Journey* from a Bibliographical Perspective." *Papers of the Bibliographical Society of Canada* 31, 2 (Fall 1993): 21–51.

–. "Samuel Hearne's Accounts of the Massacre at Bloody Fall, 17 July 1771." *ARIEL: A Review of English Literature* 22, 1 (January 1991): 25–51.

MacLeod, D. Peter. *Northern Armageddon: The Battle of the Plains of Abraham and the Making of the American Revolution.* New York: Knopf, 2016.

McLynn, Frank. *Captain Cook: Master of the Seas.* New Haven, CT: Yale University Press, 2011.

Miller, Robert J. *Native America, Discovered and Conquered: Thomas Jefferson, Lewis and Clark, and Manifest Destiny.* Westport, CT: Praeger, 2006.

Moore, Peter. *Endeavour: The Ship and the Attitude That Changed the World.* London: Vintage, 2019.

Munford, James, ed. *John Ledyard's Journal of Captain Cook's Last Voyage.* Introduction by Sinclair H. Hitchings. Corvallis: Oregon State University Press, 1963.

Murray, Stuart. "'Notwithstanding Our Signs to the Contrary': Textuality and Authority at the Endeavour River, June to August 1771." In *Captain Cook: Explorations and Reassessments,* edited by Glyndwr Williams, 59–76. Woodbridge, UK: Boydell, 2004.

Nicandri, David L. "The Rhyme of the Great Navigator: The Literature of Captain Cook and Its Influences on the Journals of Lewis and Clark; Part 2: The Grandest Sight." *We Proceeded On* 42, 2 (2016): 17–23.

Norris, John. "The Strait of Anian and British America." *BC Studies* 36 (Winter 1977–78): 3–22.

Obeyesekere, Gananath. *The Apotheosis of Captain Cook: European Mythmaking in the Pacific.* Princeton, NJ: Princeton University Press, 1992.

Robson, John. *Captain Cook's War and Peace: The Royal Navy Years, 1755–1768.* Barnsley, UK: Seaforth, 2009.

–. *Captain Cook's World: Maps of the Life and Voyages of James Cook, R.N.* Seattle: University of Washington Press, 2000.

–. "A Comparison of the Charts Produced during the Pacific Voyages of Louis-Antoine Bougainville and James Cook." In *Captain Cook: Explorations and Reassessments,* edited by Glyndwr Williams, 137–60. Woodbridge, UK: Boydell, 2004.

Sahlins, Marshall. *How "Natives" Think: About Captain Cook, for Example.* Chicago: University of Chicago Press, 1995.

Salmond, Anne. *The Trial of the Cannibal Dog: The Remarkable Story of Captain Cook's Encounters in the South Seas.* New Haven, CT: Yale University Press, 2003.

–. "Tute: The Impact of Polynesia on Captain Cook." In *Captain Cook: Explorations and Reassessments,* edited by Glyndwr Williams, 77–93. Woodbridge, UK: Boydell, 2004.

Saunt, Claudio. *West of the Revolution: An Uncommon History of 1776.* New York: Norton, 2014.

Smith, Bernard. *Imagining the Pacific: In the Wake of the Cook Voyages.* New Haven, CT: Yale University Press, 1992.

Stacey, C.P. *Quebec, 1759: The Siege and the Battle.* London: Macmillan, 1973.

Stark, Peter. *Astoria: John Jacob Astor and Thomas Jefferson's Lost Pacific Empire.* New York: HarperCollins, 2014.

Stern, Harry. "Sea Ice in the Western Portal of the Northwest Passage from 1778 to the Twenty-First Century." In *Arctic Ambitions: Captain Cook and the Northwest Passage,* edited by James K. Barnett and David L. Nicandri, 331–58. Seattle: University of Washington Press, 2015.

Suthren, Victor. *To Go upon Discovery: James Cook and Canada, from 1758 to 1779.* Toronto: Dundern, 2000.

Thomas, Nicholas. *Cook: The Extraordinary Voyages of Captain Cook.* New York: Walker, 2003.

Trevenen, James. *A Memoir of James Trevenen.* Edited by Christopher Lloyd and R.C. Anderson. London: Navy Records Society, 1959.

Tuchman, Barbara. *The First Salute: A View of the American Revolution.* New York: Knopf, 1988.

Van Orman, Richard A. *The Explorers: Nineteenth-Century Expeditions in Africa and the American West.* Albuquerque: University of New Mexico Press, 1984.

Vancouver, George. *A Voyage of Discovery to the North Pacific Ocean and Round the World, 1791–95.* Edited by W. Kaye Lamb. 4 vols. Cambridge: Cambridge University Press/ Hakluyt Society, 1984.

Watt, Sir James. "Medical Aspects and Consequences of Cook's Voyages." In *Captain James Cook and His Times,* edited by Robin Fisher and Hugh Johnston, 129–57. Seattle: University of Washington Press, 1979.

Weeks, W.F. *On Sea Ice.* Fairbanks: University of Alaska Press, 2010.

Williams, Glyndwr, ed. *Arctic Labyrinth: The Quest for the Northwest Passage.* London: Penguin, 2009.

–. "'As Befits Our Age, There Are No More Heroes': Reassessing Captain Cook." In *Captain Cook: Explorations and Reassessments,* edited by Glyndwr Williams, 230–45. Woodbridge, UK: Boydell, 2004.

–, ed. *Captain Cook: Explorations and Reassessments.* Woodbridge, UK: Boydell, 2004.

–. *The Death of Captain Cook: A Hero Made and Unmade.* London: Profile Books, 2008.

–. "Myth and Reality: James Cook and the Theoretical Geography of Northwest America." In *Captain James Cook and His Times*, edited by Robin Fisher and Hugh Johnson, 59–79. Seattle: University of Washington Press, 1979.

–. *Naturalists at Sea: Scientific Travelers from Dampier to Darwin*. New Haven, CT: Yale University Press, 2013.

–. *Voyages of Delusion: The Search for the Northwest Passage in the Age of Reason*. London: HarperCollins, 2002.

Winchester, Simon. *Pacific: Silicon Chips and Surfboards, Coral Reefs and Atom Bombs, Brutal Dictators and Fading Empires*. New York: Harper, 2015.

Withey, Lynne. *Voyages of Discovery: Captain Cook and the Exploration of the Pacific*. Berkeley: University of California Press, 1989.

Zug, James. *American Traveler: The Life and Adventures of John Ledyard*. New York: Basic Books, 2005.

Photo Credits

Page

26–27   Library and Archives Canada, Plans and Charts/NMC 21353, R12585–0–9, donated by H.R. MacMillan.

33   Thomas Jefferys, *The American Atlas* (London: R. Sayer and J. Bennett, 1775), Rare Books and Special Collections, University of British Columbia Library, Vancouver, G3435, 1775.J4.

44   National Portrait Gallery, London, NPG 5868.

66   Louis-Antoine de Bougainville, *A Voyage round the World* (London: J. Nourse and T. Davies, 1772; repr., Amsterdam: N. Israel, 1967), Rare Books and Special Collections, University of British Columbia Library, Vancouver, G.420.B823.1967.

87   Mitchell Library, State Library of New South Wales, Sydney, PXD 11, vol. 5, no. 28.

101   Mitchell Library, State Library of New South Wales, Sydney, PXD 11, vol. 5, no. 27a.

106   James Cook and James King, *A Voyage to the Pacific Ocean* (London: G. Nichol and T. Cadell, 1784), atlas, frontispiece, Rare Books and Special Collections, University of British Columbia Library, Vancouver, FC 3821.243. A1 1784.

120   National Portrait Gallery of Australia, Canberra, 2009.55.

129   James Cook, *A Voyage towards the South Pole and round the World*, vol. 2 (London: W. Strahan and T. Cadell, 1777), no. 34, Rare Books and Special Collections, University of British Columbia Library, Vancouver, FC 3821.242. A1 1777.

133   National Archives, Kew, London, ADM 55/107 f.205b.

136   James Cook, *A Voyage towards the South Pole and round the World*, vol. 2 (London: W. Strahan and T. Cadell, 1777), opposite p. 210, Rare Books and Special Collections, University of British Columbia Library, Vancouver, FC 3821.242. A1 1777.

157   National Maritime Museum, Greenwich, London, Greenwich Hospital Collection, BHC 2628.

172   Washington State Historical Society, 2003.16.19.

174–75   James Ford Bell Library, University of Minnesota.

Index

Printed and bound in Canada by Friesens

Set in New Baskerville by Artegraphica Design Co.

Copy editor: Lesley Erickson

Proofreader: Judith Earnshaw

Indexer: Margaret de Boer

Cartographer: Eric Leinberger

Cover designer: George Kirkpatrick

Cover illustration: *Foreground:* "Portrait of Captain James Cook," (detail, circa 1780). Oil on canvas by John Webber. Museum of New Zealand Te Papa Tongarewa. 1960-0013-1. *Background:* "Ice Islands: *Resolution & Adventure* among Ice Bergs," by William Hodges. Mitchell Library, State Library of New South Wales, Sydney. PXD 11, vol. 5, no 28.